MATHEMATICS IN ACTION

An Introduction to Algebraic, Graphical, and Numerical Problem Solving

The Consortium for Foundation Mathematics

Ralph Bertelle	*Columbia-Greene Community College*
Judith Bloch	*University of Rochester*
Roy Cameron	*SUNY Cobleskill*
Carolyn Curley	*Erie Community College—South Campus*
Ernie Danforth	*Corning Community College*
Brian Gray	*Howard Community College*
Arlene Kleinstein	*SUNY Farmingdale*
Kathleen Milligan	*Monroe Community College*
Patricia Pacitti	*SUNY Oswego*
Rick Patrick	*Adirondack Community College*
Renan Sezer	*LaGuardia Community College*
Patricia Shuart	*SUNY Oswego*
Sylvia Svitak	*Queensborough Community College*
Assad J. Thompson	*LaGuardia Community College*

An Imprint of Addison Wesley Longman, Inc.

Boston • San Francisco • New York • Harlow, England
Don Mills, Ontario • Sydney • Mexico City • Madrid • Amsterdam

Publisher: Jason A. Jordan

Acquisitions Editor: Jennifer Crum

Editorial Project Manager: Ruth Berry

Managing Editor: Ron Hampton

Text and Cover Design Supervision: Susan C. Raymond

Cover Designer: Leslie Haimes

Production Coordinator: Sheila Spinney

Composition: Scott Silva, Lynn Lowell

Editorial Assistant: Greg Erb

Media Producer: Lorie Reilly

Marketing Manager: Dona Kenly

Marketing Coordinator: Elan Hanson

Prepress Services Buyer: Caroline Fell

First Print Buyer: Evelyn May Beaton

Library of Congress Cataloging-in-Publication Data
Mathematics in action: an introduction to algebraic, graphical, and numerical problem solving/the Consortium for Foundation Mathematics.
p. cm.

   ISBN 0-201-66041-5 (softback)
   1. Mathematics. I. Consortium for Foundation Mathematics.

   QA39.2 .M3844 2000
   510—dc21

ISBN: 0-201-66041-5

1 2 3 4 5 6 7 8 9 10   CRW   03 02 01 00

Contents

CHAPTER 1 NUMBER SENSE

CHAPTER 3 FUNCTION SENSE

CHAPTER 4   LINEAR FUNCTIONS

Cluster 1 Introduction to Linear Functions

Cluster 2 Problem Solving with Linear Functions

Cluster 3 Systems of Linear Equations

CHAPTER 5   QUADRATIC FUNCTIONS

Cluster 1 Quadratic Functions

Preface

Our Vision

Mathematics in Action: *An Introduction to Algebraic, Graphical, and Numerical Problem Solving* is intended to empower college mathematics students for mathematical literacy in the real world and simultaneously help them build a solid foundation for future study in mathematics and other disciplines.

Our team of fourteen faculty, primarily from the State University of New York and the City University of New York systems, used the AMATYC *Crossroads* standards to develop this two-book series to serve a very large population of college students who, for whatever reason, have not yet succeeded in learning mathematics. It became apparent to us that teaching the same content in the same way to students who have not previously comprehended it is not effective, and this realization motivated us to develop a new approach.

Mathematics in Action is based on the principle that students learn mathematics best by doing mathematics within a meaningful context. In keeping with this premise, students solve problems in a series of realistic situations from which the crucial need for mathematics arises. *Mathematics in Action* guides students toward developing a sense of independence and taking responsibility for their own learning. Students are encouraged to construct, reflect on, apply, and describe their own mathematical models, which they use to solve meaningful problems. We see this as the key to bridging the gap between abstraction and application, and as the basis for transfer learning. Appropriate technology is integrated throughout the books, allowing students to interpret real-life data verbally, numerically, symbolically, and graphically.

We expect that by using the *Mathematics in Action* series, all students will be able to achieve the following goals:

- Develop mathematical intuition and a relevant base of mathematical knowledge.
- Gain experiences that connect classroom learning with real-world applications.
- Prepare effectively for further college work in mathematics and related disciplines.
- Learn to work in groups as well as independently.
- Increase knowledge of mathematics through explorations with appropriate technology.
- Develop a positive attitude about learning and using mathematics.
- Build techniques of reasoning for effective problem solving.

- Learn to apply and display knowledge through alternative means of assessment, such as mathematical portfolios and journal writing.

Pedagogical Features

The pedagogical core of *Mathematics in Action* is a series of guided discovery activities in which students work in groups to discover mathematical principles embedded in realistic situations. The key principles of each activity are highlighted and summarized at the activity's conclusion. Each activity is followed by exercises that reinforce the concepts and skills revealed in the activity.

The activities are clustered within each chapter. Each cluster contains regular activities along with project and lab activities that relate to particular topics. The lab activities require more than just paper, pencil, and calculator; they also require measurements and data collection and are ideal for in-class group work. The project activities are designed to allow students to explore specific topics in greater depth, either individually or in groups. These activities are usually self-contained and have no accompanying exercises. For specific suggestions on how to use the three types of activities, we strongly encourage instructors to refer to the *Instructor's Resource Guide* that accompanies this text.

Each cluster concludes with two sections: What Have I Learned? and How Can I Practice? The What Have I Learned? activities are designed to help students pull together the key concepts of the cluster. The How Can I Practice? activities are designed primarily to provide additional work with the numeric, graphical, and algebraic skills of the cluster. Taken as a whole, these activities give students the tools they need to bridge the gaps between abstraction, skills, and application.

Two sets of exercises called Skills Check follow Clusters 3 and 5 in Chapter 1 to provide students with additional opportunities to practice basic numerical skills. In addition, Exploring Numeracy boxes that feature realistic-problem situations appear in Chapter 1 to further help students understand how mathematical literacy is relevant to the world in which they live.

Changes from the Preliminary Edition

Instructors who have used the preliminary edition of *Mathematics in Action* will notice that the text has expanded. In response to the requests of many users, a new chapter—Chapter 5, Quadratic Functions—has been added as a brief introduction to the topic. To prepare students for the factoring in Chapter 5, multiplication of binomials has been added to Chapter 2. In Chapters 1 and 2, much of the material has been reorganized to improve the flow of topics. In Chapter 1, a new cluster on comparisons and proportional reasoning presents a conceptually unified approach to a traditional topic.

Other sections have been revised as a result of further class testing to clarify and increase emphasis on some skill areas. Specific revisions include the following:

- All exercises have been carefully reviewed to ensure an appropriate level of difficulty for this course of arithmetic and elementary algebra.

- Summaries have been added to the end of each activity.

- The exposition within the development of several topics, including exponents and order of operations, has been expanded.

- Throughout Chapters 3–5, a graphing calculator icon alerts students to the fact that a graphing calculator is recommended to solve designated problems. Students may refer to Appendix D for instruction on how to use the TI-83 graphing calculator.

- The appendices have been revised and reorganized into the following sections: Appendix A, Fractions; Appendix B, Decimals; Appendix C, Algebraic Extensions; Appendix D, The TI-83 Graphing Calculator. The appendix covering signed numbers has been eliminated, because signed numbers are covered extensively in the text itself.

- We have included appendix icons throughout the text to highlight concepts and/or skill areas that could be enhanced by the use of Appendix A, Fractions; Appendix B, Decimals; or Appendix C, Algebraic Extensions.

World Wide Web Supplement
www.mathinaction.com

A new Web site accompanies the *Mathematics in Action* series, providing valuable resources for both instructors and students. Students can use the site to access InterAct Math® tutorial exercises for each chapter of the books in the series. Instructors will find teaching tips, sample syllabi, information about alternative assessment, and links to math forum sites.

General information about the *Mathematics in Action* series can also be found at Addison Wesley Longman's Developmental Mathematics Reform Forum Web Site: **www.mathreform.com**

Instructor Supplements

Annotated Instructor's Edition
ISBN 0-201-66042-3

This special version of the text provides answers to all exercises directly beneath each problem.

Instructor's Resource Guide
ISBN 0-201-66045-8

This valuable teaching resource includes the following materials:

- Sample syllabi suggesting ways to structure the course around core and supplemental activities.

- Teaching notes for each chapter.

- Extra practice worksheets for topics with which students typically have difficulty.

- Sample chapter tests and final exams for in-class and take-home use by individual students and groups.*

- Information about incorporating technology in the classroom, including sample graphing calculator assignments.

* Electronic versions of these tests are available on disk in Microsoft Word. To receive a copy, please send your request via e-mail to math@awl.com, specifying which *Mathematics in Action* text you are using to teach your course and providing your school mailing address and phone number.

- Extensive information about group learning.
- Guidance for alternative means of assessment.
- Journal assignments.

TestGen-EQ with QuizMaster-EQ
ISBN 0-201-70509-5

Available on a dual-platform, Windows/Macintosh CD-ROM, this fully networkable software enables instructors to create, edit, and administer tests using a computerized test bank of questions organized according to the chapter content. Six question formats are available, and a built-in question editor allows the user to create graphs, import graphics, and insert mathematical symbols and templates, variable numbers, or text. An Export to HTML feature lets instructors post practice tests to the Web, and using QuizMaster-EQ, instructors can post quizzes to a local computer network so that students can take them on-line. QuizMaster-EQ automatically grades the quizzes, stores results, and lets the instructor view or print a variety of reports for individual students or for an entire class or section.

InterAct Math® Plus
Windows ISBN 0-201-63555-0
Macintosh ISBN 0-201-64805-9

This networkable software provides course-management capabilities and on-line test administration for Addison Wesley Longman's InterAct Math® Tutorial Software (see Student Supplements). InterAct Math® Plus enables instructors to create and administer on-line tests, summarize students' results, and monitor students' progress in the tutorial software, creating a valuable teaching and tracking resource.

Instructor Training Video
ISBN 0-201-70959-7

This innovative video discusses effective ways to implement the teaching pedagogy of the *Mathematics in Action* series, focusing on how to make collaborative learning, discovery learning, and alternative means of assessment work in the classroom.

Student Supplements

InterAct Math® Tutorial Software

ISBN 0-201-70507-9

Available on a dual-platform, Windows/Macintosh CD-ROM, this interactive tutorial software provides algorithmically generated practice exercises that are correlated to the chapter content of the texts. Every exercise in the program is accompanied by an example and a guided solution designed to involve students in the solution process. The software tracks students' activity and scores and can generate printed summaries of students' progress. The software also recognizes common student errors and provides appropriate feedback. Instructors can use the InterAct Math® Plus course-management software (see Instructor Supplements) to create, administer, and track on-line tests and monitor student performance during their practice sessions in InterAct Math®.

InterAct Math® XL
http://www.mathxl.com

InterAct MathXL subscription bundled with text: ISBN 0-201-71345-4

InterAct MathXL subscription purchased separately: ISBN 0-201-71630-5

InterAct MathXL Instructor Coupon: ISBN 0-201-71111-7

InterAct MathXL is a Web-based tutorial system that helps students prepare for tests by allowing them to take practice tests and receive a personalized study plan based on their results. Practice tests are correlated directly to chapter content. Once a student has taken an on-line practice test, the software scores the test and generates a study plan that identifies strengths, pinpoints topics where more review is needed, and links directly to the appropriate section(s) of the InterAct Math tutorial software for additional practice and review. Students gain access to the InterAct MathXL Web site through a password-protected subscription; subscriptions can either be bundled with new copies of the *Mathematics in Action* texts or purchased separately.

Acknowledgments

The Consortium would like to acknowledge and thank the following people for their invaluable assistance in reviewing and testing material for this text:

Russell L. Baker, *Howard Community College*

Joseph Bertorelli, *Queensborough Community College*

Vera Brennan, *Ulster County Community College*

Judy Cain, *Tompkins Cortland Community College*

Ellen Clay, *Richard Stockton College of New Jersey*

Irene Duranczyk, *Eastern Michigan University*

Janet Evert, *Erie Community College*

Mary Ann Fiore, *Queensborough Community College*

Susan Forman, *Bronx Community College*

Miguel Garcia, *Gateway Community-Technical College*

Charlotte Grossbeck, *SUNY Cobleskill*

Johanna G. Halsey, *Dutchess Community College*

Stephen Paul Hess, *Grand Rapids Community College*

Winfield Ihlow, *SUNY Oswego*

Marilyn Jacobi, *Gateway Community-Technical College*

Vijay S. Joshi, *Virginia Intermont College*

Maryann Justinger, *Erie Community College*

Kandace Kling, *Portland Community College, Sylvania*

Anne F. Landry, *Dutchess Community College*

Jennifer Laveglia, *Bellevue Community College*

Geraldine Lockard, *Queensborough Community College*

J. Robert Malena, *Community College of Allegheny County, South Campus*

Len Malinowski, *Finger Lakes Community College*

Dolores Maue, *Erie Community College*

Gabriel Melendez, *Mohawk Valley Community College*

Loretta Monk, *Fayetteville Technical Community College*

Shailaja Nagarkatte, *Queensborough Community College*

Peg Pankowski, *Community College of Allegheny County*

Amy C. Salvati, *Adirondack Community College*

We would like to acknowledge and extend a special thanks to Valerie Ann Ledford for her invaluable assistance throughout the entire project in a wide range of areas, from cheerfully providing much-needed administrative support on a sometimes daily basis to supplying feedback from using the Preliminary Edition in her classes. We would also like to thank Antonia Del Carmen Collado for her help typing solutions and assembling the manuscript and Patrice Herring for her assistance working solutions. We would also like to thank our accuracy checkers, Jeff Suzuki and Bridget Desrosiers.

Finally, a special thank you to our families for their unwavering support and sacrifice, which enabled us to make this text a reality.

The Consortium for Foundation Mathematics

To the Student

The book in your hands is most likely very different from any algebra book you have seen before. In this book, you will take an active role in developing the important ideas of algebra. You will be expected to add your own words to the text. This will be part of your daily work, both in and out of class and for homework. It is the belief of the authors that students learn mathematics best when they are actively involved in solving problems that are meaningful to them.

The text is primarily a collection of situations drawn from real life. Each situation leads to one or more problems. By answering a series of questions, and solving each part of the problem, you will be led to use one or more ideas of introductory college mathematics. Sometimes, these will be basic skills that build on your knowledge of arithmetic. Other times, they will be new concepts that are more general and far-reaching. The important point is that you won't be asked to master a skill until you see a real need for that skill as part of solving a realistic application.

Another important aspect of this text and the course you are taking is the benefit gained by collaborating with your classmates. Much of your work in class will result from being a member of a team. Working in small groups, you will help each other work through a problem situation. While you may feel uncomfortable working this way at first, there are several reasons we believe it is appropriate in this course. First, it is part of the learning-by-doing philosophy. You will be talking about mathematics, needing to express your thoughts in words. This is a key to learning. Secondly, you will be developing skills that will be very valuable when you leave the classroom. Currently, many jobs and careers require the ability to collaborate within a team environment. Your instructor will provide you with more specific information about this collaboration.

One more fundamental part of this course is that you will have access to appropriate technology at all times. You will have access to calculators and some form of graphics tool—either a calculator or computer. Technology is a part of our modern world, and learning to use technology goes hand in hand with learning mathematics. Your work in this course will help prepare you for whatever you pursue in your working life.

This course will help you develop both the mathematical and general skills necessary in today's workplace, such as organization, problem solving, communication, and collaborative skills. By keeping up with your work and following the suggested organization of the text, you will gain a valuable resource that will serve you well in the future. With hard work and dedication you will be ready for the next step.

The Consortium for Foundation Mathematics

Number Sense

Your goal in this chapter is to use the numerical mathematical skills you already have —and those you will learn or relearn—to solve problems. Chapter activities are based on practical, real-world situations that you may encounter in your daily life and work.

Before you begin the activities in Chapter 1, we ask you to think about your previous encounters with mathematics and choose one word to describe those experiences.

Enjoy solving the problems of Chapter 1!

Using Number Sense to Solve Problems

ACTIVITY 1.1

The Bookstore
Topic: *Problem Solving*

By 11:00 A.M., a line has formed outside the crowded bookstore. You ask the guard at the gate how long you can expect to wait. She provides you with the following information: She is permitted to let 6 people into the bookstore only after 6 people have left; students are leaving at the rate of 2 students per minute; she has just let 6 new students in. Also, each student spends an average of 15 minutes gathering books and supplies and 10 minutes waiting in line to check out.

Currently 38 people are ahead of you in line. You know that it is a ten-minute walk to your noon class. Can you buy your books and still expect to make it to your noon class on time?

1. What was your initial reaction after reading the problem?

2. Have you ever worked a problem such as this before?

3. Organizing the information will help you solve the problem.

    a. How many students must leave the bookstore before the guard allows more to enter?

    b. How many students per minute leave the bookstore?

    c. How many minutes are there between groups of students entering the bookstore?

    d. How long will you stand in line outside the bookstore?

    e. Now finish solving the problem and answer the question: How early or late for class will you be?

4. Write, in complete sentences, what you did to solve this problem. Then, explain your solution to a classmate.

EXERCISES

1. Think about the various approaches you and your classmates used to solve Activity 1.1, The Bookstore. Choose the approach that is best for you, and describe it in complete sentences.

2. What mathematical operations and skills did you use?

ACTIVITY 1.2

Time Management
Topics: *Problem Solving, Estimating, Tables, Bar Graphs, Using Fractions*

Your friend is a first-semester freshman carrying a full course load of 15 credit hours. Because Biology lab lasts for three hours, not one, her schedule actually consists of 17 classroom hours. Two hours of study are expected for each hour spent in the classroom or laboratory. She is also working 18 hours per week and has two children at home.

Your friend commutes to campus 1 hour each way every day that she has class, Monday through Friday. Child care and household chores consume 25 hours per week. Each day these chores consume a minimum of 1 hour before school and 2 hours after school. She averages 7 hours of sleep per night, never less than 5 hours or more than 10 hours. She tries to reserve 2 hours per day for leisure and exercise activities and 3 hours per day for personal time.

1. List each activity and the number of hours spent per week on each activity.

2. Is this schedule possible? Explain.

3. What suggestions would you make to your friend for a more workable schedule?

4. Use the information from Problems 1 and 3 to fill in the following grid with a workable schedule:

TIME	MON.	TUES.	WED.	THURS.	FRI.	SAT.	SUN.
12:00 MIDNIGHT							
1:00 A.M.							
2:00							
3:00							
4:00							
5:00							
6:00							
7:00							
8:00							
9:00	Math		Math		Math		
10:00		Biology		Biology			
11:00	English		English		English		
12:00 NOON	Psychology		Psychology		Psychology		
1:00 P.M.				Lab			
2:00	Art		Art	Lab	Art		
3:00				Lab			
4:00							
5:00							
6:00							
7:00							
8:00							
9:00							
10:00							
11:00							

5. Make a bar graph of your friend's schedule from Problem 4. Put the activities (chores, classes, etc.) on the horizontal axis and the number of hours per week on the vertical axis.

6. a. What activity does the highest bar on the graph represent? What part of the week's total of 168 hours does your friend spend doing that activity?

b. As an estimate, is this activity closest to $\frac{1}{2}$, $\frac{1}{3}$, or $\frac{1}{4}$ of the total week? Explain.

c. What advice would you give your friend about the time she spends doing this activity?

7. Your friend spends 17 hours per week in class, out of the week's total of 168 hours. That means she spends $\frac{17}{168}$ of the week in class. As an estimate, is the time spent in class closer to $\frac{1}{5}$ or $\frac{1}{10}$ of the total week?

8. Which activities consume at least $\frac{1}{4}$ of your friend's week? Explain.

9. a. Choose another activity and determine the fractional part of the total week your friend spends doing that activity.

b. What simple fraction would be a suitable estimate for your answer to part a?

EXERCISES

1. List the time commitments in your life by category as you did in Problem 1 for your friend.

2. Use the following grid to create your weekly schedule.

TIME	MON.	TUES.	WED.	THURS.	FRI.	SAT.	SUN.
12:00 MIDNIGHT							
1:00 A.M.							
2:00							
3:00							
4:00							
5:00							
6:00							
7:00							
8:00							
9:00							
10:00							
11:00							
12:00 NOON							
1:00 P.M.							
2:00							
3:00							
4:00							
5:00							
6:00							
7:00							
8:00							
9:00							
10:00							
11:00							

3. a. What part of your total weekly schedule do you spend studying outside of class?

b. What part of your total time do you spend working at a paid job?

c. What part of your total time do you spend on household chores and/or child care?

4. Make a bar graph depicting your schedule.

5. Use your bar graph to think about your schedule. Are there any changes that you could make in your schedule to help you use your time more efficiently and effectively as you pursue your educational goals? Use complete sentences to write down your thoughts.

Course Grades and Your GPA

Topics: *Problem Solving, Using Percents, Simple Averages, Weighted Averages*

Throughout your college years, different instructors may use different methods to determine your grade.

> The most common method of determining your course grade is to average exam scores that are all based on 100 points. You add all your scores and divide the sum by the number of exams. An average calculated this way is called a **simple average.** Note that a simple average is also called a **mean.**

You are a college freshman and the end of the semester is approaching. You are concerned about keeping a B– (80) average in your English literature class, where your instructor uses the simple-average method to determine averages. Your grade will be determined by averaging your exam scores. So far, you have scores of 82, 75, 85, and 93 on four exams. Each exam is based on 100 points.

1. What is your current average for the four exams?

2. What is the lowest score you could achieve on the fifth exam to maintain an 80 average?

3. Describe, in complete sentences, the procedure that you used to answer Problem 2.

4. Is it possible to achieve an A– average (90) in the course? Explain why or why not.

In some grading systems, certain graded work is counted more heavily than other work in determining your course grade. In such systems, the average is known as a **weighted average.** In one method for determining a weighted average, your instructor would assign different percent weights to different components of the course (for example, quizzes, 20%; exams, 50%; and projects, 30%). Note that the weights (given here in percent) will always sum to 100% or 1.0.

> Here is how to determine the weighted average of several components:
> 1. Calculate the simple average (based on 100) of each component.
> 2. Multiply each component's simple average by the weight assigned to that component.
> 3. Sum these weighted component averages to obtain the final weighted average.

For example, in mathematics class, your grade is determined by averaging quizzes, group activities, projects, and exams. The instructor uses the following set of weights: quizzes are worth 15% of the grade; group activities, 30%; projects, 10%; and exams, 45%. The following table categorizes your grades. Your instructor drops your lowest quiz grade when determining your quiz average. Each score is based on 100 points.

QUIZZES	GROUP ACTIVITIES	PROJECTS	EXAMS
70	95	100	81
85	90	95	85
40	75		72
90	60		
75			

5. Determine the simple average for each of the categories and list it in the given table. Also list the weight of each component.

COMPONENT	SIMPLE AVERAGE	WEIGHT
QUIZZES		0.15
GROUP ACTIVITIES		
PROJECTS		
EXAMS		

6. Multiply the simple average for each category (from Problem 5) by the weight assigned to that category. Then sum these weighted values to determine your weighted average in mathematics class.

7. Suppose that the grading scale used at your college is as given in the following chart. What letter grade would you earn in mathematics class? Note that "90 to 93" means to include 90 but not 93.

CLASS AVERAGE	93 to 100	90 to 93	87 to 90	83 to 87	80 to 83	77 to 80	73 to 77	70 to 73	67 to 70	63 to 67	60 to 63	below 60
LETTER GRADES	A	A–	B+	B	B–	C+	C	C–	D+	D	D–	F

Your college grade point average (GPA) is another example of a weighted average.

The weight of a course is the fractional portion of your total credit load that the course represents. It is determined by dividing the number of credit hours of the course by your total credit hours for the semester. Therefore, the larger the number of credits carried by the course, the more strongly it will be weighted in your GPA.

For example, in a 12-credit load, a 3-credit course would have a weight of $\frac{3}{12} = 0.250$ and a 2-credit course would have a weight of $\frac{2}{12} = 0.167$.

8. This semester, your friend is taking two 3-credit courses, one in English and the other in Spanish, a 4-credit course in mathematics, and a 5-credit chemistry course. What is the weight of each course?

COURSE	English	Spanish	Mathematics	Chemistry
WEIGHT				

Your letter grade for a course is translated to a numerical equivalent by the following table:

LETTER GRADES	A	A–	B+	B	B–	C+	C	C–	D+	D	D–	F
NUMERICAL EQUIVALENTS	4.00	3.67	3.33	3.00	2.67	2.33	2.00	1.67	1.33	1.00	0.67	0.00

Suppose you took 17 credit hours this past semester, your third semester in college. You earned an A– in psychology (3 hours); a C+ in economics (3 hours); a B+ in chemistry (4 hours); a B in English (3 hours), and a B– in mathematics (4 hours).

9. Your semester GPA is the weighted average of the numerical equivalents of your course grades. Use the following procedure to complete the table and calculate your semester GPA. (Use a spreadsheet program if you have access to a computer.)

   a. Use the first four columns to record the information regarding the courses you took. As a guide, the information for your psychology course has been recorded for you.

1 COURSE	2 LETTER GRADE	3 NUMERICAL EQUIVALENT	4 CREDIT HOURS	5 WEIGHT	6 CONTRIBUTION TO GPA
Psychology	A–	3.67	3	3/17 = 0.176	0.176 · 3.67 = 0.646

   b. Calculate the weight for each course and enter it in column 5.

c. For each course, multiply the course's weight (column 5) by your numerical grade (column 3). Round to three decimal places and enter this product in column 6, the course's contribution to your GPA. You can now calculate your semester GPA by summing the contributions of all your courses. What is your semester GPA?

EXERCISES

1. A grade of W is given if you withdraw from a course before a certain date. The W appears on your transcript but is not included in your grade point average. Suppose that instead of a C+ in economics, you receive a W. Recalculate your GPA.

2. Now suppose that you earn an F in economics. The F is included in your grade point average. Recalculate your GPA.

3. In your first semester in college, you took 13 credit hours and earned a GPA of 2.13. The following semester your GPA of 2.34 was based on 12 credit hours. You calculated this past semester's GPA in Problem 9.

   a. Explain why the calculation of your overall GPA for the three semesters requires a weighted average.

   b. Calculate your overall GPA for the three semesters.

4. You are concerned about passing your economics class with a C– (70) average. Your grade is determined by averaging your exam scores. So far, you have scores of 78, 66, 87, and 59 on four exams. Each exam is based on 100 points. Your economics instructor uses the simple-average method to determine your average.

   a. What is your current average for the four exams?

   b. What is the lowest score you could achieve on the fifth exam to have at least a 70 average?

5. In chemistry class, your grade is determined by calculating the weighted average of your quiz, lab, and exam grades. The instructor uses the following set of weights: quizzes are worth 20% of the grade; labs, 30%; and exams, 50%. The following table categorizes your grades. Your instructor drops your lowest quiz grade when determining your quiz average. Each score is based on 100 points. Calculate your average for chemistry.

QUIZZES	LABS	EXAMS
78	90	81
85	80	85
55	85	93
90	95	
80	75	
	80	

6. Suppose you took 15 credit hours last semester. You earned an A− in English (3 hours); a B in mathematics (4 hours); a C+ in chemistry (3 hours); a B+ in health (2 hours); and a B− in history (3 hours). Calculate your GPA for the semester.

7. Suppose your history professor discovers an error in his calculation of your grade from last semester. Your newly computed history grade is a B+. Use this new grade and the information in Exercise 6 to recalculate your GPA.

8. Have you encountered any other grading systems used by your instructors? Describe them. How do they work? Give an example.

The Electric Bill

Topics: *Problem Solving, Estimating, Percents, Bar Graphs*

Suppose you receive the following electric bill.

Electric Service

Niagara Mohawk buys low-cost energy that includes hydroelectric power purchased from the New York Power Authority. These hydroelectric purchases have a savings value of $8.88.

This meter reading, Feb. 24, 00 (actual)............   18052
Last meter reading, Dec. 23, 99 (actual)............ − 17307
Amount of electricity used........................kWh   745

Current charges for 63 days—residential service (Rate 1)

Basic service charge (not including usage)......... $14.66
Charge for 745 kWh @ 10.1530¢ each kWh... + 75.64
Electric adjustments.. +   3.64
Sales tax (2.000%).. +   1.88

Total cost for electric service........................... $95.82

Your energy use and its cost

☐ = Actual reading
▨ = Estimated reading
☐ = Average customer

This chart shows your energy-use pattern over the last 13 months. It also shows the current month's usage by our average residential customer.

Electric Meter 87-099-474

DAILY AVERAGES		
	Last year	This period
Temp	13°	23°
kWh	10.6	11.8
Cost $	1.38	1.52

1. This bill is filled with numerical information, such as percents and rates. What other examples of numerical information can you find on this electric bill?

2. What time period does this bill represent?

3. Explain how the amount of electricity used for the time period of this bill was calculated.

4. What does each bar in the bar graph represent?

5. a. Which billing period had the most usage of electricity?

   b. Estimate the number of kilowatt-hours (kWh) used in that period.

6. a. Which billing period had the least usage?

   b. Estimate the number of kilowatt-hours used in that period.

7. How does your current electricity usage compare to that of the average residential customer? Express this comparison as a fraction.

8. Explain how the usage charge of $75.64 for this billing period was calculated.

9. Explain how the sales tax of $1.88 was calculated.

10. a. How is the daily average usage for this billing period calculated?

b. Where is this amount indicated on the bill?

11. On a budget plan, you would pay the same amount each billing period for electricity. To do this, the utility company calculates the average usage over all billing periods during the previous year.

    a. If you choose this option for next year, calculate your average usage per billing period.

    b. Based on the current rate, calculate your usage charge for each billing period on the budget plan.

E X E R C I S E S

1. List the different mathematical operations used in this activity.

2. Use the appropriate ideas you listed in Exercise 1 to calculate the following: You buy three CDs priced at $12.99, $17.99, and $19.99. If the sales tax rate is 7.5%, what is the total cost of your purchase?

3. You spend $13.48, $18.90, and $21.20 for your dinners on a three-day vacation and leave a 15% tip for each meal (calculated on the subtotal before taxes). The local sales tax is 7%.

    a. How much money do you spend on your meals, before tax and tip?

b. What is the total tax for all three meals?

c. How much do you leave in tips for all three meals?

d. What is the total cost of all three meals?

e. What is the average cost of a meal, tax and tips included?

4. At the bookstore you buy books and supplies for the semester. The five text-books that you need cost $42.00, $56.00, $34.50, $79.10, and $68.00. You also need to buy five spiral notebooks at $2.89 each, a journal book that costs $2.19, and a $94 graphics calculator for your mathematics class. The local sales tax is 7%. Your financial aid package provides you with a $400 bookstore account for the semester.

   a. What is the average price of your textbooks?

   b. What is your total bookstore bill?

   c. Is there enough money in your bookstore account to pay the total bill? Explain.

5. Summer weather can vary, depending on the jet stream. The following chart gives daily high and low temperatures for a week in July.

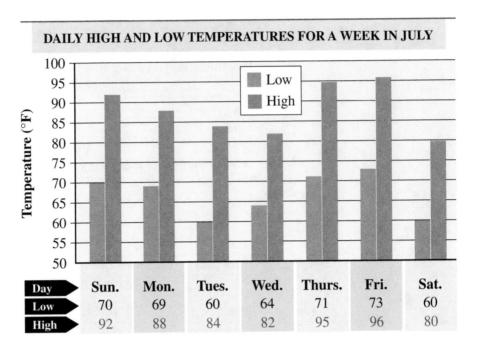

DAILY HIGH AND LOW TEMPERATURES FOR A WEEK IN JULY

Day	Sun.	Mon.	Tues.	Wed.	Thurs.	Fri.	Sat.
Low	70	69	60	64	71	73	60
High	92	88	84	82	95	96	80

a. Determine the average high temperature for the week.

b. Determine the average low temperature for the week.

c. Determine the average difference between the daily high and low temperatures for the week.

ACTIVITY 1.5

You and Your Calculator

Topics: *Properties of Real Numbers, Whole-Number Exponents, Scientific Notation with Large Numbers, Order of Operations, Grouping*

A calculator is a powerful tool for problem solving. A quick look at the desks of those around you should show you that calculators come in many sizes and shapes and with varying capabilities. Some calculators perform only basic operations such as addition, subtraction, multiplication, division, and square roots. Others also handle operations with exponents, perform operations with fractions, and do trigonometry and statistics. There are calculators that graph equations and generate tables of values; some even manipulate algebraic symbols.

Unlike people, however, calculators do not think for themselves and can only perform tasks in the way that you instruct them (or program them). Therefore, you need to understand the properties of numbers and become familiar with the way your calculator operates with numbers. In particular, you will learn how your calculator deals with very large numbers and the order in which it performs the operations you request.

Appendix

NOTE: Detailed information about using the TI-83 calculator for more complex tasks appears in Appendix D.

1. Use your calculator to determine the sum $146 + 875$.

2. **a.** Now, input $875 + 146$ into your calculator and evaluate. How does this sum compare to the sum in Problem 1?

   b. If you use numbers other than 146 and 875, does reversing the order of the numbers change the result? Explain by giving examples.

   If the order in which two numbers are added is reversed, the sum remains the same. This property is called the **commutative property** of addition and can be written symbolically as
   $$a + b = b + a$$

3. Is the commutative property true for the operation of subtraction? Multiplication? Division? Explain by giving examples for each operation.

It is sometimes convenient to do mental arithmetic (i.e., without the aid of your cal-culator). For example, to evaluate $4 \cdot 27$ without the aid of your calculator, think about the multiplication as follows: 27 can be written as $20 + 7$. Therefore, $4 \cdot 27$ can be written as $4 \cdot (20 + 7)$, which can be evaluated as $4 \cdot 20 + 4 \cdot 7$. The product $4 \cdot 27$ can now be thought of as $80 + 28$, or 108. To summarize,

$$4 \cdot 27 = \underline{4 \cdot (20 + 7) = 4 \cdot 20 + 4 \cdot 7} = 80 + 28 = 108$$

The bracketed step involves a very important property called the *distributive property*. In particular, multiplication is distributed over addition or subtraction.

The **distributive property** is written symbolically as:

$$c \cdot (a + b) = c \cdot a + c \cdot b \text{ for addition,}$$

or

$$c \cdot (a - b) = c \cdot a - c \cdot b \text{ for subtraction.}$$

Note that $c \cdot (a + b)$ can also be written as $c(a + b)$ and $c \cdot (a - b)$ can also be written as $c(a - b)$. When a number is immediately followed by parentheses, the operation of multiplication is implied.

4. Another way to express 27 is $25 + 2$ or $30 - 3$.

  a. Express 27 as $25 + 2$ and use the distributive property to multiply $4 \cdot 27$.

  b. Express 27 as $30 - 3$ and use the distributive property to multiply $4 \cdot 27$.

5. Mentally evaluate the following multiplication problems using the distributive property. Verify your answer using your calculator.

  a. $7 \cdot 82$

  b. $5 \cdot 108$

6. a. Evaluate $5 + 6 \cdot 2$ in your head and record the result. Verify using your calculator.

  b. What operations are involved in this problem?

c. In what order did you and your calculator perform the operations to get the answer?

d. Evaluate $(5 + 6) \cdot 2$ and record the result. Verify using your calculator.

e. Why is the result in part d different from the result in part a?

Scientific and graphing calculators are programmed to perform operations in a universally accepted order as illustrated by the previous problems. Part of the **order of operations** priority convention is as follows:

1. Perform multiplication and division before addition and subtraction.
2. If both multiplication and division are present, perform the operations in order, from left to right.
3. If both addition and subtraction are present, perform the operations in order, from left to right.

For example: $10 - 2 \cdot 4 + 7 = 10 - 8 + 7 = 2 + 7 = 9$

7. Perform the following calculations without a calculator. Then use your calculator to verify your result.

a. $42 \div 3 + 4$

b. $42 \div (3 + 4)$

c. $(4 + 8) \div 2 \cdot 3 - 9$

d. $4 + 8 \div 2 \cdot 3 - 9$

8. a. Perform the calculation $\frac{24}{2 + 6}$ without your calculator.

b. Now use your calculator to evaluate $\frac{24}{2 + 6}$. Did you obtain 3 as the result? If not, then perhaps you entered the expression as $24 \div 2 + 6$ and your answer is 18.

c. Explain why the result of $24 \div 2 + 6$ is 18.

Note that $\frac{24}{2+6}$ is the same as the quotient $\frac{24}{8}$. Therefore, the addition in the denominator is done first, followed by the division. To write $\frac{24}{2+6}$ in a horizontal format, you must use parentheses to group the expression in the denominator to indicate that the addition is performed first. That is, write $24 \div (2 + 6)$.

9. Enter $24 \div (2 + 6)$ into your calculator using the parenthesis keys and verify that the result is 3.

Parentheses are grouping symbols that are used to override the standard order of operations. Operations contained in parentheses are performed first.

For example, $2 \cdot (3 + 4 \cdot 5) = 2 \cdot (3 + 20) = 2 \cdot 23 = 46$.

10. Evaluate the following mentally and verify on your calculator.

a. $\dfrac{6}{3+3}$

b. $\dfrac{2+8}{4-2}$

c. $5 + 2 \cdot (4 \div 2 + 3)$ d. $10 - (12 - 3 \cdot 2) \div 3$

Calculators can easily perform repeated multiplication. Recall that $5 \cdot 5$ can be written as 5^2 (5 squared). There are two ways to square a number on your calculator.

11. a. One method to evaluate 5^2 is to use the $\boxed{x^2}$ key. Input $\boxed{5}$ and then press the $\boxed{x^2}$ key. Do this now and record your answer.

b. Another way you can evaluate 5^2 is by using the exponent key. Depending on your calculator, the exponent key may be $\boxed{x^y}$, $\boxed{y^x}$, or $\boxed{\wedge}$. To calculate 5^2, input $\boxed{5}$, press the exponent key, then enter the exponent as $\boxed{2}$ and press $\boxed{\text{ENTER}}$. Do this now and record your answer.

c. The exponent key can be used with any exponent. For example, $5 \cdot 5 \cdot 5$ can be written as 5^3. Evaluate $5 \cdot 5 \cdot 5$ as written, and evaluate 5^3 using the exponent key.

An expression such as 5^3 is called a **power** of 5. The base is 5 and the exponent is 3. Note that the exponent indicates how many times the base is multiplied by itself. Note also that 5^3 is read as "five raised to the third power." When a power (also known as an *exponential expression*) is contained in an expression, it is evaluated *before* any multiplication or division.

For example, to evaluate the expression $20 - 2 \cdot 3^2$, you can proceed as follows:

$$20 - 2 \cdot 3^2 =$$
$$20 - 2 \cdot 9 =$$
$$20 - 18 =$$
$$2$$

12. Enter the expression $20 - 2 \cdot 3^2$ into your calculator and verify the result.

13. Evaluate the following.

    a. $6 + 3 \cdot 4^3$ **b.** $2 \cdot 3^4 - 5^3$

14. a. Use the exponent key to evaluate the following powers: 3^0, 8^0, 23^0, 526^0.

    b. Evaluate other nonzero numbers with a zero exponent.

    c. Evaluate 0^0. What is the result?

    d. Write a sentence describing the result of raising a nonzero number to a zero exponent.

15. a. Now use the exponent key to evaluate the following powers of 10: 10^2, 10^3, 10^4, and 10^6. What relationship do you notice between the exponent and the number of zeros in the result?

    b. Evaluate 10^5. Is the result what you expected? How many zeros are in the result?

c. Evaluate 10^0. Is the result what you expected? How many zeros are in the result?

16. a. Now use the exponent key to evaluate the following powers of 10: 10^{-1}, 10^{-2}, 10^{-3}, 10^{-4}, and 10^{-6}. What relationship do you notice between the exponent and the number of zeros in the result?

b. Evaluate 10^{-5}. Is the result what you expected? How many zeros are in the result?

17. a. Because 5000 is 5×1000, the number 5000 can be written as 5×10^3. Use your calculator to verify this by evaluating 5×10^3.

b. There is another way to evaluate a number times a power of 10, such as 5×10^3, on your calculator. Find the key labeled (EE) or (E) or (EXP) ; sometimes it is a second function. This key takes the place of the (10) (^) keys. To evaluate 5×10^3, enter (5), press the (EE) key, then enter (3). Try this now and verify the result.

> When a number is written as the product of a decimal number between 1 and 10 and a power of 10, it is expressed in **scientific notation.**

For example, 423 is expressed in scientific notation as 4.23×10^2.

The (EE) or (E) key on your calculator is used to enter a number in scientific notation. For example,

$$423 = 4.23 \times 100 = 4.23 \times 10^2 \text{ and can be entered as } 4.23 \text{(EE)}\, 02$$

$$147{,}000 = 1.47 \times 100{,}000 = 1.47 \times 10^5 \text{ and can be entered as } 1.47 \text{(EE)}\, 05$$

18. a. Input 4.23 (EE) 2 into your calculator, press the (ENTER) or (=) key, and see if you obtain the result that you expect.

b. Input 1.47 (EE) 5 into your calculator, press the (ENTER) or (=) key, and see if you obtain the result that you expect.

One advantage of using scientific notation becomes evident when you have to work with very large numbers or very small numbers.

19. a. The average distance from Earth to the Sun is 93,000,000 miles. Write this number in words and in scientific notation. Then enter it into your calculator using the (EE) key.

b. Estimate 5 times the average distance from Earth to the Sun.

c. Now use your calculator to verify your estimate from part a. Write the answer in standard form, in scientific notation, and in words.

20. There are heavenly bodies that are thousands of times farther away from Earth than the Sun is. Multiply 93,000,000 miles by 1000. Write the result in scientific notation, in standard form, and in words.

21. a. You can approximate the distance that Earth travels in one orbit around the sun by multiplying 93,000,000 miles by $2 \times \pi$. (The Greek letter π, pronounced "pie," represents a specific number that, when rounded to the nearest hundredth, is 3.14). Use the (π) button on your calculator to calculate this distance.

b. In part a, you calculated the circumference of a circle whose radius is 93,000,000 miles. The calculator gives a result with many digits. Do you think it makes sense to report the distance with that much accuracy? Explain.

22. Use your calculator to perform each of the following calculations.

   a. $12 - 2 \cdot (8 - 2 \cdot 3) + 3^2$ **b.** $5^4 + 2 \cdot 4^0$

c. $\dfrac{128}{16 - 2^3}$

d. $3.26 \times 10^4 \cdot 5.87 \times 10^3$.

Enter the numbers in scientific notation.

e. $10\pi \div (0.25\pi)$

f. $25 \times 10^{-2} + 750 \times 10^{-3}$

SUMMARY

1. If the order in which two numbers are added is reversed, the sum remains the same. This property is called the **commutative property** of addition and can be written symbolically as

$$a + b = b + a.$$

2. The **distributive property** of multiplication over addition or subtraction is written symbolically as

$$c \cdot (a + b) = c \cdot a + c \cdot b \text{ for addition}$$

or

$$c \cdot (a - b) = c \cdot a - c \cdot b \text{ for subtraction.}$$

3. The **Order of Operations** priority convention for parentheses, addition, subtraction, multiplication, division and exponentiation is:

   a. First priority: operations contained within parentheses (performed according to the accepted priority convention).

   b. Second priority: exponentiation.

   c. Third priority: multiplication or division (whichever comes first when read from left to right).

   d. Fourth priority: addition or subtraction (whichever comes first when read from left to right).

4. Any number, except zero, raised to the zero power equals 1. The symbol 0^0 is undefined.

5. A number is expressed in **scientific notation** when it is written as the product of a decimal number between 1 and 10 and a power of 10.

EXERCISES

1. The following numbers are written in standard notation. Convert each number to scientific notation.

   a. 213 040 000 000

   b. 0.000041324

    c. 555 140 500 000 000　　　　　　**d.** 0.00000000000213749

2. The following numbers are written in scientific notation. Convert each number to standard notation.

    a. 4.532×10^{11}　　　　　　　　**b.** 2.162×10^{-3}

    c. 4.532×10^{7}　　　　　　　　　**d.** 4.532×10^{-7}

3. Mentally evaluate each of the following expressions by performing the operations in the appropriate order. Use your calculator to check your results.

    a. $6 + 18 \div 3 \cdot 4$　　　　　　　　**b.** $5 \cdot 2^{3} - 6 \cdot 2 + 5$

    c. $48 \div 6 + 2 \cdot 2$　　　　　　　　**d.** $5 + 5 \cdot 3 - 2 \cdot 2^{2}$

    e. $5^{3} \cdot 5^{5}$　　　　　　　　　　　**f.** $500 \div 25 \cdot 2 - 3 \cdot 2$

    g. $\dfrac{7 - 3 \cdot 2}{5}$　　　　　　　　　**h.** $\dfrac{6}{8 - 2 \cdot 3}$

4. Solve the following problems by first changing the numbers to scientific notation and then performing the appropriate operations.

   a. The distance that light travels in 1 second is 186,000 miles. How far will light
   travel in 1 year? This distance is called a light-year. (There are approximately 31,500,000 seconds in 1 year.)

   b. The total area of the oceans of the world is about 140,000,000 square miles.
   An acre is about 0.00156 square mile. Determine the number of acres in
   the oceans of the world by dividing 140,000,000 by 0.00156. Express your
   answer in standard form and in words.

Target Heart Rate
Topics: *Algorithms, Percents*

NOTE: In preparation for this activity, you need to find your resting heart rate by taking your pulse for 60 seconds before getting out of bed in the morning.

You have joined a fitness center to get into shape and to lose a few pounds. You have chosen to use the Nautilus equipment as well as the aerobic machines. To maximize the benefit from your workout, you must increase and maintain your heart rate within a targeted range for a period of 10 to 20 minutes. Your instructor explains two different methods for determining this range.

1. Use method 1 to compute your target heart-rate range.

DETERMINING YOUR TARGET HEART RATE

Method 1

   a. Calculate your resting heart rate by taking your pulse for 60 seconds before getting out of bed in the morning:

   Resting heart rate = _____ beats per minute

   b. Calculate your approximate maximum heart rate:

   $220 - \text{age}$ = _____ maximum heart rate

   c. Calculate your approximate working heart rate:

   Maximum heart rate − resting heart rate = _____ working heart rate

   d. Calculate your approximate lower heart-rate limit by multiplying your approximate working heart rate by 60% and then adding the result to your resting heart rate:

   Working heart rate · 60% + resting
   heart rate = _____ lower heart-rate limit

   e. Calculate your approximate upper heart-rate limit by multiplying your approximate working heart rate by 80%, then add the result to your resting heart rate.

   Working heart rate · 80% + resting
   heart rate = _____ upper heart-rate limit

   f. Your target heart-rate range during aerobic exercise is defined by your lower and upper heart-rate limits:

   Target heart-rate range is from _____ (lower heart-rate limit)
   to _____ (upper heart-rate limit)

Source: YMCA

2. Use method 2 to compute your target heart-rate range.

DETERMINING YOUR TARGET HEART RATE

Method 2

   a. Calculate your approximate maximum heart rate:

   $220 - \text{age} = $ _____ beats per minute   maximum heart rate

   b. Calculate your lower heart-rate limit by multiplying your maximum heart rate by 70%

   Maximum heart rate $\cdot$ 70% = _____   lower heart-rate limit

   c. Calculate your upper heart-rate limit by multiplying your maximum heart rate by 85%:

   Maximum heart rate $\cdot$ 85% = _____   upper heart-rate limit

   d. Your target heart-rate range during aerobic exercise is defined by your lower and upper heart-rate limits:

   Target heart-rate range is from _____ (lower heart-rate limit)

   to _____ (upper heart-rate limit)

Source: YMCA

---

3. Compare the two methods. Which one do you think is more accurate? Explain in complete sentences.

4. When exercising, you take your pulse for only 10 seconds, not an entire minute. Use method 1 to compute your 10-second heart-rate range.

5. Why do the calculations in method 2 suggest that a person's target heart-rate range decreases as he or she gets older?

Income and Expenses
Topics: *Tables, Percents,*
Bar Graphs

You have just completed your two-year degree and are starting your first job as a salesperson at a sporting goods store. Your income consists of a base salary of $8.00 per hour, with time and a half for more than 40 hours per week and double time for holidays. You also receive a 3% commission on total sales.

1. Based on a 5-day week, 8-hour day, and 52-week year, determine your annual gross base salary.

2. You are paid biweekly. What is your gross base salary per pay period?

3. To determine your net (take-home) biweekly pay, the following must be deducted from your gross income: Social Security, Medicare, federal and state taxes, and union dues.

   a. If Social Security is 6.2% of gross income, determine the amount of your Social Security deduction.

   b. If Medicare is 1.45% of gross income, determine your Medicare deduction.

   c. Use the charts on pages 32–34 to find the federal and state biweekly deductions for a single person (claiming zero for withholding allowances).

   d. Ten dollars per pay period for union dues is also deducted. Determine the total of all deductions.

   e. What is your net base salary per pay period?

4. During the first pay period of a certain month, you work 10 days. Two of these days are holidays. During the second pay period, you work 10 days, and on four occasions you work overtime accumulating $2\frac{1}{4}$, $4\frac{2}{3}$, $5\frac{1}{2}$, and $3\frac{3}{4}$ hours overtime. Calculate your net pay for these pay periods by filling in the accompanying table.

SINGLE Persons—BIWEEKLY Payroll Period   FEDERAL

If the wages are—		And the number of withholding allowances claimed is—										
At least	But less than	0	1	2	3	4	5	6	7	8	9	10
		The amount of income tax to be withheld is—										
$0	$105	0	0	0	0	0	0	0	0	0	0	0
105	110	1	0	0	0	0	0	0	0	0	0	0
110	115	2	0	0	0	0	0	0	0	0	0	0
115	120	2	0	0	0	0	0	0	0	0	0	0
120	125	3	0	0	0	0	0	0	0	0	0	0
125	130	4	0	0	0	0	0	0	0	0	0	0
130	135	5	0	0	0	0	0	0	0	0	0	0
135	140	5	0	0	0	0	0	0	0	0	0	0
140	145	6	0	0	0	0	0	0	0	0	0	0
145	150	7	0	0	0	0	0	0	0	0	0	0
150	155	8	0	0	0	0	0	0	0	0	0	0
155	160	8	0	0	0	0	0	0	0	0	0	0
160	165	9	0	0	0	0	0	0	0	0	0	0
165	170	10	0	0	0	0	0	0	0	0	0	0
170	175	11	0	0	0	0	0	0	0	0	0	0
175	180	11	0	0	0	0	0	0	0	0	0	0
180	185	12	0	0	0	0	0	0	0	0	0	0
185	190	13	0	0	0	0	0	0	0	0	0	0
190	195	14	0	0	0	0	0	0	0	0	0	0
195	200	14	0	0	0	0	0	0	0	0	0	0
200	205	15	0	0	0	0	0	0	0	0	0	0
205	210	16	0	0	0	0	0	0	0	0	0	0
210	215	17	0	0	0	0	0	0	0	0	0	0
215	220	17	1	0	0	0	0	0	0	0	0	0
220	225	18	2	0	0	0	0	0	0	0	0	0
225	230	19	3	0	0	0	0	0	0	0	0	0
230	235	20	3	0	0	0	0	0	0	0	0	0
235	240	20	4	0	0	0	0	0	0	0	0	0
240	245	21	5	0	0	0	0	0	0	0	0	0
245	250	22	6	0	0	0	0	0	0	0	0	0
250	260	23	7	0	0	0	0	0	0	0	0	0
260	270	24	8	0	0	0	0	0	0	0	0	0
270	280	26	10	0	0	0	0	0	0	0	0	0
280	290	27	11	0	0	0	0	0	0	0	0	0
290	300	29	13	0	0	0	0	0	0	0	0	0
300	310	30	14	0	0	0	0	0	0	0	0	0
310	320	32	16	0	0	0	0	0	0	0	0	0
320	330	33	17	1	0	0	0	0	0	0	0	0
330	340	35	19	3	0	0	0	0	0	0	0	0
340	350	36	20	4	0	0	0	0	0	0	0	0
350	360	38	22	6	0	0	0	0	0	0	0	0
360	370	39	23	7	0	0	0	0	0	0	0	0
370	380	41	25	9	0	0	0	0	0	0	0	0
380	390	42	26	10	0	0	0	0	0	0	0	0
390	400	44	28	12	0	0	0	0	0	0	0	0
400	410	45	29	13	0	0	0	0	0	0	0	0
410	420	47	31	15	0	0	0	0	0	0	0	0
420	430	48	32	16	0	0	0	0	0	0	0	0
430	440	50	34	18	2	0	0	0	0	0	0	0
440	450	51	35	19	3	0	0	0	0	0	0	0
450	460	53	37	21	5	0	0	0	0	0	0	0
460	470	54	38	22	6	0	0	0	0	0	0	0
470	480	56	40	24	8	0	0	0	0	0	0	0
480	490	57	41	25	9	0	0	0	0	0	0	0
490	500	59	43	27	11	0	0	0	0	0	0	0
500	520	61	45	29	13	0	0	0	0	0	0	0
520	540	64	48	32	16	0	0	0	0	0	0	0
540	560	67	51	35	19	3	0	0	0	0	0	0
560	580	70	54	38	22	6	0	0	0	0	0	0
580	600	73	57	41	25	9	0	0	0	0	0	0
600	620	76	60	44	28	12	0	0	0	0	0	0
620	640	79	63	47	31	15	0	0	0	0	0	0
640	660	82	66	50	34	18	1	0	0	0	0	0
660	680	85	69	53	37	21	4	0	0	0	0	0
680	700	88	72	56	40	24	7	0	0	0	0	0
700	720	91	75	59	43	27	10	0	0	0	0	0
720	740	94	78	62	46	30	13	0	0	0	0	0
740	760	97	81	65	49	33	16	0	0	0	0	0
760	780	100	84	68	52	36	19	3	0	0	0	0
780	800	103	87	71	55	39	22	6	0	0	0	0

SINGLE Persons—BIWEEKLY Payroll Period   FEDERAL

If the wages are–		And the number of withholding allowances claimed is—										
At least	But less than	0	1	2	3	4	5	6	7	8	9	10
		The amount of income tax to be withheld is—										
$800	$820	106	90	74	58	42	25	9	0	0	0	0
820	840	109	93	77	61	45	28	12	0	0	0	0
840	860	112	96	80	64	48	31	15	0	0	0	0
860	880	115	99	83	67	51	34	18	2	0	0	0
880	900	118	102	86	70	54	37	21	5	0	0	0
900	920	121	105	89	73	57	40	24	8	0	0	0
920	940	124	108	92	76	60	43	27	11	0	0	0
940	960	127	111	95	79	63	46	30	14	0	0	0
960	980	130	114	98	82	66	49	33	17	1	0	0
980	1,000	133	117	101	85	69	52	36	20	4	0	0
1,000	1,020	136	120	104	88	72	55	39	23	7	0	0
1,020	1,040	139	123	107	91	75	58	42	26	10	0	0
1,040	1,060	142	126	110	94	78	61	45	29	13	0	0
1,060	1,080	145	129	113	97	81	64	48	32	16	0	0
1,080	1,100	151	132	116	100	84	67	51	35	19	3	0
1,100	1,120	156	135	119	103	87	70	54	38	22	6	0
1,120	1,140	162	138	122	106	90	73	57	41	25	9	0
1,140	1,160	167	141	125	109	93	76	60	44	28	12	0
1,160	1,180	173	144	128	112	96	79	63	47	31	15	0
1,180	1,200	179	149	131	115	99	82	66	50	34	18	2
1,200	1,220	184	154	134	118	102	85	69	53	37	21	5
1,220	1,240	190	160	137	121	105	88	72	56	40	24	8
1,240	1,260	195	165	140	124	108	91	75	59	43	27	11
1,260	1,280	201	171	143	127	111	94	78	62	46	30	14
1,280	1,300	207	177	146	130	114	97	81	65	49	33	17
1,300	1,320	212	182	152	133	117	100	84	68	52	36	20
1,320	1,340	218	188	158	136	120	103	87	71	55	39	23
1,340	1,360	223	193	163	139	123	106	90	74	58	42	26
1,360	1,380	229	199	169	142	126	109	93	77	61	45	29
1,380	1,400	235	205	174	145	129	112	96	80	64	48	32
1,400	1,420	240	210	180	150	132	115	99	83	67	51	35
1,420	1,440	246	216	186	155	135	118	102	86	70	54	38
1,440	1,460	251	221	191	161	138	121	105	89	73	57	41
1,460	1,480	257	227	197	167	141	124	108	92	76	60	44
1,480	1,500	263	233	202	172	144	127	111	95	79	63	47
1,500	1,520	268	238	208	178	148	130	114	98	82	66	50
1,520	1,540	274	244	214	183	153	133	117	101	85	69	53
1,540	1,560	279	249	219	189	159	136	120	104	88	72	56
1,560	1,580	285	255	225	195	164	139	123	107	91	75	59
1,580	1,600	291	261	230	200	170	142	126	110	94	78	62
1,600	1,620	296	266	236	206	176	145	129	113	97	81	65
1,620	1,640	302	272	242	211	181	151	132	116	100	84	68
1,640	1,660	307	277	247	217	187	157	135	119	103	87	71
1,660	1,680	313	283	253	223	192	162	138	122	106	90	74
1,680	1,700	319	289	258	228	198	168	141	125	109	93	77
1,700	1,720	324	294	264	234	204	173	144	128	112	96	80
1,720	1,740	330	300	270	239	209	179	149	131	115	99	83
1,740	1,760	335	305	275	245	215	185	155	134	118	102	86
1,760	1,780	341	311	281	251	220	190	160	137	121	105	89
1,780	1,800	347	317	286	256	226	196	166	140	124	108	92
1,800	1,820	352	322	292	262	232	201	171	143	127	111	95
1,820	1,840	358	328	298	267	237	207	177	147	130	114	98
1,840	1,860	363	333	303	273	243	213	183	152	133	117	101
1,860	1,880	369	339	309	279	248	218	188	158	136	120	104
1,880	1,900	375	345	314	284	254	224	194	164	139	123	107
1,900	1,920	380	350	320	290	260	229	199	169	142	126	110
1,920	1,940	386	356	326	295	265	235	205	175	145	129	113
1,940	1,960	391	361	331	301	271	241	211	180	150	132	116
1,960	1,980	397	367	337	307	276	246	216	186	156	135	119
1,980	2,000	403	373	342	312	282	252	222	192	161	138	122
2,000	2,020	408	378	348	318	288	257	227	197	167	141	125
2,020	2,040	414	384	354	323	293	263	233	203	173	144	128
2,040	2,060	419	389	359	329	299	269	239	208	178	148	131
2,060	2,080	425	395	365	335	304	274	244	214	184	154	134
2,080	2,100	431	401	370	340	310	280	250	220	189	159	137

$2,100 and over   Use Table 2(a) for a **SINGLE person** on page 34. Also see the instructions on page 32.

NY State—Single Persons—Biweekly Payroll Period

WAGES		EXEMPTIONS CLAIMED										10
At Least	But Less Than	0	1	2	3	4	5	6	7	8	9	or more
		TAX TO BE WITHHELD										
$0	$200	$0.00										
200	210	0.00										
210	220	0.00										
220	230	0.00	$0.00									
230	240	0.00	0.00									
240	250	0.00	0.00									
250	260	0.00	0.00									
260	270	0.00	0.00	$0.00								
270	280	0.30	0.00	0.00								
280	290	0.70	0.00	0.00								
290	300	1.10	0.00	0.00								
300	320	1.70	0.10	0.00	$0.00							
320	340	2.50	0.90	0.00	0.00							
340	360	3.30	1.70	0.20	0.00	$0.00						
360	380	4.10	2.50	1.00	0.00	0.00						
380	400	4.90	3.30	1.80	0.30	0.00	$0.00					
400	420	5.70	4.10	2.60	1.10	0.00	0.00					
420	440	6.50	4.90	3.40	1.90	0.30	0.00	$0.00				
440	460	7.30	5.70	4.20	2.70	1.10	0.00	0.00	$0.00			
460	480	8.10	6.50	5.00	3.50	1.90	0.40	0.00	0.00			
480	500	8.90	7.30	5.80	4.30	2.70	1.20	0.00	0.00	$0.00		
500	520	9.70	8.10	6.60	5.10	3.50	2.00	0.40	0.00	0.00		
520	540	10.50	8.90	7.40	5.90	4.30	2.80	1.20	0.00	0.00	$0.00	
540	560	11.30	9.70	8.20	6.70	5.10	3.60	2.00	0.50	0.00	0.00	
560	580	12.10	10.50	9.00	7.50	5.90	4.40	2.80	1.30	0.00	0.00	$0.00
580	600	12.90	11.30	9.80	8.30	6.70	5.20	3.60	2.10	0.60	0.00	0.00
600	620	13.80	12.10	10.60	9.10	7.50	6.00	4.40	2.90	1.40	0.00	0.00
620	640	14.70	13.00	11.40	9.90	8.30	6.80	5.20	3.70	2.20	0.60	0.00
640	660	15.60	13.90	12.20	10.70	9.10	7.60	6.00	4.50	3.00	1.40	0.00
660	680	16.50	14.80	13.10	11.50	9.90	8.40	6.80	5.30	3.80	2.20	0.70
680	700	17.40	15.70	14.00	12.30	10.70	9.20	7.60	6.10	4.60	3.00	1.50
700	720	18.50	16.60	14.90	13.10	11.50	10.00	8.40	6.90	5.40	3.80	2.30
720	740	19.50	17.50	15.80	14.00	12.30	10.80	9.20	7.70	6.20	4.60	3.10
740	760	20.60	18.60	16.70	14.90	13.20	11.60	10.00	8.50	7.00	5.40	3.90
760	780	21.60	19.60	17.60	15.80	14.10	12.40	10.80	9.30	7.80	6.20	4.70
780	800	22.80	20.70	18.60	16.70	15.00	13.30	11.60	10.10	8.60	7.00	5.50
800	820	24.00	21.70	19.70	17.70	15.90	14.20	12.50	10.90	9.40	7.80	6.30
820	840	25.20	22.90	20.70	18.70	16.80	15.10	13.40	11.70	10.20	8.60	7.10
840	860	26.40	24.10	21.80	19.80	17.80	16.00	14.30	12.50	11.00	9.40	7.90
860	880	27.50	25.30	23.00	20.80	18.80	16.90	15.20	13.40	11.80	10.20	8.70
880	900	28.70	26.50	24.20	21.90	19.90	17.80	16.10	14.30	12.60	11.00	9.50
900	920	29.90	27.60	25.40	23.10	20.90	18.90	17.00	15.20	13.50	11.80	10.30
920	940	31.10	28.80	26.50	24.30	22.00	19.90	17.90	16.10	14.40	12.70	11.10
940	960	32.30	30.00	27.70	25.50	23.20	21.00	19.00	17.00	15.30	13.60	11.90
960	980	33.40	31.20	28.90	26.60	24.40	22.10	20.00	18.00	16.20	14.50	12.70
980	1,000	34.60	32.40	30.10	27.80	25.50	23.30	21.10	19.00	17.10	15.40	13.60
1,000	1,020	35.80	33.50	31.30	29.00	26.70	24.50	22.20	20.10	18.10	16.30	14.50
1,020	1,040	37.00	34.70	32.40	30.20	27.90	25.60	23.40	21.10	19.10	17.20	15.40
1,040	1,060	38.30	35.90	33.60	31.40	29.10	26.80	24.50	22.30	20.20	18.20	16.30
1,060	1,080	39.60	37.10	34.80	32.50	30.30	28.00	25.70	23.50	21.20	19.20	17.20
1,080	1,100	41.00	38.40	36.00	33.70	31.40	29.20	26.90	24.60	22.40	20.30	18.20
1,100	1,120	42.40	39.80	37.20	34.90	32.60	30.40	28.10	25.80	23.50	21.30	19.30
1,120	1,140	43.80	41.10	38.50	36.10	33.80	31.50	29.30	27.00	24.70	22.50	20.30
1,140	1,160	45.10	42.50	39.90	37.30	35.00	32.70	30.40	28.20	25.90	23.60	21.40
1,160	1,180	46.50	43.90	41.20	38.60	36.20	33.90	31.60	29.40	27.10	24.80	22.50
1,180	1,200	47.90	45.20	42.60	40.00	37.30	35.10	32.80	30.50	28.30	26.00	23.70
1,200	1,220	49.20	46.60	44.00	41.30	38.70	36.30	34.00	31.70	29.40	27.20	24.90
1,220	1,240	50.60	48.00	45.30	42.70	40.10	37.40	35.20	32.90	30.60	28.40	26.10
1,240	1,260	52.00	49.30	46.70	44.10	41.40	38.80	36.30	34.10	31.80	29.50	27.30
1,260	1,280	53.30	50.70	48.10	45.40	42.80	40.20	37.50	35.30	33.00	30.70	28.40
1,280	1,300	54.70	52.10	49.50	46.80	44.20	41.50	38.90	36.40	34.20	31.90	29.60
1,300	3,460	6.85% (.0685) of the excess over $1,300 plus:										
		55.40	52.80	50.10	47.50	44.90	42.20	39.60	37.00	34.80	32.50	30.20
$3,460 & OVER		Use Method II, "Exact Calculation Method," on page T-13 of this booklet										

REMINDER: Use 6.2% for Social Security, 1.45% for Medicare, and the tables for federal and state taxes. Don't forget the union dues.

	FIRST PAY PERIOD	SECOND PAY PERIOD
GROSS BASE		
GROSS OT/HOL.		
TOTAL GROSS		
FEDERAL INCOME TAX		
STATE INCOME TAX		
SOCIAL SECURITY TAX		
MEDICARE		
UNION DUES		
TOTAL NET		

5. You sell $10,000 worth of merchandise during each pay period.

   a. What is your total commission?

   b. Your commission is paid to you in a separate paycheck. Calculate your take-home pay for commission for this month. (Remember that federal and state tax tables are based on *biweekly* payroll periods.)

   c. What is your total take-home pay for the month, including commission?

   d. What percent of your gross income this month is from commission?

   e. What percent of your gross income this month comes from overtime/holiday hours?

f. Create a bar graph to show a graphical view of your month's total gross income. Include these three categories: gross overtime/holiday, gross commission, and gross base salary.

6. a. You want to purchase a new TV that costs $549 plus 7% sales tax. You plan to save your overtime pay to purchase the TV outright and not pay monthly installments on the bill. Estimate the number of hours you need to work to yield enough net overtime pay to purchase the TV.

b. Your overtime hours for the month in Problem 4 are a reasonable estimate of the hours you work each month. How many months of saving will it take for you to purchase the TV?

7. You are trying not to go any further into debt, and you want to pay off some of your college debts. Each month you analyze how you spend your money and categorize your expenses under the following headings: (1) household, (2) medical, (3) entertainment, (4) loans, (5) insurance, (6) personal, and (7) miscellaneous.

Your actual expenses incurred for one month are shown in the following list. The number in parentheses indicates the category of the expense.

car payment (4), $125	rent (1), $300
car insurance (5), $72	phone (1), $60
parking (7), $50	movies (3), $20
bowling (3), $60	towels (1), $22
shoes (6), $43	student loan (4), $60
dentist (2), $30	film (7), $5
groceries (1), $150	savings (6), $100
credit-card debt (4), $50	medicines (2), $47
bank charges (7), $7	clothes (6), $110
gas/car (7), $80	

If computers are accessible, create a spreadsheet; otherwise, use the accompanying table to group the information. List the given expenses in the appropriate columns in the table.

HOUSEHOLD (1)	MEDICAL (2)	ENTERTAINMENT (3)	LOANS (4)	INSURANCE (5)	PERSONAL (6)	MISC. (7)

8. Do you think that the expenses you incurred this month will be consistent with other months' expenses? Explain.

9. What are the total expenses for the month?

10. What percent of the total is used for household expenses? What percent is for entertainment? Create a bar graph showing the percentages for all of the categories.

11. Will you have enough income to cover your expenses this month? Explain.

12. You are considering a move to another apartment that has two bedrooms. How much more do you think you can afford to pay, based on your calculated income and expenses? Will the move be possible?

What Have I Learned?

The activities in this cluster gave you an opportunity to use mathematics to solve problems in several different contexts. Fractions, decimals, and percents were involved, as well as tables, bar graphs, and calculators. If you need to brush up on some of these skills, you can find many practice problems in the Appendices. Apply the skills you used in this cluster to solve the following problem.

1. You are shopping for new clothes for the semester and decide to buy a sweater priced at $63, shoes for $74, and pants priced at $39. All items in the store are on sale at 30% off the tag price. If the total cost of your purchase after the discount is more than $100, you may deduct an additional $25. The 7% sales tax is computed on the final cost after all the discounts have been taken. You have $100 to spend.

   a. Organize these data in a way that makes sense to you.

   b. Will $100 cover your purchases? Explain in complete sentences. Include the calculations that you used to determine your answer.

2. George Polya's book *How to Solve It* outlines a four-step process for solving problems.

   1. Understand the problem (see clearly what is involved).
   2. Make a plan (look for connections to obtain the idea of a solution).
   3. Carry out the plan.
   4. Look back at the completed solution (review and discuss it).

Exercise numbers appearing in color are answered in the Selected Answers section of this book.

Describe how your procedures in Problem 1 correspond with Polya's suggestions.

How Can I Practice?

1. **a.** Estimate the average temperature in your hometown for each month from January to December.

MONTH	AVERAGE TEMPERATURE
January	
February	
March	
April	
May	
June	
July	
August	
September	
October	
November	
December	

b. Draw a bar graph to represent your data.

2. a. Write four million three hundred thousand forty-two in standard form.

b. Write 12,578 in words.

3. Use pencil and paper to perform the following calculations. Then use your calculator to verify your results.

a. $4 \cdot (6 + 3) - 9 \cdot 2$

b. $5 \cdot 9 \div 3 - 3 \cdot 4 \div 6$

c. $2 + 3 \cdot 4^3$

d. $\dfrac{256}{8 + 6^2}$

4. a. Write 214,000,000,000 in scientific notation.

b. Write 7.83×10^4 in standard notation.

5. The sales tax in Erie County, New York, is 8%. Determine the cost of a new car, including sales tax, if the sticker price is $12,073.

6. The summer of 1999 was exceptionally hot and dry throughout the eastern seaboard and Midwest of the United States. The bad news was that crop yields for corn were poor, but the good news was that mosquitoes were scarce. During July 1999, scientists in Monmouth County, New Jersey, counted 1711 mosquitoes in their traps, which was only about one-fifth of the typical July catch they averaged in the previous six years. About how many mosquitoes were trapped in Monmouth County during a typical July?

7. A newly discovered binary star, Shuart1, is located 185 light-years from Earth. One light-year is 9,460,000,000,000 kilometers.

   a. Express the distance to Shuart1 in kilometers. Write the answer in scientific notation.

   b. The speed of light in a vacuum is approximately 300,000 kilometers per second. Approximately how many years does it take light to travel to Earth from Shuart1? Assume that there are 365 days per year. Write your result in scientific notation.

8. A professional softball player was injured partway through the season. She played 75% of her games before the injury and 25% of them after she recuperated and returned to play. Her batting average was .420 in the first part of the season and .360 when she returned. What was her average for the entire season?

CLUSTER 2

Problem Solving with Fractions

ACTIVITY 1.8

How Much Do I Owe?
Topics: *Equivalent Fractions, Addition, Subtraction, and Multiplication of Fractions*

You decide to have friends over to watch some videos. After the first movie, you call the local sub shop to order three giant submarine sandwiches, which are on sale for $9.95 each (tax included, free delivery). When the subs are delivered, you pay the bill of $29.85, plus a $4.00 tip, and everyone agrees to reimburse you, depending on how much they eat.

Because some friends are hungrier than others, you cut one sub into three equal (large) parts, a second sub into six equal (medium) parts, and the third sub into twelve equal (small) parts.

Sub 1: Large pieces

Sub 2: Medium pieces

Sub 3: Small pieces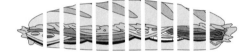

Because the first sub is divided into three equal parts, each part (large piece) represents $\frac{1}{3}$ of a giant sub.

1. **a.** What fraction of giant sub 2 does each medium piece represent? Explain.

b. What fraction of giant sub 3 does each small piece represent? Explain.

The following table represents the number and size of the portion(s) that each friend consumes. You will complete the remaining columns as you proceed through this activity.

NAME	LARGE PIECES ($\frac{1}{3}$ SUB)	MEDIUM PIECES ($\frac{1}{6}$ SUB)	SMALL PIECES ($\frac{1}{12}$ SUB)	FRACTIONAL PART OF SUB	AMOUNT OWED
Pete	1	0	0		
Joaquin	0	0	4		
Halima	0	2	0		
Leah	1	0	1		
Pat	0	2	0		
Marty	0	1	2		
Jennifer	0	1	1		
You	1	0	0		

2. What fractional part of a giant sub does Pete eat? Record your result in column 5 of the table.

3. In computing how much Halima owes, you notice that two medium pieces placed end to end measure the same as one large piece. What single fraction of a sub represents Halima's combined portion? Explain how you obtained your answer. Compare your method with those used by your classmates.

You could have used either of two different approaches to obtain your answer in Problem 3:

Add the fractions: $\frac{1}{6} + \frac{1}{6} = \frac{1+1}{6} = \frac{2}{6}$

Use multiplication to do repeated addition: $\frac{1}{6} + \frac{1}{6} = 2 \cdot \frac{1}{6} = \frac{2}{1} \cdot \frac{1}{6} = \frac{2 \cdot 1}{1 \cdot 6} = \frac{2}{6}$

Appendix

If you find you need some review with fractions and operations with fractions, refer to Appendix A, which contains several examples with solutions and practice exercises.

4. Because two medium-sized pieces equal one large piece, $\frac{2}{6}$ should be equivalent to $\frac{1}{3}$. Describe a procedure to write $\frac{2}{6}$ equivalently as $\frac{1}{3}$.

5. Use the procedure you described in Problem 4 to compare Joaquin's portion with Pete's piece. Use the following diagrams to help support your answer.

Pete's portion

Joaquin's portion

In Problems 4 and 5, you determined that the fractions $\frac{1}{3}$, $\frac{2}{6}$, and $\frac{4}{12}$ are **equivalent.** These fractional parts of a sub all represent the same portion. When $\frac{2}{6}$ and $\frac{4}{12}$ are written as $\frac{1}{3}$, the fractions $\frac{2}{6}$ and $\frac{4}{12}$ are said to be written in **reduced form** (lowest terms).

PROPERTY OF EQUIVALENT FRACTIONS

If the numerator and denominator of a fraction are both multiplied (or divided) by the same nonzero number, an **equivalent fraction** is obtained.

6. Write each fraction as an equivalent fraction in lowest terms.

   a. $\frac{10}{12}$ b. $\frac{18}{24}$

7. There are many fractions that are equivalent to $\frac{1}{3}$. For each of the following, determine the value of the missing numerator or denominator so that the resulting fraction is equivalent to $\frac{1}{3}$.

   a. $\dfrac{1}{3} = \dfrac{?}{15}$ b. $\dfrac{1}{3} = \dfrac{6}{?}$

   c. $\dfrac{1}{3} = \dfrac{?}{33}$ d. $\dfrac{1}{3} = \dfrac{90}{?}$

8. Leah has one large piece and one small piece of a sub. What single fraction of a sub represents Leah's combined portions? *Recall:* To add fractions, each fraction must have the same denominator.

9. You have expressed Pete's, Halima's, Joaquin's, and Leah's portions as a single fraction of a sub. Compute the fractional part for your remaining friends (and yourself!) and record your results in column 5 of the table following Problem 1. Show your work below.

10. a. Determine how much Pete owes. Remember to include the tip in the total cost of the order before calculating the cost of one sub. Round your final answer to the nearest cent.

b. Determine the amount each additional person owes, and record your answers in the last column of the table. Round your answers to the nearest cent.

c. Do you have enough money to reimburse yourself? Explain.

d. What fractional part of a sub is left over?

11. What total fractional part of the sandwich order did Pete, Joaquin, Halima and Marty eat?

12. Use the sandwich diagrams on page 42 to help answer the following.

a. How many sandwiches do Leah, Pete, Joaquin, and Marty consume altogether? Express this answer in twelfths.

b. The fraction that is your answer to part a is called an **improper fraction**. What characterizes improper fractions?

Appendix

c. You can also express your answer to part a as a **mixed number**. Explain how to convert between an improper fraction and a mixed number.

d. Express the number of sandwiches consumed by Marty, Leah, Jennifer, Halima, and Pete as a mixed number.

13. Suppose that you and Pete are the only ones who show up to eat the subs and you each eat the portions reported in the table. You divide the remaining portions into two equal parts to take home. How much do you each take home?

Appendix

14. Leah likes this method of cutting sandwiches and orders the same three subs the next day for her sorority gathering.

   a. The sorority sisters eat $1\frac{5}{12}$ subs. How much is left?

   b. Some friends arrive later and eat $\frac{5}{6}$ more. How much remains to feed to the dog?

E X E R C I S E S

1. Your stock goes up $\frac{1}{8}$ of a point (1 point = $1). If you own 220 shares, how much money do you make?

2. The winnings from a horse race are distributed among the owners. You own $\frac{4}{25}$ share of the horse, Mulligan, who won $235,000 at Rolling Hills Raceway. How much money will you receive for your share?

3. At the end of the semester, the bookstore will buy back books that will be used again in courses the next semester. Usually, they will give you $\frac{1}{6}$ the original cost of the book. If you spend $243 on books this semester and the bookstore will buy back all your books, how much money can you expect to receive?

4. You use $\frac{1}{3}$ of your take-home pay each week for rent, $\frac{1}{5}$ for food, and $\frac{1}{10}$ for insurance. What part of your paycheck is left for your other expenses and savings?

5. List at least five fractions that are equivalent to $\frac{3}{5}$. Explain or show why they are equivalent.

6. Perform the indicated operations.

   a. $\dfrac{4}{9} + \dfrac{7}{3}$

   b. $\dfrac{6}{7} - \dfrac{1}{4}$

   c. $\dfrac{3}{8} \cdot \dfrac{4}{9}$

   d. $\dfrac{7}{3} + \dfrac{4}{5}$

7. Your youngest brother, who is in elementary school, asks you why $\frac{3}{4}$ and $\frac{6}{8}$ are equivalent. How would you answer his question?

ACTIVITY 1.9

Fractions Invade Campus Life

Topics: *Adding, Subtracting, Multiplying, and Dividing Fractions*

Fractions are a part of everyday life, and many calculators support operations with fractions. In this activity, you will calculate with fractions that you are likely to encounter during your college career and beyond. Your goal is to become comfortable doing arithmetic operations with fractions manually, as well as with the aid of your calculator. Both ways of dealing with fractions are important for your future success in using math to solve problems.

1. In the course of a typical day (24 hours), you, as a student, expect to spend $\frac{1}{4}$ of the time sleeping, $\frac{1}{6}$ of the time in class, $\frac{1}{3}$ of the time studying, and $\frac{1}{8}$ of the time eating. What fraction of your day is left as "free time"? Explain how you arrived at your result.

2. As part of your life as a college student, you decide to try some baking. Your favorite muffin recipe calls for $2\frac{2}{3}$ cups of flour, 1 cup of sugar, $\frac{1}{2}$ cup of crushed cashews, and $\frac{5}{8}$ cup of milk, plus assorted spices. How many cups of mixture do you have?

Appendix

Go to Appendix A if you need help to review mixed numbers.

3. The syllabus for your history course contains the following information about the fractional parts that will be used to calculate your course grade.

Quiz average	$\frac{1}{5}$
Exam average	$\frac{1}{2}$
Final exam	$\frac{1}{5}$

The rest of your grade is based on in-class participation. What fraction of your course grade does in-class participation represent?

a. Show your manual calculations.

b. Explain how you used your calculator to verify your result.

4. You correctly answer $\frac{2}{3}$ of the 75 questions on your first psychology exam. How many questions do you answer incorrectly? Show your manual work. Explain the steps you followed using your calculator.

5. You are taking five 3-credit courses this semester. On the average, you spend 35 hours per week outside of class doing course work. You use about 10 of these 35 hours to work on math.

   a. What fraction of your study time is devoted to your math course?

   b. If you spend equal time on each course, what fraction of your study time should you allot to the math course?

6. You must take medicine in four equal doses each day. Each day's medicine comes in a single container and measures $3\frac{1}{5}$ tablespoons. How much medicine is in each dose?

7. The wall space for bookshelves in your dorm room is $4\frac{1}{2}$ feet across. A board you have measures $12\frac{2}{3}$ feet in length.

   a. Without calculating, roughly estimate how many shelves you can cut from the board.

   b. Check your estimate by calculating $12\frac{2}{3} \div 4\frac{1}{2}$.

   c. Did the calculation verify your estimate? Explain why or why not.

EXERCISES

Appendix

Recall that Appendix A contains examples of operations with fractions, including solutions and practice exercises.

1. The year that you enter college, your freshman class consists of 760 students. According to statistical studies, about $\frac{4}{7}$ of these students will actually graduate. Approximately how many of your classmates will receive their degree?

2. You rent an apartment for the academic year (two semesters) with three of your college friends. The rent for the entire academic year is $10,000. Each semester you receive a bill for your share of the rent. If you and your friends divide the rent equally, how much must you pay each semester?

3. Your residence hall has been designated a quiet building. This means that there is a no-noise rule from 10:00 P.M. every night to noon the next day. During what fraction of a 24-hour period is one allowed to make noise?

4. You would like to learn to play the harp but are concerned with time constraints. A friend of yours plays and for three consecutive days before a recital, she practices for $1\frac{1}{4}$ hours, $2\frac{1}{2}$ hours, and $3\frac{2}{3}$ hours. What is her total practice time before a recital?

5. You are planning a summer cookout and decide to serve $\frac{1}{4}$-pound hamburgers. If you buy $5\frac{1}{2}$ pounds of hamburger meat, how many burgers can you make?

6. Perform the indicated operations.

   a. $4\frac{2}{3} - 1\frac{6}{7}$

   b. $5\frac{1}{2} + 2\frac{1}{3}$

   c. $2\frac{1}{6} \cdot 4\frac{1}{2}$

   d. $2\frac{3}{7} + \frac{14}{5}$

   e. $\dfrac{4}{5} \div \dfrac{8}{3}$

   f. $4\frac{1}{5} \div \frac{10}{3}$

ACTIVITY 1.10

Delicious Recipes

Topics: *Adding, Subtracting, Multiplying, and Dividing Fractions*

The recipes in this activity are for foods to be served at a party. Use these recipes to help plan the party by answering the questions following the recipes. Reduce each fraction, using mixed numbers when appropriate.

NOTE: You may not have worked extensively with fraction operations for some time. Appendix A contains many practice problems involving adding, subtracting, multiplying, and dividing fractions.

Crab Supreme

4 small (6 oz) cans crab meat	2 dashes of Tabasco sauce
1 egg, hard-boiled and mashed	$3\frac{1}{2}$ tbsp chopped fresh chives
$\frac{1}{2}$ cup mayonnaise	$\frac{1}{4}$ tsp salt
$2\frac{1}{2}$ tbsp chopped onion	$\frac{1}{2}$ tsp garlic powder
$3\frac{2}{3}$ tbsp plain yogurt	1 tsp lemon juice

Drain and rinse crab in cold water. Mash crab and egg together. Add all remaining ingredients except chives. Stir well. Chill, top with chives, and serve with chips or crackers. SERVES 6.

---

1. Determine the ingredients for one-half of this recipe. Fill in the blanks below.

   _____ small (6 oz) cans crab   _____ dashes Tabasco

   _____ egg, hard boiled and mashed   _____ tbsp chives

   _____ cup(s) mayonnaise   _____ tsp salt

   _____ tbsp, chopped onion   _____ tsp garlic powder

   _____ tbsp, plain yogurt   _____ tsp lemon juice

2. List the ingredients needed for the crab recipe if 18 people attend the party.

   _____ small (6 oz) cans crab   _____ dashes Tabasco

   _____ egg, hard-boiled and mashed   _____ tbsp chives

   _____ cup(s) mayonnaise   _____ tsp salt

   _____ tbsp chopped onion   _____ tsp garlic powder

   _____ tbsp plain yogurt   _____ tsp lemon juice

3. If a container of yogurt holds 1 cup, how many batches of crab appetizer can you make with one container? (1 cup = 16 tbsp)

4. If each person drinks $2\frac{2}{3}$ cups of soda, how many cups of soda will be needed for 18 people?

Apple Crisp

4 cups tart apples $\frac{1}{3}$ cup softened butter
   peeled, cored, and sliced $\frac{1}{2}$ tsp salt
$\frac{2}{3}$ cup packed brown sugar $\frac{3}{4}$ tsp cinnamon
$\frac{1}{4}$ cup rolled oats $\frac{1}{8}$ tsp allspice or nutmeg
$\frac{1}{2}$ cup flour

Preheat oven to 375°. Place apples in a greased 8-inch square pan. Blend remaining ingredients until crumbly, and spread over the apples. Bake approximately 30 minutes uncovered, until the topping is golden and the apples are tender. SERVES 4.

---

5. List the ingredients needed for the apple crisp recipe if 18 people attend the party.

6. How many times would you need to fill a $\frac{2}{3}$-cup container to measure 4 cups of apples?

7. If it takes $\frac{3}{4}$ tsp of cinnamon to make one batch of apple crisp and you have only 6 tsp of cinnamon left in the cupboard, how many batches can you make?

EXERCISES

Use the two preceding recipes in this activity, as well as the following recipe for potato pancakes, to answer the questions.

Potato Pancakes

6 cups potato	$\frac{1}{3}$ cup flour
(pared and grated)	$3\frac{3}{8}$ tsp salt
9 eggs	$2\frac{1}{4}$ tbsp grated onion

Drain the potatoes well. Beat eggs and stir into the potatoes. Combine and sift the flour and salt, then stir in the onions. Add to the potato mixture. Shape into patties and sauté in hot fat. Best served hot with applesauce. MAKES 36 3-inch pancakes.

---

1. If you were to make one batch of each of the three recipes, how much salt would you need? How much flour? How much onion?

2. If you have 2 cups of flour in the cupboard before you start cooking for the party and you make one batch of each recipe, how much flour will be left in the cupboard?

What Have I Learned?

Spend some time to reflect on operations with fractions by answering the following questions.

1. To add or subtract fractions, they must be written in equivalent form with common denominators. However, to multiply or divide fractions you do not need a common denominator. Why is this reasonable?

2. The operation of division can be viewed from several different points of view. For example, $24 \div 3$ has at least two meanings:

   - 24 can be written as the sum of how many 3s?
   - If 24 is divided into 3 equal-sized parts, how large is each part?

   These interpretations can be applied to fractions as well as to whole numbers.

   a. Calculate $2 \div \frac{1}{2}$ by answering this question: 2 can be written as the sum of how many $\frac{1}{2}$s?

   b. Calculate $\frac{1}{5} \div 2$ by answering this question: If you divide $\frac{1}{5}$ into 2 equal parts, how large is each part?

   c. Do your answers to parts a and b agree with the results you would obtain by using the procedures for dividing fractions reviewed in this cluster? Explain.

How Can I Practice?

1. You are in a golf tournament and there is a prize for driving the green on the sixth hole, with the drive closest to the hole winning. You drive the ball to within 4 feet $2\frac{3}{8}$ inches of the hole and your nearest competitor is 4 feet $5\frac{1}{4}$ inches from the hole. By how many inches did you win?

2. One of your jobs as the assistant to a weather reporter is to determine the average thickness of the ice in a bay on the St. Lawrence River. Ice fishermen use this report to determine if the ice is safe for fishing. You must chop holes in five different areas, measure the thickness, and take the average. During the first week in January, you record the following measurements: $2\frac{3}{8}$, $5\frac{1}{2}$, $6\frac{3}{4}$, 4, and $5\frac{7}{8}$ inches. What do you report as the average thickness of the ice in this area? Do you think the ice is safe?

3. You and two others in your family will divide 120 shares of a computer stock left by a relative who died. The stock is worth $10\frac{9}{16}$ per share. If you decide to sell your share of the stock, how much money will you receive?

4. You are about to purchase a rug for your college dorm room. The rug's length is perfect for your room. The width of the rug you want to purchase is $6\frac{1}{2}$ feet. If you center the rug in the middle of your 10-foot-wide room, how much floor space will you have on each side of the rug?

5. A plumber has $12\frac{1}{2}$ feet of plastic pipe. He uses $3\frac{2}{3}$ feet for the sink line and $5\frac{3}{4}$ feet for the washing machine. How much does he have left? He needs approximately $3\frac{1}{2}$ feet for a disposal. Does he have enough pipe left for a disposal?

6. Perform the following operations.

   a. $\frac{5}{7} + \frac{2}{7}$

   b. $\frac{3}{4} + \frac{3}{8}$

   c. $\frac{3}{8} + \frac{1}{12}$

   d. $\frac{4}{5} + \frac{5}{6}$

   e. $\frac{1}{2} + \frac{3}{5} + \frac{4}{15}$

   f. $\frac{11}{12} - \frac{5}{12}$

   g. $\frac{7}{9} - \frac{5}{12}$

   h. $\frac{2}{3} - \frac{1}{4}$

   i. $\frac{7}{30} - \frac{3}{20}$

j. $\frac{4}{5} - \frac{3}{4} + \frac{1}{2}$

k. $\frac{3}{5} \cdot \frac{1}{2}$

l. $\frac{2}{3} \cdot \frac{7}{8}$

m. $\frac{15}{8} \cdot \frac{24}{5}$

n. $5 \cdot \frac{3}{10}$

o. $\frac{3}{8} \div \frac{3}{4}$

p. $8 \div \frac{1}{2}$

q. $\frac{5}{7} \div \frac{20}{21}$

r. $4\frac{5}{6} + 3\frac{2}{9}$

s. $12\frac{5}{12} - 4\frac{1}{6}$

t. $6\frac{2}{13} - 4\frac{7}{26}$

u. $2\frac{1}{4} \cdot 5\frac{2}{3}$

v. $6\frac{3}{4} \div 1\frac{2}{7}$

EXPLORING NUMERACY

Suppose that each of the 263,814,032 residents of the United States skipped one meal per week for a year. Estimate the number of hungry people in the world who could be fed three meals per day for a year.

ACTIVITY 1.11

Which One Is Better?

Topics: *Ratios, Equivalent Ratios, Proportions, Fractions, Decimals, Percent*

Comparisons and Proportional Reasoning

The following table summarizes Michael Jordan's statistics during the six games of the 1996 National Basketball Association (NBA) championship series.

GAME	POINTS	FIELD GOALS	FREE THROWS
1	28	9 out of 18	9 out of 10
2	29	9 out of 22	10 out of 16
3	36	11 out of 23	11 out of 11
4	23	6 out of 19	11 out of 13
5	26	11 out of 22	4 out of 5
6	22	5 out of 19	11 out of 12

1. What was his points-per-game average over the six-game series?

2. In which game did he score the most points?

3. In which game(s) did he score the most field goals? The most free throws?

Problem 3 focused on the *actual* number of Jordan's successful field goals and free throws in these six games. Another way of assessing Jordan's performance is to *compare* the number of successful shots to the total number of attempts for each game. This comparison gives you information on the *relative* success of his shooting. For example, in each of games 1 and 2, Jordan made 9 field goals. Relatively speaking, you could argue that he was more successful in game 1 because he made 9 out of 18 attempts; in game 2, he only made 9 out of 22 attempts.

4. Use the free-throw data from the six games to express Jordan's *relative* performance in the given comparison formats (verbal, fraction, division, and decimal). The data from the first game have been entered for you.

JORDAN'S RELATIVE FREE-THROW PERFORMANCE

	VERBAL	FRACTION	DIVISION	DECIMAL
GAME 1	9 out of 10	$\frac{9}{10}$	$9 \div 10$ or $10\overline{)9}$	0.90
GAME 2				
GAME 3				
GAME 4				
GAME 5				
GAME 6				

5. a. For which of the six games was his relative free-throw performance highest?

b. Which comparison format did you use to answer part a? Why?

6. For which of the six games was Jordan's *actual* free-throw performance the lowest?

7. For which of the six games was Jordan's *relative* free-throw performance the lowest?

When relative comparisons using quotients are made between different values or quantities of the same kind (e.g., number of baskets to number of baskets), the comparison is called a **ratio**. Ratios can be expressed in any of several forms—verbal, fraction, division, or decimal, as you saw in Problem 4.

Proportional reasoning, a critically important quantitative skill, is the ability to recognize when two ratios are equivalent, that is, when equivalent ratios represent the same relative performance level.

Two ratios are said to be **equivalent** if the ratios have equal numerical (e.g., decimal or fraction) values. The mathematical statement that two ratios are equivalent is called a **proportion.** In fraction form, the proportion is written $\frac{a}{b} = \frac{c}{d}$.

You can find *equivalent* ratios the same way you find equivalent fractions. For example, 3 out of 4 is equivalent to 6 out of 8, because $\frac{3}{4} \cdot \frac{2}{2} = \frac{6}{8}$.

8. Fill in the blanks in each of the following proportions.

a. 3 out of 4 is equivalent to _____ out of 12

b. 3 out of 4 is equivalent to _____ out of 32

c. 3 out of 4 is equivalent to _____ out of 100

d. Write the resulting proportion from part c using a fraction format.

9. a. Which of the following ratios are equivalent?

      i. 28 out of 40   **ii.** 175 out of 250   **iii.** 75 out of 100

  b. Explain the method you used to answer part a.

  c. Write each of the equivalent ratios from part a in fraction form.

  d. Determine the "reduced" form of the equivalent fractions from part c. What do you observe?

10. a. Explain why the following ratios are equivalent.

      i. 27 out of 75   **ii.** 63 out of 175   **iii.** 36 out of 100

  b. Write each ratio in fraction form.

  c. Determine the "reduced" form of the equivalent fractions from part b.

  d. With which of the equivalent ratios in part a do you feel most comfortable? Explain.

The number 100 is a very familiar quantity of comparison: There are 100 cents in a dollar and frequently 100 points on a test. Therefore, people feel most comfortable with a ratio such as 70 out of 100 or 36 out of 100.

> The phrase "out of 100" is commonly referred to by its Latin equivalent, *percent*. Per means "division" and cent means "100," so **percent** means "divide by 100."

Therefore, 70 out of 100 can be rephrased as 70 percent and written in the familiar notation 70%, which equals $70 \div 100 = \frac{70}{100} = 0.70$. Similarly, 36 out of 100 can be rephrased as 36 percent and written in the familiar notation $36\% = 36 \div 100 = \frac{36}{100} = 0.36$.

11. Complete the following table using Michael Jordan's field goal data from the beginning of the activity.

JORDAN'S RELATIVE FIELD GOAL PERFORMANCE

	VERBAL	FRACTION	DECIMAL	PERCENT
GAME 1	9 out of 18	$\frac{9}{18}$	0.50	50%
GAME 2				
GAME 3				
GAME 4				
GAME 5				
GAME 6				

SUMMARY

When comparisons using quotients are made between different quantities of the same kind, the comparison is called a **ratio.**

Ratios can be expressed verbally (4 out of 5), as a fraction $\left(\frac{4}{5}\right)$, as a division $(5\overline{)4})$, as a decimal (0.80), or as a percent (80%).

Two ratios are said to be if the ratios have equal numerical values. The mathematical statement that two ratios are equivalent is called a **proportion.**

Finding equivalent ratios is often accomplished by finding equivalent fractions.

EXERCISES

1. Complete the following table by representing each comparison in all four formats.

VERBAL	REDUCED FRACTION	DECIMAL	PERCENT
	$\frac{1}{3}$		
	$\frac{4}{5}$		
	$\frac{16}{25}$		
5 out of 8			
250 out of 600			
144 out of 48			
		0.75	
		0.375	
		0.6	
			40%
			0.25%
			500%

2. There are 1240 females out of 2200 freshmen at the local community college. Compare the number of females to the total number of freshmen in the following formats.

   a. As a fraction   **b.** As a decimal   **c.** As a percent

3. At the state university near the community college in Exercise 2, the freshman class consists of 1480 males and 1620 females. In which freshman class, the community college or the university, is the relative number of females larger? Explain your reasoning.

4. At competitive colleges, the admissions office often compares the number of students accepted to the total number of applications received. This comparison is known as the *selectivity index*. The admissions office also compares the number of students who actually attend to the number of students who have been accepted for admission. This comparison is known as the *yield*. Complete the following table to determine the selectivity index and yield (in percent format) for colleges A, B, and C.

	NUMBER OF APPLICANTS	NUMBER ACCEPTED	NUMBER ATTENDING	SELECTIVITY INDEX	YIELD (AS A %)
COLLEGE A	5500	3500	1000		
COLLEGE B	8500	4800	2100		
COLLEGE C	4200	3200	900		

Which college do you think is the most competitive? The least competitive? Explain.

5. Here are your scores on three graded assignments. On which assignment did you perform best?

   a. 25 out of 30   **b.** 30 out of 40   **c.** 18 out of 25

6. In a typical year, 3884 million kilograms of the plastic PVC is produced in the United States and 9 million kilograms is recycled. What percent of the PVC manufactured in this country is recycled?

7. Baseball batting averages are the ratio of hits to at bats. They are reported as three-digit decimals. Compare the batting average of three players with the given records.

   a. 24 hits out of 70 at bats   b. 35 hits out of 124 at bats

   c. 87 hits out of 273 at bats

8. In their championship 1999 season, the New York Yankees won 98 and lost 64 of their regular season games. What percent of the games played did they win?

9. The win-loss records of three pitchers are given below. Use the data to rank the three pitchers by their relative performances. Explain how you determined your answer. Compare your results and methods with those of your learning partners. Did you arrive at the same conclusion? Why or why not?

	WINS	LOSSES
BOB	14	14
DAN	26	11
TOM	17	7

ACTIVITY 1.12

Social Issues
Topics: *Fractions, Percent, Ratio and Proportion, Contingency Tables, Relative Frequency, Probability*

In an effort to increase the education level of their police officers, many municipalities are requiring new recruits to have at least a two-year college degree. A recent survey indicated that 1 out of 5 New York City (NYPD) police officers holds a four-year college degree.

1. There are approximately 41,000 NYPD officers. Based on the ratio above, estimate (i.e., without making an actual calculation) how many NYPD officers hold a four-year college degree. Explain how you made the estimation.

2. To calculate the number of four-year college degree holders more precisely, you can start with the proportion statement

$$1 \text{ out of } 5 \ = \ \underline{\ ?\ } \text{ out of } 41{,}000$$

   a. Rewrite this proportion in fraction form.

Appendix

   b. Solve the proportion using equivalent fractions. (Recall your work in Activity 1.11.)

There are several ways to solve proportion equations.

For example, consider the proportion

$$2 \text{ out of } 5 = \underline{\hspace{1cm}} \text{ out of } 6000$$

As in Problem 2, the proportion can be written in fraction form, this time as

$$\frac{2}{5} = \frac{n}{6000},$$

where n is the unknown quantity of the proportion. You can determine the value of the unknown numerator n by writing $\frac{2}{5}$ as an equivalent fraction with denominator 6000:

$$6000 \div 5 = 1200,$$

so,

$$\frac{2}{5} \cdot \frac{1200}{1200} = \frac{2400}{6000}.$$

Therefore, the unknown value is 2400.

Another method of solving proportions uses the fact that the two mathematical statements

$$\frac{a}{b} = \frac{c}{d} \quad \text{and} \quad a \cdot d = b \cdot c$$

Appendix

are equivalent.

The act of transforming the statement on the left (containing two fractions) into the statement on the right is customarily called **cross-multiplication** because the numerator of the first fraction is multiplied by the denominator of the second and the numerator of the second fraction is multiplied by the denominator of the first.

3. Use cross-multiplication to determine which pair(s) of fractions are equal.

  a. $\frac{11}{15}$ and $\frac{5}{8}$ **b.** $\frac{5}{9}$ and $\frac{65}{117}$ **c.** $\frac{3}{10}$ and $\frac{14}{45}$

You can use cross-multiplication to solve proportions. The proportion $\frac{2}{5} = \frac{n}{6000}$ is equivalent to $2 \cdot 6000 = 5 \cdot n$. You can then view the resulting equation, $12,000 = 5n$, as a scale whose arms are in balance. The equal sign can be thought of as the balance point.

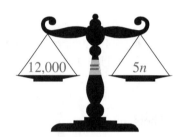

If you divide both sides of the equation by 5, the balance is maintained. Then the unknown number is isolated on one side of the equal sign and its value appears on the other side. The entire process can be written as follows:

Original proportion: $\dfrac{2}{5} = \dfrac{n}{6000}$

Cross-multiply: $2 \cdot 6000 = 5 \cdot n$

Divide both sides by 5: $\dfrac{2 \cdot 6000}{5} = \dfrac{\cancel{5} \cdot n}{\cancel{5}}$

Simplify: $2400 = n$

4. Use the cross-multiplication method to verify your result in Problem 2b.

In the original statement 2 out of 5 = _____ out of 6000, you are given a known ratio, $\frac{2}{5}$, and asked to determine what "part" of the total, 6000, will represent the same relative ratio, $\frac{2}{5}$. This "part" can be expressed mathematically as $\frac{2}{5}$ of 6000, and is calculated as

$$\frac{2}{5} \text{ times } 6000 = \frac{2}{5} \cdot 6000 = 2400.$$

5. Redo Problem 2b using the multiplication approach just described. Compare your results from all three methods. Which method do you prefer for Problem 2? Why?

6. Solve the following proportions.

    a. 2 out of 3 = _____ out of 36

    b. $\dfrac{2}{3} = \dfrac{n}{45}$

New York State has taken a leading position in raising the standards of its high school graduates. By the year 2003, every graduate will need to have passed a series of rigorous subject-matter tests called regents exams. Currently, in your cousin's county, only 6 out of 10 graduates receive regents diplomas.

7. If 5400 students in your cousin's county earned a regents diploma last year, estimate (without actually doing a calculation) the total number of high school graduates in that county last year.

This situation differs from the NYPD situation at the beginning of the activity because the total number is not known. The 5400 represents that part of the total number of high school graduates who earned a regents diploma. Written as a proportion,

$$6 \text{ out of } 10 = 5400 \text{ out of } \underline{\hspace{1cm}}$$

8. **a.** Rewrite this proportion in fraction form.

    b. To determine the total number of students, use cross-multiplication to calculate the unknown denominator in part a.

9. Solve these proportions.

   a. 2 out of 3 = 80 out of _____.

   b. $\frac{2}{3} = \frac{216}{n}$

   c. Tuition at a local community college is $1250 per term for a full-time student. This is 53% of the estimated cost of attending classes (tuition, books, transportation, lunch) published in the college catalog. What would be the total estimated cost for attending classes at the college?

Data are collected every day by agencies large and small. The Internal Revenue Service and the Congressional Budget Office collect data, as do college registrars. Data are usually organized into tables to help answer questions and make decisions. When conjectures are made based on data, it is assumed that the sample is representative of the entire population.

The following table of data was generated as a result of a survey you were asked to conduct in your new job at the county social services agency.

	MALE	FEMALE
SMOKER	154	139
NONSMOKER	238	201

10. a. How many males are represented in this table?

   b. How many females?

   c. How many smokers?

   d. How many nonsmokers?

e. How many female nonsmokers?

f. How many male smokers?

g. How many people were surveyed?

11. The table provides *actual* numbers for your answers in Problem 10. In many applications, *relative* numbers provide more information. Find the following *relative* numbers, expressed as a percent. Round your answer to the nearest percent.

a. What percent of the people surveyed are smokers?

b. What percent of the people surveyed are female?

c. What percent of the people surveyed are male smokers?

d. What percent of the males surveyed are smokers?

e. Explain why you should get different relative numbers in parts c and d.

The percents you determined in Problem 11 are sometimes referred to as **relative frequencies** (for example, the frequency of women relative to all people in the survey). A relative frequency is often used to estimate the probability (chance) of something occurring. For example, the probability that a randomly chosen person in your county is a nonsmoking female is expressed as the ratio $\frac{201}{732}$, which may also be represented as 0.27 when rounded to the nearest hundredth or as 27% when rounded to the nearest whole percent.

12. Estimate the following probabilities by finding the relative frequency. Express as a decimal rounded to the nearest hundredth and then as a percent.

a. Determine the probability that a randomly chosen person in your county is a smoker.

b. Determine the probability that a randomly chosen person in your county is a male.

    c. Determine the probability that a randomly chosen male in your county is a smoker.

    d. Determine the probability that a randomly chosen female in your county is a smoker.

    e. Who apparently smokes more in your county, men or women? Use the data from the survey to support your conclusion.

13. Assume that your survey responses are representative of the 125,300 adult residents of your county. Use proportional reasoning to approximate (to the nearest hundred) the number of people in your county who are

    a. Female

    b. Smokers

    c. Female smokers

SUMMARY

1. A proportion expressed in fraction form, $\frac{a}{b} = \frac{c}{d}$, is equivalent to the statement $ad = bc$.

2. Problems that involve **proportional reasoning** usually include a known ratio $\frac{a}{b}$ and a given piece of information, either a "part" or a "total" value, resulting in the proportion:

$$\frac{a}{b} = \frac{part}{total}$$

The missing value can be determined by constructing equivalent fractions or by cross-multiplying and solving the resulting equation.

EXERCISES

1. Determine the value of each of the following.

   a. $\frac{3}{2} \cdot 32$ b. $\frac{5}{4}$ of 200   c. $45\% \cdot 40$

   d. 15% of 24   e. $\frac{3}{8}$ of 40   f. $\frac{5}{8} \cdot 56$

   g. 7.5% of 80   h. $0.3\% \cdot 60,000$

2. Solve the following proportions.

   a. 3 out of 5 = _____ out of 20   b. $\frac{3}{5} = \frac{n}{765}$

   c. 3 out of 5 = 27 out of _____   d. 3 out of 5 = 1134 out of _____

In Exercises 3–19, write a proportion that represents the situation and then determine the unknown value in the proportion.

3. You correctly answered two-thirds of the questions on your psychology exam. There were 75 questions on the exam. How many questions did you answer correctly?

4. At the end of the semester, the bookstore will buy back books that will be used again in the courses the next semester. The bookstore will usually pay 15% of the original cost of the book. If you spend $246 on books this semester and the bookstore will buy back all your books, how much money can you expect to receive? (Recall that 15% can be thought of as the ratio $\frac{15}{100}$.)

5. During a recent downsizing, 20% of a company's workforce received pink slips. If this represents 500 job losses, how many people had been employed by this company?

6. On a recent mathematics exam, 80% of the class received a grade above 70. If 28 students performed at this C level and above, how many students are in the class?

7. The 8% sales tax on your cousin's new car is $1250. What is the actual price of the car?

8. During a recent infestation by beetles, $\frac{2}{3}$ of the ash trees in a local park were destroyed. If this represents 120 trees, how many ash trees were originally in the park?

9. In the 1996 merger of the Boeing and McDonnell Douglas corporations, McDonnell Douglas shareholders received a $\frac{2}{3}$ share of Boeing stock for each share of McDonnell Douglas stock they owned. If you owned 240 shares of McDonnell Douglas stock, how many shares of Boeing did you receive?

10. If 35% of the 2200 employees brown-bag their lunch at their desk, how many employees does this represent?

11. The sales tax rate on taxable items in Nassau County, New York, is 8.5%. Determine the tax on a new $25,000 car.

12. The expected tip on waiter service in a restaurant is now approximately 20% of the food and beverage cost. What is the customary tip on a dinner for two costing $45?

13. You invest $2500 in an account paying an annual interest of 5%. How much interest will you earn at the end of one year?

14. Your freshman class consists of 760 students. In recent years, only 4 out of 7 students actually graduated in four years. Approximately how many of your classmates are expected to graduate in four years?

15. In a very disappointing season, your softball team won only 30% of its games. If this represents 6 games, how many games were played?

16. You were given a job stuffing envelopes. After completing a box of 440 envelopes, you are told that you are only $\frac{2}{5}$ done. How many envelopes (total) are you expected to stuff?

17. A local businessman contributes $63,000 to a candidate's political campaign. This will cover 15% of the candidate's expenses. What are the total expenses in running this campaign?

18. In a recent classroom survey, the following table of data was generated. Students were categorized by their major—general studies or other (all other majors).

	MALES	FEMALES	TOTAL
GENERAL STUDIES	10	8	
OTHER	14	18	
TOTAL			

a. What percent of the people surveyed were general studies majors?

b. What percent of the males surveyed were in the category of Other?

c. What percent of the people surveyed were female?

Suppose your survey is representative of the whole college. There are 3230 students at this campus.

d. Approximate the number of people at the college who are general studies majors.

e. Approximate the number of males at the college.

f. What is the relative frequency of male general studies majors?

g. Based on the table, what is the probability of selecting at random, from the entire student body, a female general studies major?

h. What is the probability of selecting a male?

ACTIVITY 1.13

Did You Buy Enough Paint?

Topics: *Rates, Proportions, Unit Analysis*

Did you know that 1 gallon of paint covers approximately 400 square feet of wall? You plan to paint all the walls and ceilings in your new home. The total surface area that needs to be painted is 6000 square feet.

1. a. Assuming that you apply just one coat of paint, will 10 gallons be enough? 20 gallons?

b. Estimate (without actually calculating) the number of gallons of paint you will need to purchase.

To answer Problem 1 more precisely, you need to apply the same proportional reasoning you used in Activity 1.12. In this case, however, "number of gallons of paint" to "number of square feet of surface to be painted" is properly called a **rate**, since the units of measurement are different. In fact, such a rate is usually expressed as number of gallons of paint *per* square foot of surface, where *per* signifies division.

2. Determine the amount of paint you need by solving the following proportion:

$$\frac{1 \text{ gal}}{400 \text{ sq ft}} = \frac{n \text{ gal}}{6000 \text{ sq ft}}$$

Notice that when setting up a proportion involving rates, the two numerators must have the same units of measurement (in this case, number of gallons). Likewise, the denominators must also have the same units of measurement (in this case, the number of square feet).

3. Any rate can also be expressed in its reciprocal form, in this case $\frac{400 \text{ sq ft}}{1 \text{ gal}}$. Given your total surface area of 6000 square feet, set up and solve the proportion with this reciprocal rate. Compare your answer to the number of gallons of paint calculated in Problem 2.

4. Now suppose that you are given 5 gallons of a superior paint, which covers better because it is much thicker. However, 1 gallon only covers 250 square feet of wall surface.

   a. How much wall space will your 5 gallons cover?

   b. Solve a proportion to determine how much of this superior paint you need to cover your 6000 square feet of wall surface.

To solve the proportion in the original paint problem, you may have used cross-multiplication.

$$\frac{1 \text{ gal}}{400 \text{ sq ft}} = \frac{n \text{ gal}}{6000 \text{ sq ft}}$$

Cross multiply:
$$1 \text{ gal} \cdot 6000 \text{ sq ft} = n \text{ gal} \cdot 400 \text{ sq ft}$$

Divide both sides by 400 sq ft:
$$\frac{6000 \cancel{(\text{sq ft})} (\text{gal})}{400 \cancel{(\text{sq ft})}} = \frac{n \cdot 400 \cancel{(\text{sq ft})}(\text{gal})}{400 \cancel{(\text{sq ft})}}$$

Simplify:
$$15 \text{ gal} = n \text{ gal}$$

Notice that you can treat the units of measurement as factors in the fractions, dividing out the common units where possible and leaving the desired units of measurement. This technique is the key to **unit analysis** (sometimes called *dimensional analysis*). Pay close attention to the units of measurement in a problem and apply the strategy of unit analysis to more effectively find correct solutions.

Unit analysis provides a convenient shortcut to the paint problem. To determine the number of gallons of paint corresponding to 6000 square feet of wall space, you need only to multiply 6000 square feet by the appropriate fractional form of the rate:

$$6000 \text{ sq ft} \cdot \frac{1 \text{ gal}}{400 \text{ sq ft}} = \frac{6000 \cancel{(\text{sq ft})}(\text{gal})}{400 \cancel{(\text{sq ft})}} = 15 \text{ gal}$$

Notice here that the rate $\frac{1 \text{ gal}}{400 \text{ sq ft}}$ was chosen rather than $\frac{400 \text{ sq ft}}{1 \text{ gal}}$ because the unit of measurement square feet divides out, leaving the correct unit of measurement, namely gallons.

5. Your car averages 24 miles to the gallon. The 180-mile route you are planning to take across the mountain has no gas stations along the way. How many gallons should be in your gas tank to allow you to make it safely across? (In each part, show how the units of measurement divide out.)

   a. Solve by setting up a proportion, using the rate 24 miles per 1 gallon.

   b. Solve by setting up a proportion, using the rate 1 gallon per 24 miles.

   c. Solve by multiplying 180 miles by the appropriate rate.

   d. Which method from parts a–c do you prefer? Why?

6. You are still driving your trusty 24-mile-per-gallon car, planning for a 320-mile trip. Gasoline currently costs $1.12 per gallon. How much will you expect to pay on your trip for gas? *Hint:* You will need to multiply the 320 miles by two appropriate rates. Write down your calculation, and show how the units of measurement divide out.

7. As a nurse in the county hospital, you have received an order to administer 50 milligrams of a drug. The drug is available at a strength of 15 milligrams per milliliter. How many milliliters would you administer?

8. A Ford Taurus averages 380 miles on a 14-gallon tank of gas. You run out of gas on a deserted highway, but have $\frac{1}{2}$ gallon of lawn mower gas with you. Will you be able to reach the nearest gas station 15 miles away?

SUMMARY

- A **ratio** is a comparison between two quantities with the same unit of measurement (or perhaps no units).
- A **rate** is a comparison between two quantities with different units of measurement.
- **Unit analysis** is a strategy for performing calculations by treating units of measurement as factors in fractions. Common units in the fractions are divided out (cancelled) to leave desired units for the result.

EXERCISES

In the following exercises, first estimate the answer by taking an educated guess. Then solve precisely, recording your calculations with the units of measurement. Check your answer against your original estimate, making sure your final answer is reasonable.

1. As part of your job as a quality-control worker in a factory you can check 16 parts in 3 minutes. How long will it take you to check 80 parts?

2. Your car averages about 27 miles per gallon on highways. With gasoline priced at $1.25 per gallon, how much will you expect to spend on gasoline during your 500-mile trip?

3. You currently earn $11.50 per hour. Assuming that you work 40-hour weeks with no raises, what total gross salary will you earn in the next five years?

4. You have an order to administer 0.2 milligram per kilogram of a drug to a patient who weighs 28 kilograms. How many milligrams would you administer?

5. You are traveling at 75 miles per hour on a straight stretch of highway in Nevada. It is noon now. When will you arrive at the next town 120 miles away?

ACTIVITY 1.14

Uncle Sam's Place

Topics: *Unit Analysis, Metric System, U.S. System, Unit Conversion*

According to the USDA Natural Resources Conservation Service, federally owned land totaled 408 million acres in 1992. This was 21% of the total area of the United States.

FEDERALLY PROTECTED LAND

States with the most federal acreage (in millions):

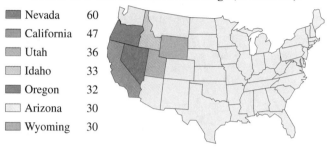

■	Nevada	60
■	California	47
■	Utah	36
■	Idaho	33
■	Oregon	32
■	Arizona	30
■	Wyoming	30

1. What percent of the total federally owned acreage is contained in the top seven states? (Round your answer to the nearest tenth of a percent.)

2. Nevada alone contains what percent of the total federally owned acreage? (Round your answer to the nearest tenth of a percent.)

The total area of the state of Nevada is 110,567 square miles. You will use this fact plus the information in the chart above to determine what percent of the total area of the state of Nevada is federally owned.

Note that to correctly compare the area of federally owned land to the total area of Nevada, both measurements must be expressed in the same units, either acres or square miles. You can use unit analysis to convert either measurement.

The conversion fact that you need is 1 square mile = 640 acres. As a rate, this fact can be expressed in two ways:

$$\frac{1 \text{ sq mi}}{640 \text{ acres}} \quad \text{or} \quad \frac{640 \text{ acres}}{1 \text{ sq mi}}$$

To convert a measurement from square miles to acres, multiply by the appropriate conversion factor. For example, to convert 20 square miles to acres, multiply by the conversion factor that will result in square miles dividing out, leaving acres in the numerator:

$$20 \text{ sq mi} \cdot \frac{640 \text{ acres}}{1 \text{ sq mi}} = 12,800 \text{ acres}$$

3. The total area of the state of Nevada is 110,567 square miles. What percent of the total acreage of the state of Nevada is federally owned?

   a. Use unit analysis to express both measurements in acres, and then calculate the requested percent.

   b. Use unit analysis to express both measurements in square miles, and then calculate the requested percent.

Refer to the U.S. system of measurement inside the front cover to determine the conversion facts needed for Problems 4–7. Write down your calculations, and show how the units of measurement divide out.

4. A 52-foot-long string is more than 600 inches long. What conversion fact is used to convert feet to inches? Determine the exact length of the string in inches.

5. How many feet are there in 3.2 miles?

6. How many ounces are there in 5 tons? (You will need to multiply by two conversion factors.)

7. Find the number of gallons in 360 fluid ounces.

Refer to the metric system of measurement inside the front cover to determine the conversion facts needed for Problems 8–11. Write down your calculations and show how the units of measurement divide out.

8. Convert 5.6 kilometers to meters.

9. Convert 5,250,000 milligrams to kilograms.

10. How many milliliters are there in 7.35 liters?

11. Describe the difference between the metric conversions and the U.S. conversions. In which system do you find the conversions easier to perform?

Unit analysis is used extensively to convert between U.S. system measurements and metric measurements. Refer to the conversion facts inside the front cover to solve the following problems.

12. Your car weighs 2504 pounds. What is its mass in kilograms?

13. You need to mail a box to Venezuela, but to be processed at the Venezuelan post office, the box must have no single dimension longer than 120 centimeters, and the sum of the three dimensions must not exceed 200 centimeters. All you can find is a tape measure in inches, and your box is 24 by 18 by 40 inches. Will your package be accepted by the post office in Venezuela?

Unit analysis is especially useful when the required conversion does not have a direct equivalent in a table. For example, suppose you want to convert 500 fluid ounces of soda into liters (abbreviated as ℓ). The table on the inside cover does not directly give conversion facts for fluid ounces and liters. However, you can relate fluid ounces to cups, cups to pints, pints to quarts and quarts to liters. Thus, the problem requires four conversion factors. They can be used all at once as follows:

$$500 \text{ fl oz} \cdot \frac{1 \text{ cup}}{8 \text{ fl oz}} \cdot \frac{1 \text{ pt}}{2 \text{ cups}} \cdot \frac{1 \text{ qt}}{2 \text{ pt}} \cdot \frac{1 \ell}{1.06 \text{ qt}} = \frac{500}{8 \cdot 2 \cdot 2 \cdot 1.06} \ell = 14.74 \ell$$

Note how all of the units "divided out" leaving you with the desired unit of liters. Each successive unit fraction was chosen to eliminate the units remaining from the previous fraction.

14. How many grams are there in 1 ton?

15. You are traveling in Canada at 70 miles per hour. The speed limit is given metrically as 100 kilometers per hour. Are you exceeding the speed limit?

16. How many millimeters are there in a length of 0.0045 inch?

17. If you are traveling at 90 feet per second, will you be exceeding the 100 km/hr Canadian speed limit?

18. Assume that the distance across a flat United States is approximately 3000 miles. How many pennies laid edge to edge would it take to span the country? What measurement of a penny must you use here?

E X E R C I S E S

In Exercises 1–4, refer to the graphic of federally owned land on page 79.

  1. How many acres of land are there in the United States? (State your answer using scientific notation.)

2. There are 43,560 square feet in 1 acre. Determine the number of square feet of land in the United States. (State your answer using scientific notation.)

3. Assume that there are about 250 million people in the United States. If everyone spread out uniformly across the country and claimed his or her own personal space, how many square feet would each person have?

4. Under the assumptions given in Exercise 3, how many people would there be in each square mile?

5. The distance between Earth and the Sun is 92,960,000 miles. Convert this distance to kilometers.

6. The height of the Empire State Building is 1414 feet. The height of the Eiffel Tower in Paris is 300.12 meters. Which is taller?

7. Compare the height of the Empire State Building to the height of the Oriental Pearl Television Tower in China, which is 468.18 meters tall.

8. a. How tall are you (in feet)?

b. Convert your height to inches.

c. What is your height in centimeters?

9. You buy a 2-liter bottle of diet cola.

   a. How many quarts do you have?

   b. Convert the quarts to pints.

   c. Do you have enough to give 8-ounce cups to each of your six companions? Explain.

10. The equatorial diameter of Earth is 12,756 kilometers. Convert this length to miles.

ACTIVITY 1.15

**Grade Point Averages:
Who Improved More?**
Topics: *Actual Change,
Relative Change, Growth
Factor, Decay Factor*

You and your friend are discussing your grade point averages (GPAs) over the past two semesters. The GPAs are summarized in the following table.

	YOUR GPA	YOUR FRIEND'S GPA
FALL	2.10	3.00
SPRING	2.55	3.50

1. Determine your actual change in GPA from the fall to the spring.

2. Determine your friend's actual change in GPA.

You see that your friend has improved more than you have and you congratulate her. She thanks you for the compliment, but says that based on the fall semester, you in fact improved more than she did. Who is right?

It is possible that you are both correct. You and your friend are discussing two different types of change: actual change and relative change. **Actual change** is the actual numerical difference by which a quantity has changed. When a quantity *increases* in value, the actual change is *positive*. When the quantity *decreases,* the actual change is *negative.* For example, if a quantity has decreased by 10, you can describe this as an actual change of -10: the magnitude, 10 (absolute value), states the size of the change, and the negative sign denotes the direction of the change.

The word *relative* used in this situation indicates that two quantities are being compared in terms of the ratio of their values. Recall that a ratio is commonly written in fraction or percent format.

> The ratio formed when calculating **relative change** always compares the actual change (numerator) to the original (or earlier) amount (denominator).

Your GPA during the fall semester, 2.10, is your original amount. Your actual change is 0.45. Therefore, your relative change is

$$\frac{0.45}{2.10} \approx 0.214 = 21.4\%$$

3. Determine the relative change in your friend's GPA. Was she justified in her claim that you have improved more than she has?

4. Determine the actual change in each of the following.

   a. The price of a share of stock was $24 last week and is now $30.

b. The number of violent crimes reported in your precinct was 40 in 1997 and 35 in 1998.

c. Miscellaneous household expenses last month were $250. This month they were $150.

5. Determine the relative change in the quantities from Problem 4. Express this ratio in both fraction and percent format.

a.   **b.**

c.

6. Suppose that the market value of your house has increased $10,000 since you purchased it. How significant is this actual change?

a. What is the relative change (expressed as a percent) if the original purchase price was $50,000?

b. What is the relative change (expressed as a percent) if your purchase price was $500,000?

7. Suppose that inflation is running at a fixed rate of 8% per year.

a. By how much will the cost of a $22,000 car increase in the next year?

b. What will the $22,000 car cost next year?

In Problem 7, you determined the new cost of the car by first computing the actual increase (8% of $22,000) and then adding this increase to the original cost of the car. It is often useful to compute the new value directly, bypassing the intermediate step of calculating the actual increase.

8. a. If a quantity increases by 50%, how does its new value compare to its original value? That is, what is the ratio of new value to original value? Complete the following table to discover/confirm your answer.

ORIGINAL VALUE	NEW VALUE (INCREASED BY 50%)	RATIO OF NEW VALUE TO ORIGINAL VALUE		
		FRACTION FORMAT	PERCENT FORMAT	DECIMAL FORMAT
20	30	$\frac{30}{20} = \frac{3}{2} = 1\frac{1}{2}$	150%	1.50
50				
100				

b. What is the ratio of new value to original value of any quantity that increases by 50%? Express this ratio in reduced fraction and decimal formats.

The ratio of a new value to an original value, which depends on a specified percent increase, is called the **growth factor.** A growth factor is obtained by adding the percent increase to 100%. For example, in Problem 8, the quantities increase by 50%, so the growth factor is 100% (the original amount) plus 50% (the increase) or 150% = 1.50.

Multiplying an original value by a growth factor results in the new value. For example, in Problem 8, if an original value, 20, increases by 50%, then the new value is $20 \cdot 1.50 = 30$.

9. a. What is the growth factor of any quantity that increases by 25%? Use this growth factor to determine the new value when an original value, 60, increases by 25%.

b. What is the growth factor of any quantity that increases by 8%? Use this growth factor to answer the question in Problem 7b.

10. It is often useful to determine the percent increase represented by a given growth factor. That is, if the growth factor is 1.40, then what percent increase of the original quantity does this represent?

11. Complete the following table.

PERCENT INCREASE	5%		7.3%	
GROWTH FACTOR		1.45		1.027

When a quantity increases in value by a fixed percent, say 10%, multiply its original value by the growth factor 1.10 to obtain its new value.

$$\text{Growth Factor} \cdot \text{Original Value} = \text{New Value}$$

Example: Acme Corporation is planning to expand its current workforce of 1500 by 20%. What is the anticipated size of its new workforce?

$$1.20\,(1500) = 1800 \text{ employees}$$

If a quantity has already increased by a fixed percent, say 10%, divide its new (larger) value by the growth factor (1.10) to obtain the original value.

$$\frac{\text{New Value}}{\text{Growth Factor}} = \text{Original Value}$$

Example: Acme's chief competitor, Arco, has already increased its workforce by 25% and currently employs 2400 workers. What was the previous size of its workforce?

$$\frac{2400}{1.25} = 1920 \text{ employees}$$

An alternate method to solve the above problems is to set up and solve a percent proportion. The growth factor represents the ratio of the new value to the original value. Thus, if the growth factor is $1.10 = 110\% = \frac{110}{100}$, then the percent proportion is

$$\frac{110}{100} = \frac{\text{New Value}}{\text{Original Value}}.$$

12. Since last year, housing prices have appreciated (increased) by 25% in your neighborhood. A house just sold for $300,000.

   a. What is the growth factor associated with a 25% increase?

   b. Is $300,000 the new value or the original value?

   c. What was last year's market value?

13. The past year has seen tremendous growth in the stock market. Your stock's price has increased 35% since you bought it last January, and the value of your investment is now $6000. How much did you invest in this stock last January?

In this next section, you will examine the process of calculating percent decrease.

14. A suit, originally priced at $400, is on sale for 30% off.

    a. Determine the amount of discount on the suit. This represents the actual decrease in the cost of the suit.

    b. What is the new ticketed price of the suit?

In Problem 14, you first determined the actual decrease (30% of 400) in the cost of the suit and then subtracted this decrease from the original value. As you did in the growth factor calculations done earlier, you can bypass the intermediate step of calculating the actual decrease by multiplying the original value by the **decay factor.**

15. a. Complete the following table to determine the decay factor of a quantity that *decreases* by 20%.

ORIGINAL VALUE	NEW VALUE	RATIO OF NEW VALUE TO ORIGINAL VALUE: DECAY FACTOR		
		FRACTION FORMAT	PERCENT FORMAT	DECIMAL FORMAT
20	16	$\frac{16}{20} = \frac{4}{5}$	80%	0.80
50				
100				

    b. What is the decay factor of any quantity that decreases by 75%? Express this ratio in reduced fraction and decimal formats.

    c. If the percent decrease is 5%, what is the decay factor?

It is important to note that although a percent decrease usually describes a portion that has been removed, the corresponding decay factor represents the percent remaining. Therefore, a 20% decrease is represented by a decay factor of 80%, or 0.80.

Multiplying the original value by a decay factor always produces the value that remains, not the amount that has been removed.

When a quantity decreases in value by a fixed percent, say 10%, multiply its original value by the decay factor 0.90 to obtain its new (smaller) value.

Decay Factor · Original Value = New Value

If a quantity has already decreased by 10%, divide its new (smaller) value by 0.90 to obtain the original value.

$$\frac{\text{New Value}}{\text{Decay Factor}} = \text{Original Value}$$

Note that you can also solve problems involving a decay factor by setting up and solving a percent proportion. The decay factor represents the ratio of the new value to the original value. Thus, if the decay factor is $0.90 = 90\% = \frac{90}{100}$, then the percent proportion is

$$\frac{90}{100} = \frac{\text{New Value}}{\text{Original Value}}$$

16. Complete the following table.

PERCENT DECREASE	45%			6%	15%		3.2%	
NEW TO OLD RATIO: DECAY FACTOR		.55	.75			.34		.986

17. A suit, originally priced at $400, is on sale for 30% off.

   a. Determine the decay factor.

   b. Use this decay factor to calculate the new ticketed price of the suit. Compare your results with your answer to Problem 14b.

18. You have been able to trim this year's budget by 22% over last year's expenses. Last year's budget was $170,000. Determine the decay factor, and use this factor to calculate this year's budget.

19. After several years of downsizing, a company now employs 1500 people. This represents a decrease of 35% from the 1997 level. How many employees worked for the company back in 1997?

SUMMARY

Actual change is the actual numerical difference by which a quantity has changed. It may be positive or negative and has the same unit of measurement as the quantity itself.

Relative change is the ratio that compares the actual change (numerator) to the original, or earlier, amount (denominator). This ratio has no unit of measure associated with it.

If the ratio of a new value to an original value is greater than 1, it is called a **growth factor.**

If the ratio of a new value to an original value is less than 1, it is called a **decay factor.**

EXERCISES

1. A house that cost $175,000 in 1990 was priced at $300,000 in 1999.

   a. Determine the actual increase in price.

   b. Calculate the percent increase in price.

2. Your school's enrollment was 8250 last year. This year the enrollment went down to 7650.

   a. Determine the actual change in enrollment.

   b. Find the percent decrease in enrollment.

3. In 1998, the average price of gasoline in your neighborhood was $1.12 per gallon. By the end of 1999, the average price rose to $1.26 per gallon. Determine the percent increase in the price of gasoline.

Exercise numbers appearing in color are answered in the Selected Answers section of this book.

4. Your hourly wage at your part-time job is $6.30 per hour, up from last year's wage of $5.95 per hour. What percent raise did you receive?

5. The number of applications to the local state college have soared by 40% since 1995. At that time, there were 4500 applicants. How many applications are anticipated this year?

6. The size of this year's graduating class represents an increase of 25% over the 1995 graduating class. This year, 2200 students are receiving their diplomas. How many graduates were there in 1995?

7. The number of homicides in your city has dramatically decreased in the past five years, down by approximately 80%. If there were 42 homicides on record for the last calendar year, approximately how many homicides were committed five years ago?

8. At the end of the season, your favorite label jacket is finally on sale for 70% off the original list price of $120. What is its current ticketed price?

9. Last year's rental cost of a power saw was $16.20 per hour. This year, the rental fee was increased by 5%. How much will you pay this summer for a 4-hour rental?

10. You have been burning 420 calories during each session on the StairMaster. Your trainer claims that you will burn off 30% more calories on the treadmill. How many calories do you expect to burn off on the treadmill?

PROJECT
ACTIVITY 1.16

**Take an Additional
10% Off**

Topics: *Fractions, Decimals,
Percent, Ratio, Growth
Factor, Decay Factor,
Consecutive Percent Change*

Homer was walking home from work one day when he noticed a $5 bill on the sidewalk. There was no one nearby, so he picked it up and placed it in his pocket. His other pocket already contained a $10 bill. Homer grinned and thought to himself, "My wealth has just increased by 50%."

Unfortunately, Homer was unaware that the pocket that held the $5 bill had a large hole in it. When he arrived home, he sadly discovered that the $5 was missing. "That's not so bad," he explained to his disappointed wife. "Earlier our wealth increased by 50%, but now it has decreased by only $33\frac{1}{3}\%$. We're still ahead by nearly 17%!"

1. **a.** Show how Homer calculated his increase of 50%. Was he correct?

    b. Explain how Homer calculated his decrease of $33\frac{1}{3}\%$. Was he correct?

    c. Explain how Homer calculated his net gain of nearly 17%. Was he correct?

When a sequence of consecutive changes occurs in the value of a quantity, such as Homer's short-lived wealth, you often need to calculate the overall change.

> The total change is the *sum* of the individual actual changes that occur.

2. Suppose you track the price of a stock over two weeks. The starting price is $24, and the changes in price are an increase of $6 the first week and $6 the second week.

    a. Determine the actual change in stock price over these two weeks.

    b. Now, calculate the relative change in stock price over the *first* week. Express this relative change in both fraction and percent format.

    c. At the end of the first week, the price has risen to _____. Calculate the relative change in stock price over the *second* week. Express this relative change in both fraction and percent format.

d. Determine the relative change in stock price over the full two-week period. That is, compare the actual two-week change to the starting price. Express this relative change in both fraction and percent format.

e. Do the relative changes over each week sum to the relative change over the two-week period?

> The calculations you have just done demonstrate that the numerical values (in percent, decimal, or fraction format) of a sequence of relative changes do *not* sum.

The following example should be familiar to anyone who shops during storewide sales.

3. You have just clipped an "additional 10% off" coupon from the newspaper. Your favorite shirt is already on sale for 30% off the original price of $36.00. The store will apply these discounts consecutively.

a. Determine the decay factor corresponding to each discount.

b. Apply these decay factors one at a time to calculate the final sale price if the 30% discount is taken first, followed by the 10% discount.

c. Apply these decay factors one at a time to calculate the final sale price if the 10% discount is taken first, followed by the 30% discount.

d. In which order would you prefer these discounts to be taken? Explain.

e. Calculate the ratio of the final sale price to the original price and write it in decimal form. Compare it with the product of the two decay factors you determined in part a. What do you observe?

4. The average score on a fourth-grade reading test in your district was 65 in 1980. By 1990, the average score had fallen by 60%. After a decade of extensive curriculum changes, the superintendent proudly announced that the average 1999 score was up 70% since 1990.

   a. What is the growth or decay factor from 1980 to 1990? What was the 1990 average reading score?

   b. What is the growth or decay factor from 1990 to 1999? What is the 1999 average reading score?

   c. What was the relative (percent) change from 1980 to 1999?

   d. What percent of the 1980 score is the 1999 score?

   e. Compare your result from part d with the product of the two factors you found in parts a and b.

> The previous problems have demonstrated that the cumulative effect of a sequence of relative changes is the **product** of the associated growth or decay factors.

To give another example, if a value is increased by 50%, followed by an increase of 20%, then the cumulative effect is given by the product $1.50 \cdot 1.20 = 1.80$. Therefore, the net effect of consecutively applying the 50% and 20% increases is an increase of 80%. Notice that the order in which these increases are applied has no effect on the cumulative growth factor. Why?

5. Suppose an item's value is decreased by 50%, followed by a decrease of 20%. What is the cumulative percent decrease?

6. Suppose an item's value is decreased by 30%, followed by an increase of 30%.

   a. Does its value return to the original level? Explain by giving an example.

b. Suppose instead that the 30% increase were taken first, followed by the 30% decrease. Would the item's value return to the original level? Explain.

7. Homer's wealth increased by 50%, only to decrease by $33\frac{1}{3}\%$. What is the cumulative result of these changes? Show how you obtained the result mathematically. Does the result you obtain here agree with your initial reaction to Homer's plight?

8. Homer was eyeing a beautiful diamond pendant for his wife, but the $2000 list price was far too high for his modest budget. During the next several weeks, he rejoiced as he witnessed the following successive discounts on the pendant:

   20% off list price

   30% off marked price

   an additional 50% off marked price of every item in the display case

   At this point, he rushed into the store, expecting to purchase the pendant for 100% off!

   a. How do you think Homer calculated the total discount?

   b. For what price is the store actually selling the pendant?

   c. How would the final price differ if the discounts had been taken in the reverse order (50% off, 30% off, and then 20% off)?

9. The accompanying graph is adapted from *USA Today* (January 5, 1996). Answer the following questions concerning the graph.

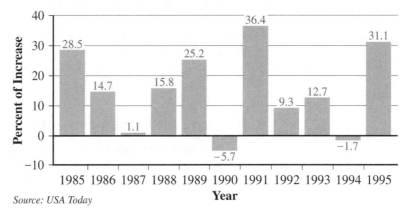

Mutual Funds Performance
Average Total Return for General Stock Funds

Source: USA Today

a. Describe what the graph represents.

b. Which year had the highest average percent return?

c. Which year had the lowest average percent return?

d. If Homer had invested money in 1990, does the graph imply that he would have lost money on his investment for that year? Explain.

e. Based on the past performance of the mutual funds, is it possible for Homer to make any predictions for 1996?

f. In 1994, Homer invested $1200 in mutual funds. Using the average return for that year, what is his net gain or loss on this investment for 1994?

g. Homer invested $1000 at the beginning of 1989. Assuming that he received the average return as indicated on the graph, what was the value of his investment by the end of 1990?

h. If Homer invested $3000 in January 1988 and left his earnings or losses in the same mutual fund until December 1991, what would be his expected average total percent return? Assume that he received the percents shown on the graph.

What Have I Learned?

1. On a 75-question practice exam, you were able to answer 61 questions correctly. On the exam itself you managed to answer 39 out of 45 correctly. Explain the distinction between your actual performance and relative performance on each of these exams.

2. Fifteen out of 25 students in your mathematics class commute to school by car.

   a. Express this ratio in reduced fraction form, as a decimal, and as a percent.

   b. Suppose this ratio accurately represents the commuting habits of the entire student body. Explain the strategy you would use to answer the following questions.

      i. Suppose there are 4500 students at your college. How many students commute by car?

      ii. Suppose there are 4500 students on campus who commute by car. What is the total enrollment at your college?

   c. Determine the answers to questions i and ii.

3. In converting from inches to centimeters, do you multiply or divide by 2.54? Use an example to illustrate how you can be sure of your answer.

4. a. Explain how to determine the growth factor associated with any percent increase.

b. Explain how to determine the decay factor associated with any percent decrease.

5. Which numbers represent possible growth factors and which represent possible decay factors? Explain. Determine the percent increase or decrease corresponding to each of these factors.

    a. 1.35   **b.** 107%   **c.** 0.97

    d. 86%   **e.** 2.00

6. You deposited $1000 in a special bank account for your child. The yearly rate of interest is 5.6% as long as you do not remove any money from the account for three years. How much is in your account at the end of the third year? Show how you obtained your answer.

How Can I Practice?

1. In a class of 27 students, 16 are female. What percent of students are female? (Round your answer to the nearest whole number.)

2. If a student answered 67 questions correctly out of 80 questions on an exam, what percent of the questions did the student answer correctly? (Round your answer to the nearest whole number.)

3. A total of 2365 students on your campus are commuters. The remaining 1325 live in the college dormitories. What percent of students are commuters?

4. If the sales tax in your county is 7.5%, how much tax will you pay on books and supplies costing $185?

5. You are planning a weekend trip to New York City. You estimate that the round trip is 400 miles. Your car gets 24 miles per gallon and gas costs $1.26 per gallon. Compute your gas costs for the round trip.

6. The blueprint for your new home is on a scale of 0.25 in. : 1 ft. This means that every $\frac{1}{4}$ inch on the blueprint represents 1 foot of actual space. You have just purchased a $6\frac{1}{2}$ foot sofa that you will place against the long wall in your family room. On the blueprint, the wall measures $3\frac{1}{2}$ inches. How much space is left along the wall for end tables?

7. The taxes on a house assessed at $22,000 are $900. At the same tax rate, what are the taxes (to the nearest dollar) on a house assessed at $30,000?

8. In Suffolk County on Long Island, 6152 mosquitoes were trapped in July 1999, compared with 27,161 the previous July. If the number of trapped mosquitoes accurately represents the mosquito population in the county, determine the percent decrease in the mosquito population from 1998 to 1999.

9. Ample storage space is an important factor in the design and building of modern structures. It seems as if there are never enough closets to keep all those "essential" items we all love to store. Museums are not immune to these problems.

Did you ever wander through a museum and wonder how much of the museum's collection you are viewing? According to *New York* magazine (July 10, 1995), many New York City museums display only a small percent of their total number of holdings.

MUSEUM	TOTAL NUMBER OF ITEMS IN THE COLLECTION	ITEMS ON DISPLAY
Whitney Museum	10,000	200
Museum of Modern Art	78,000	600
New York Historical Society Museum	150,000	2000
Guggenheim Museum	5000	75
Brooklyn Museum	1,500,000	6000
Jewish Museum	2700	1000
Frick Museum	1500	1425

a. For each museum, express the number of items on display as a percent of the total number of items in the collection.

b. Change these percents to decimals, and order them from smallest to largest.

c. Which museum displays the smallest percent? The largest?

d. If the collections from all seven museums are combined, express the number of items on display as a percent of the total number of items in the collection.

e. Is the percent in part d the same as the average of the percents in part a? Explain in complete sentences.

f. If you were writing a newspaper article about these museums, which of the two numbers discussed in part e best expresses the average percent of the number of items on display in the seven museums?

10. You have a friend in graduate school who is completing a master's thesis. You decide to help her out by offering to type her paper for a small fee.

a. You make 8 mistakes in typing on the first 6 pages. If the thesis is 324 pages long, how many mistakes can you expect to make?

b. If you can type 11 pages per hour, how long will it take you to finish the 324-page thesis?

c. If you charge your friend 50 cents per page, what is the total your friend owes you?

d. Your friend's thesis is due in three days. Is there enough time to finish the typing? Explain.

11. You decide to buy a new beach ball for the summer, but you cannot remember whether the diameter is 60 centimeters, or 60 kilometers. Which diameter is reasonable? Explain.

12. A friend from Canada is visiting for Thanksgiving. Your friend wants to know how much the 18-pound turkey you bought weighs in kilograms. You also want to tell her the amount in liters of the 3 quarts of eggnog you have made. Calculate these values.

13. Your patient is supposed to receive 500 milligrams of medication in an elixir at a strength of 875 milligrams per 5 milliliters of solution. How many milliliters would you administer?

14. A stamping machine punches out 10 parts per minute. How many parts are stamped out during a continuous 8-hour shift?

15. a. A sweater that sold for $32 before Christmas is selling for 30% off during an after-Christmas sale. What is the sale price of the sweater?

b. In January, the sweater is still available but has been reduced an additional 60%. What will it cost to buy the sweater in January? What is the total percent savings from the original cost of the sweater?

16. You used to drink 67.6 ounces (2 liters) of diet cola each day. Now you drink four 12-ounce cans a day. By what percent have you decreased your diet cola consumption?

17. **a.** You used to maintain your weight at 130 pounds. This year you put on some weight and tip the scale at 138 pounds. By what percent has your weight increased?

b. If you lose 1% of what you weigh each week while following a diet and exercise program, how many weeks will it take you to reduce back to 130 pounds?

18. Your new diet requires you to decrease your daily caloric intake from 1800 to 1400 calories. By what percent do you need to decrease your caloric intake?

19. It is customary for the tax and tip at a restaurant to come to 25% of the actual food and beverage charge. What is the total cost, including tax and tip, of a meal whose food and beverage charge is $48.50?

20. A generous diner left 30% to cover the tax and tip on his meal. If his total expense came to $110, what was his food and beverage charge?

1. Subtract 187 from 406.

2. Multiply 68 by 79.

3. Write 16.0709 in words.

4. Write three thousand four hundred two and twenty-nine thousandths in standard form.

5. Round 567.0468 to the nearest hundredth.

6. Round 2.59945 to the nearest thousandth.

7. Add $48.2 + 36 + 2.97 + 0.743$.

8. Subtract 0.48 from 29.3.

9. Multiply 2.003 by 0.36.

10. Divide 28.71 by 0.3.

11. Change 0.12 to a percent.

12. Change 3 to a percent.

13. The sales tax rate in some states is 6.5%. Write this percent as a decimal.

14. Write 360% in decimal form.

15. Write $\frac{2}{15}$ as a decimal.

16. Write $\frac{3}{7}$ as a percent. Round the percent to the nearest tenth.

17. The membership in the Nautilus Club increased by 12.5%. Write this percent as a fraction.

18. Write the following numbers in order from smallest to largest:
    3.027   3.27   3.0027   3.0207

Exercise numbers appearing in color are answered in the Selected Answers section of this book.

19. Find the average of 43, 25, 37, and 58.

20. In a recent survey, 14 of the 27 people questioned preferred Coke to Pepsi. What percent of the people preferred Coke?

21. If the sales tax is 7%, what is the cost, including tax, of a new baseball cap that is priced at $8.10?

22. If you spend $63 a week for food and you earn $800 per month, what percent of your monthly income is spent on food?

23. The enrollment at a local college has increased 5.5%. Last year's enrollment was 9500 students. How many students are expected this year?

24. You decide to decrease the number of calories in your diet from 2400 to 1500. Determine the percent decrease.

25. The area of the land masses of the world is about 57,000,000 square miles. If the area of 1 acre is about 0.00156 square mile, determine the number of acres of land in the world.

26. You and a friend have been waiting for the price on a winter coat to come down sufficiently from the manufacturer's suggested retail price (MSRP) of $500 so you each could afford to buy one. The retailer always discounts the MSRP by 10%. Between Thanksgiving and New Year's, the price was further reduced by 40%, and in January, it was again reduced by 50%.

    a. Your friend remarked that it looks like you can get the coat for free in January. Is that true? Explain.

    b. What does the coat cost before Thanksgiving, in December, and in January?

c. How would the final price differ if the discounts had been taken in the reverse order (50% off, 40% off, and 10% off)? How would the intermediate prices be affected?

27. A local education official proudly declares that although in the 1980s math scores fell by nearly 55%, they have since rebounded over 65%. This sounds like great news for your district, doesn't it? Determine how the current math scores actually compare with the pre-1980 scores.

28. Last year, the MIA Corporation showed a 10.8% increase in retail sales in January, a 4.7% decrease in February, and a 12.4% increase in March. If retail sales at the beginning of January were approximately 6 million dollars, what were the sales at the beginning of April?

29. Each year, Social Security payments are adjusted for cost of living. Your grandmother relies on Social Security for a significant part of her retirement income. In 1996, her Social Security income was approximately $1100 per month. If cost of living increases were 1.7% for 1997, 1.6% for 1998, and 2.6% for 1999, what was the amount of your grandmother's monthly Social Security checks in 2000?

Problem Solving with Signed Numbers

ACTIVITY 1.17

Celsius Thermometers

Topic: *Adding and Subtracting Signed Numbers*

So far, the activities in this book have mostly involved positive numbers and zero. In the real world, you also encounter negative numbers.

1. What are some situations in which you have encountered negative numbers?

> The collection of positive counting numbers, negative counting numbers, and zero is basic to our number system. For easy referral, this collection is called the set of **integers.**

A good technique for visualizing the relationship between positive and negative numbers is to use a number line, scaled with integers, much like a thermometer's scale.

On a number line, 0 separates the positive and negative numbers. Recall that numbers increase in value as you move to the right. Thus, -1 has a value greater than -4, which makes sense in the thermometer model, since $-1°$ is a warmer temperature than $-4°$.

In this activity, a thermometer model illustrates addition and subtraction of signed numbers. On a Celsius thermometer, $0°$ represents the temperature at which water freezes, and $100°$ represents the temperature at which water boils. The following thermometers show temperatures from $-10°C$ to $+10°C$.

2. a. What is the practical meaning of positive numbers in the Celsius thermometer model?

   b. What is the practical meaning of negative numbers in the Celsius thermometer model?

You can use signed numbers to represent a change in temperature. A rise in temperature is indicated by a positive number, and a drop in temperature is indicated by a negative number. As shown on the following thermometers, a rise of 6° from −2° results in a temperature of 4°, symbolically written as −2° + 6° = 4°.

3. Answer the following questions, using the thermometer models.

a. What is the result if a temperature starts at 5° and rises 3°? Symbolically, you are calculating 5° + 3°. Use the thermometer below to demonstrate your calculation.

b. If a temperature starts at −5° and rises 3°, what is the result? Write the calculation symbolically. Use the thermometer below to demonstrate your calculation.

c. If a temperature starts at 5° and drops 3° (−3), the resulting temperature is 2°. Symbolically, you are calculating 5° + (−3°). Use the thermometer below to demonstrate your calculation.

d. What is the result when a temperature drops 3° from −5°? Write the calculation symbolically. Use the thermometer below to demonstrate your calculation.

4. a. In what direction did you move on the thermometer when you added positive degrees to a starting temperature in Problem 3?

b. In each case, was the result greater or less than the starting number?

5. a. When you added negative degrees to a starting temperature in Problem 3, in what direction did you move on the thermometer?

b. In each case, was the result greater than or less than the starting number?

6. a. Evaluate each of the following mentally. Check your result using a calculator.

$$4 + 6 = \qquad\qquad 6 + 8 = \qquad\qquad -7 + (-2) =$$

$$-16 + (-10) = \qquad\qquad (-.5) + (-1.4) =$$

$$-\frac{5}{2} + \left(-\frac{3}{2}\right) =$$

b. In each calculation in part a, what do you notice about the signs of the two numbers being added?

c. How do the signs of the numbers being added determine the sign of the result?

d. How do you calculate the numerical part of the result from the numerical parts of the numbers being added?

7. a. Evaluate each of the following mentally. Check your result using a calculator.

$$4 + (-6) \qquad\qquad -6 + 8 \qquad\qquad 7 + (-2)$$

$$-16 + 10 \qquad\qquad -\tfrac{6}{2} + \tfrac{7}{2} = \qquad\qquad .5 + (-2.8)$$

b. In each calculation in part a, what do you notice about the signs of the numbers being added?

c. How do the signs of the numbers being added determine the sign of the result?

d. How do you calculate the numerical part of the result from the numerical parts of the numbers being added?

8. Evaluate each of the following. Check your result using a calculator.

a. $-8 + (-6)$ **b.** $9 + (-12)$ **c.** $6 + (-8)$ **d.** $-7 + (-9)$

e. $16 + 10$ **f.** $16 + (-10)$ **g.** $-9 + 8$ **h.** $-2 + (-3)$

i. $9 + (-5)$ **j.** $3 + (-6)$ **k.** $-\tfrac{5}{8} + \tfrac{1}{8}$ **l.** $-\tfrac{5}{6} + \left(-\tfrac{1}{6}\right)$

m. $\tfrac{3}{4} + \left(-\tfrac{1}{4}\right)$ **n.** $\tfrac{1}{2} + \tfrac{1}{3}$ **o.** $-5.9 + (-4.7)$ **p.** $0.50 + 0.06$

q. $(-5.75) + 1.25$ **r.** $-12.1 + 8.3$ **s.** $-6 + \left(-\tfrac{2}{3}\right)$ **t.** $5 + \left(-1\tfrac{3}{4}\right)$

Suppose you know the starting and ending temperatures for a certain period of time and that you are interested in the *change* in temperature over that period.

> **Change in value** is calculated by subtracting the initial (original) value from the final value. That is,
>
> Final number − initial number = change in value (difference).

For example, the change in value from 2° to 9° is 9° − 2° = 7°. Here, 9° is the final temperature, 2° is the original, or initial, temperature, and 7° is the change in temperature. Notice that the temperature has risen; therefore, the change in temperature of 7° is positive, as shown on the thermometer on the left.

Suppose that −6° is the final temperature and 5° is the original, or initial temperature.

Symbolically, this is written −6° −5° producing a result of −11°, as indicated on the thermometer on the right. The significance of a negative change is that the temperature has fallen.

9. a. What is the change in temperature from 3° to 5°? That is, what is the difference between the final temperature, 5°, and the initial temperature 3°? Symbolically you are calculating $+5° - (+3°)$. Use the thermometer below to demonstrate this calculation. Has the temperature risen or fallen?

b. The change in temperature from 3° to $-5°$ is $-8°$. Symbolically, you write $-5° - (3°) = -8°$. Use the thermometer below to demonstrate this calculation. Has the temperature risen or fallen?

c. A temperature in Montreal last winter rose from −3° to 5°. What was the change in temperature? Write the calculation symbolically and determine the result. Use the thermometer below to demonstrate your calculation.

d. The temperature on a March day in Montana was −3° at noon and −5° at 5 P.M. What was the change in temperature? Write the calculation symbolically and determine the result. Use the thermometer below to demonstrate your calculation. Has the temperature risen or fallen?

10. When calculating a change in value, in which position in the subtraction calculation must the initial value be placed—to the left or to the right of the subtraction symbol?

Problem 9 showed that you use subtraction to calculate the *change* or *difference* in value between two numbers and Problem 3 demonstrated that addition is used to move from an initial number to another number.

You can use the concept of opposites to relate the operations of addition and subtraction of integers.

Opposites are numbers that when added together give a sum of zero.

For example, 10 and −10 are opposites because $10 + (-10) = 0$.

11. a. What is the opposite of 5?

b. What is the opposite of −8?

c. $0 - 5 =$ $0 + (-5) =$

d. $0 - 12 =$ $0 + (-12) =$

e. $0 - (-8) =$ $0 + 8 =$

f. $0 - (-6) =$ $0 + 6 =$

g. From your experience in parts c–f, what can you conclude about the result of subtracting a number from zero? about adding a number to zero?

h. From your experience with parts c–f, is it reasonable to believe that subtracting a number gives the same result as adding its opposite? Explain.

> Subtracting a signed number is equivalent to adding its opposite.

For example, $-4 - 6$ becomes $(-4) + (-6)$ and equals -10 by the addition of signed numbers.

12. Convert each of the following to an equivalent addition problem and evaluate. Check your result using a calculator.

a. $4 - 11$ **b.** $-8 - (-6)$ **c.** $9 - (-2)$

d. $-7 - 1$ **e.** $10 - 15$ **f.** $6 - (-4)$

g. $-8 - 10$ **h.** $-2 - (-5)$ **i.** $-\frac{5}{8} - \frac{1}{8}$

j. $\frac{5}{6} - \left(-\frac{1}{6}\right)$ **k.** $-\frac{3}{4} - \left(-\frac{1}{4}\right)$ **l.** $\frac{1}{2} - \frac{1}{3}$

m. $5.9 - (-4.7)$ **n.** $-3.75 - 1.25$

o. $-6 - \left(-\frac{2}{3}\right)$ **p.** $5 - \left(-1\frac{3}{4}\right)$

Number lines and models such as thermometers provide a good visual approach to adding and subtracting signed numbers. However, you will find it more efficient to use a general method when dealing with signed numbers in applications. A convenient way of adding and subtracting signed numbers is to use the concept of absolute value.

The **absolute value** of a number represents the distance that the number is from zero on the number line. For example, $+9$ and -9 are both 9 units from 0 but in opposite directions so they have the same absolute value, 9. Symbolically we write $|9| = 9$, and $|-9| = 9$ and $|0| = 0$. Notice that opposites have the same absolute value. More generally, the absolute value of a number represents the size or magnitude of the number.

The procedures you have developed in this activity for adding and subtracting signed numbers can be restated using absolute value.

ADDING AND SUBTRACTING SIGNED NUMBERS

When *adding* two numbers with the *same* sign, add the absolute values of the numbers. The sign of the sum is the same as the sign of the numbers being added.

When *adding* two numbers with *opposite* signs, find their absolute values and then subtract the smaller from the larger. The sign of the sum is the same as the sign of the number with the larger absolute value.

Subtracting a signed number is equivalent to *adding the opposite* of the signed number. Two changes must be made. The subtraction symbol is changed to an addition symbol, and the number following the subtraction symbol is replaced by its opposite. The new addition problem is evaluated using the addition rule.

EXERCISES

1. Another illustration of adding signed numbers can be found on the gridiron—that is, on the football field. On the number line, a gain is represented by a move to the right, while a loss results in a move to the left. For example, a 7-yard loss and a 3-yard gain results in a 4-yard loss. This can be written symbolically as $-7 + 3 = -4$.

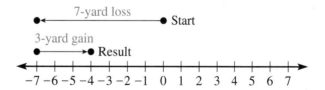

Use the number line to complete the following.

a. A 6-yard gain and a 3-yard gain results in a _____ .

b. A 5-yard gain and a 2-yard loss results in a _____ .

c. A 4-yard loss and a 3-yard loss results in a _____ .

d. A 1-yard gain and an 8-yard loss results in a _____ .

2. Write each of the situations in Problem 1 as an addition problem

    a.   **b.**

    c.   **d.**

3. The San Francisco 49ers are famous for selecting their first ten plays before the game begins. If the yardage gained or lost from each play is totaled, the sum represents the total yards gained (or lost) for those plays. One play sequence proceeded as follows:

PLAY	1	2	3	4	5	6	7	8	9	10
YARDS	Lost 5	Gained 7	Gained 9	Lost 5	Gained 15	Gained 4	Lost 6	Gained 20	Lost 1	Gained 5

What was the total yards gained (or lost) for these ten plays?

4. Evaluate each of the following. Use a calculator to verify your result.

    a. $-3 + (-7)$ **b.** $-6 + 2$ **c.** $4 + (-10)$ **d.** $16 - 5$

    e. $7 - 12$ **f.** $-2 - 8$ **g.** $-5 + (-10)$ **h.** $(-6) + 9$

    i. $-8 - 10$ **j.** $-4 - (-5)$ **k.** $0 - (-6)$ **l.** $-3 + 0$

    m. $9 - 9$ **n.** $9 - (-9)$ **o.** $0 + (-7)$ **p.** $-4 + (-5)$

    q. $-8 - (-3)$ **r.** $(-9) - (-12)$ **s.** $5 - 8$ **t.** $(-7) - 4$

    u. $-\frac{5}{9} + \frac{4}{9}$ **v.** $2.56 - 3.14$ **w.** $-75 - 30.5$ **x.** $33\frac{1}{3} - 66\frac{2}{3}$

5. The current temperature is $-12°C$. If the forecast is that it will be 5° warmer tomorrow, write a symbolic expression and evaluate it to determine tomorrow's predicted temperature.

6. The current temperature is −4°C. If the forecast is that it will be 7° colder tomorrow, write a symbolic expression and evaluate it to determine tomorrow's predicted temperature.

7. On a cold day in April, the temperature rose from −5°C to 7°C. Write a symbolic expression and evaluate it to determine the change in temperature.

8. A heat wave hits New York City, increasing the temperature an average of 3°C per day for five days, starting with a high of 20°C on June 6. What is the approximate high temperature on June 11?

ACTIVITY 1.18

Gains and Losses in Business

Topic: *Multiplying and Dividing Signed Numbers*

Situation 1

Stock values are reported daily in some newspapers. Reports usually state the values at both the opening and the closing of a business day. Suppose you own 220 shares of Corning Incorporated stock.

1. Your stock opens today valued at $31.00 per share, but goes down to $30.50 per share by the end of the day. What is the total change in the value of your shares today? State your answer in words and as a signed number.

2. Your stock opens another day at $42.00 per share and goes up $1.25 per share by the end of the day. What is the total value of your shares at the end of the day?

Situation 2

You are one of three equal partners in a company.

3. Your company made a profit of $300,000 in 1999. How much money did you and your partners make each? Write the answer in words and as a signed number.

4. Your company experienced a loss of $150,000 in 1998. What was your share of the loss? Write the answer in words and as a signed number.

5. Over the two-year period from January 1, 1998, to December 31, 1999, what was the net profit for each partner? Express your answer in words and as a signed number.

6. Suppose that in 2000, your corporation suffered a further loss of $180,000. Over the three-year period from January 1, 1998, to December 31, 2000, what was the net change for each partner? Express your answer in words and as a signed number.

Situation 3

Each individual account in your small business has a current dollar balance, which can be either positive (a credit) or negative (a debit). To find your current total account balance, you simply add all the individual balances.

7. Suppose your records show the following individual balances.

-220	-220	350	350	-220	-220	-220	350

a. What is the total of the individual positive balances?

b. What is the total of the individual negative balances?

c. In part b, did you sum the five negative balances? Or did you multiply -220 by 5? Which method is more efficient? What is the sign of the result of your calculation?

d. What is the total balance for these accounts?

8. From past experience, you know that multiplication is repeated addition. (See Problem 7c.) Represent the following as repeated addition.

For example, $2(-3) = $
$$\begin{array}{r} -3 \\ + \ -3 \\ \hline -6 \end{array}$$

a. $4(-12)$ **b.** $3(9)$

You can use the concept of opposites when representing multiplication as repeated addition. Think of $-2(3)$ as the opposite of $2(3) = 6$. Then $-2(3) = -6$.

9. Use your calculator to determine the following. Compare the results of parts a–d.

    a. $9 \cdot (-5) =$ **b.** $-5 \cdot 9 =$

    c. $-9 \cdot 5 =$ **d.** $5 \cdot (-9) =$

10. Evaluate the following **mentally,** then check using your calculator.

    a. $-2 \cdot 6$ **b.** $(-4)(5)$ **c.** $6 \cdot (-6)$

    d. $3(-9)$ **e.** $-8 \cdot 11$ **f.** $(-5.1) \cdot 7$

    g. From your experience in this activity, what is the sign of the product of a positive number and a negative number?

11. a. Fill in the blanks to complete the pattern begun in the first few equations.

$$3 \cdot -2 = -6$$
$$2 \cdot -2 = -4$$
$$1 \cdot -2 = \underline{\hspace{3em}}$$
$$0 \cdot -2 = \underline{\hspace{3em}}$$
$$-1 \cdot -2 = 2$$
$$-2 \cdot -2 = \underline{\hspace{3em}}$$
$$-3 \cdot -2 = \underline{\hspace{3em}}$$

    b. What do you conclude is the sign of the product of two negative numbers?

The rule you discovered in Problem 11 can also be obtained using the ideas of repeated addition and opposites. Because $3(-2) = -6$, we know that $-3(-2)$ is the opposite of -6, which is 6. Thus, $-3(-2) = 6$.

12. Evaluate the following mentally. Then check using your calculator.

    a. $-5 \cdot (-7)$ **b.** $-3 \cdot (-6)$ **c.** $-4 \cdot 6$

    d. $-8(-2)$ **e.** $-1 \cdot 2.718$ **f.** $-3.14 \cdot (-1)$

    g. $-6(0)$ **h.** $4(-7)$ **i.** $(-3)(-8)$

Once you know how to multiply signed numbers, division of signed numbers follows naturally. Recall that division is "undoing" multiplication. For example, $12 \div 4 = 3$ because $4 \cdot 3 = 12$. Similarly, $12 \div -4 = -3$ because $(-4) \cdot (-3) = 12$. (Check this on your calculator.)

13. Evaluate the following mentally. Then check using your calculator.

a. $25 \div (-5)$ **b.** $32 \div (-8)$ **c.** $(-12) \div 6$

d. $16 \div (-2)$ **e.** $-48 \div 6$ **f.** $-36 \div (-9)$

g. $-11 \cdot (-3)$ **h.** $-6 \div 3$ **i.** $0 \div (-5)$

j. $-6 \div 0$ **k.** $-63 \div (-1)$ **l.** $-7 \cdot 4$

SIGNED NUMBER MULTIPLICATION AND DIVISION

The product of two numbers with the same sign is positive.

The product of two numbers with different signs is negative.

The quotient of two numbers with the same sign is positive.

The quotient of two numbers with different signs is negative.

EXERCISES

1. Calculate mentally or use your calculator. Compare the results for similarities and differences.

a. $12 \cdot (-6)$ **b.** $12 - 6$ **c.** $12 \div (-6)$ **d.** $-6 \div 12$

2. Calculate mentally and check using your calculator. Compare the results for similarities and differences.

a. $\frac{1}{2} \div 2$ **b.** $\frac{1}{2} \div (-2)$ **c.** $-2 \div \frac{1}{2}$ **d.** $2 \div \left(-\frac{1}{2}\right)$

3. Evaluate each of the following. Use a calculator to verify your answers.

    a. $2.1(-8)$ **b.** $-8 \cdot (-7)$ **c.** $11(-3)(-4)$ **d.** $-125 \div 25$

    e. $10,000 \div (-250)$ **f.** $-6 \cdot 9$ **g.** $-24 \div (-3)$

    h. $-12 \div 4$ **i.** $0.2 \cdot (-0.3)$ **j.** $-1.6 \div (-4)$

    k. $\left(-\frac{3}{4}\right)\left(-\frac{2}{9}\right)$ **l.** $-4\frac{2}{3} \div \left(-\frac{7}{9}\right)$

4. Yesterday, you wrote five checks for \$25 each.

    a. What was the total amount of the checks?

    b. You originally had \$98 in your checking account. What was your balance after you wrote the five checks?

5. A cold front moves through your hometown in January, dropping the temperature an average of 3° per day. If the temperature is 7° on January 1, what will the temperature be five days later?

6. You own 180 shares of Bell Atlantic stock. The following table lists the changes in a given week for a single share. Did you earn a profit or suffer a loss on the value of your shares during the given week? How much profit or loss?

DAY	MON.	TUES.	WED.	THURS.	FRI.
DAILY CHANGE	+0.38	−0.75	−1.25	−1.13	2.55

ACTIVITY 1.19

Order of Operations Revisited

Topics: *Order of Operations with Signed Numbers, Negative Exponents, Scientific Notation*

You can combine the procedures for adding and subtracting signed numbers with those for multiplying and dividing signed numbers by using the order of operations you learned in Activity 1.5, You and Your Calculator. That order of operations is valid for *all* numbers, positive and negative, including decimals, fractions, and numerical expressions with exponents. Your ability to follow these procedures correctly with all numbers will help you use the formulas that you will encounter in applications.

1. Calculate the following expressions by hand. Then check your answer using your calculator.

   a. $6 + 4 \cdot (-2)$ **b.** $-6 + 2 - 3$ **c.** $(6 - 10) \div 2$

   d. $-3 + (2 - 5)$ **e.** $(2 - 7) \cdot 5$ **f.** $-3 \cdot (-3 + 4)$

   g. $-2 \cdot 3^2 - 15$ **h.** $(2 + 3)^2 - 10$ **i.** $-7 + 8 \div (5 - 7)$

   j. $2.5 - (5.2 - 2.2)^2 + 8$ **k.** $4.2 \div 0.7 - (-5.6 + 8.7)$

   l. $\frac{1}{4} - \left(\frac{2}{3} \cdot \frac{9}{8}\right)$ **m.** $\frac{5}{6} \div (-10) + \frac{7}{12}$

2. Evaluate $5 - 3^2$ by hand and then check using your calculator. Which operation must be performed first?

3. Evaluate $0 - 3^2$ by hand and then check using your calculator.

4. Evaluate -3^2 by hand. Did you obtain the same answer as in Problem 3?

In Problem 4, two operations are performed on the number 3: exponentiation and negation. This means that, by order of operations, you first square 3—that is, $3^2 = \underline{\quad 9 \quad}$. Then you write the opposite of 9 (negate the 9), which is _____.

Problems 3 and 4 together show that you can interpret a leading negative sign as subtraction by simply subtracting from zero. For example: $-16 = 0 - 16$. Therefore,

$$-4^2 = 0 - 4^2$$
$$= 0 - 16$$
$$= -16.$$

5. Evaluate -7^2 by using this technique.

6. a. Evaluate $(-3)^2$ by hand and then check with your calculator.

   b. Is $(-3)^2$ the same as -3^2?

The base for the exponent in the expression $(-3)^2$ is -3. You are calculating the square of -3. In the expression -3^2, the base for the exponent is 3. You are calculating the opposite of the square of 3.

7. Evaluate the following expressions by hand. Check each one using your calculator before going on to the next.

   a. -5^2 **b.** $(-5)^2$ **c.** $(-3)^3$

   d. -1^4 **e.** $2 - 4^2$ **f.** $(2 - 4)^2$

   g. $-5^2 - (-5)^2$ **h.** $(-1)^8$ **i.** $-5^2 + (-5)^2$

   j. $-\left(\frac{1}{2}\right)^2 + \left(\frac{1}{2}\right)^2$ **k.** $-1.5^2 - (1.5)^2$ **l.** $(-1.5)^2 - (1.5)^2$

So far in this book, you have been using only zero or positive integers as exponents. Are numbers with negative exponents meaningful? How would you calculate an expression such as 10^{-2}? In the following problems, you will discover answers to these questions.

8. a. Complete the following table.

EXPONENTIAL FORM	EXPANDED FORM	VALUE
10^5	$10 \times 10 \times 10 \times 10 \times 10$	100,000
10^4	$10 \times 10 \times 10 \times 10$	10,000
10^3	$10 \times 10 \times 10$	1000
10^2		
10^1		
10^0		
10^{-1}		
10^{-2}	$\frac{1}{10} \times \frac{1}{10}$	
10^{-3}		$\frac{1}{1000}$ or 0.001
10^{-4}		

b. What is the relationship between each negative exponent in column 1 and the number of zeros in the denominator of the corresponding fraction in column 3? In the decimal result? List and explain everything you observe.

9. What is the meaning of the expression, 10^{-6}? That is, write 10^{-6} as a fraction and as a decimal number.

10. Evaluate each of the following expressions.

    a. 10^3 **b.** 1×10^3 **c.** 5×10^3 **d.** 2.72×10^3

    e. $10 - {}^{-3}$ **f.** 1×10^{-3} **g.** 5×10^{-3} **h.** 2.72×10^{-3}

11. a. Recall from Activity 1.5 that you used the $\boxed{\text{EE}}$ key on your calculator to enter numbers written in scientific notation. For example, $10^2 = 1 \times 10^2$ is entered as $\boxed{1}$ $\boxed{\text{EE}}$ $\boxed{2}$. Try it.

b. Now enter $10^{-2} = 1 \times 10^{-2}$ as $\boxed{1}$ $\boxed{\text{EE}}$ $\boxed{-}$ $\boxed{2}$. What is the result written as a decimal? Written as a fraction?

c. Use the $\boxed{\text{EE}}$ key to enter the numbers in Problem 10 into your calculator. Record your results and compare them with your results in Problem 10.

12. Pluto takes 247.7 years to circumnavigate the sun.

    a. How many seconds does it take Pluto to circumnavigate the sun?

    b. Round your result to the nearest million seconds.

    c. Write the rounded result in standard scientific notation.

13. Tests show that the unaided eye can detect objects that have a diameter of 0.1 millimeter. What is the diameter in inches (1 in. = 25.4 mm)? Write your result in decimal form and in scientific notation.

14. The wavelength of x-rays is 1×10^{-8} centimeter, and that of ultraviolet light is 2×10^{-5} centimeter.

    a. Which wavelength is shorter?

    b. How many times shorter? Write your answer in standard form and in scientific notation.

15. Use your calculator to evaluate the following expressions.

    a. $\dfrac{16}{4 \times 10^{-2} - 2 \times 10^{-3}}$
        b. $\dfrac{3.2 \times 10^{3}}{(8.2 \times 10^{-2})(3.0 \times 10^{2})}$

EXERCISES

1. Complete the following table and compare with the table in Problem 8 of the activity.

EXPONENTIAL FORM	EXPANDED FORM	VALUE
2^5	$2 \times 2 \times 2 \times 2 \times 2$	32
2^4	$2 \times 2 \times 2 \times 2$	16
2^3		8
2^2		
2^1		
2^0		
2^{-1}		
2^{-2}	$\frac{1}{2} \times \frac{1}{2}$	
2^{-3}		$\frac{1}{8}$
2^{-4}		

2. Evaluate the following expressions by hand or by calculator as you see fit. Estimate your results to see if your answers are reasonable.

    a. $(8 - 17) \div 3 + 6$ **b.** $-7 + 3(1 - 5)$ **c.** $-3^2 \cdot 2^2 + 25$

    d. $1.6 - \left(1.2 + 2.8\right)^2$ **e.** $\frac{5}{16} - 3\left(\frac{3}{16} - \frac{7}{16}\right)$ **f.** $\frac{3}{4} \div \left(\frac{1}{2} - \frac{5}{12}\right)$

    g. -1^{-2} **h.** $4^3 - \left(-4\right)^3$ **i.** $\left(\frac{2}{3}\right)^2 - \left(\frac{2}{3}\right)^2$

    j. $\dfrac{-3 \times 10^2}{3 \times 10^{-2} - 2 \times 10^{-2}}$ **k.** $\dfrac{1.5 \times 10^3}{\left(-5.0 \times 10^{-1}\right)\left(2.6 \times 10^2\right)}$

3. In 1994, the U.S. government paid 296.3 billion dollars in interest on the national debt.

   a. Write this number in standard notation and in scientific notation.

   b. Assume that there were approximately 250 million people in the United States in 1994. How much of the debt could we say each person owed on the interest?

4. You have found 0.4 gram of gold while panning for gold on your vacation.

   a. How many pounds of gold do you have $(1 \text{ g} = 2.2 \times 10^{-3} \text{ lb})$? Write your result in decimal form and in scientific notation.

   b. If you were to tell a friend how much gold you have, would you state the quantity in grams or in pounds? Explain.

5. You are doing an experiment in a chemistry research lab and have to calculate the molecular weight of the compound you have isolated from the other compounds in the experiment. This leads you to the following numerical expression:

$$\text{Molecular weight of the compound (grams per mole)} = \frac{(0.134)(0.082)(371)}{(0.9697)(0.0532)}$$

Use your calculator to evaluate the expression, and write your result to the nearest thousandth.

What Have I Learned?

Your study group is preparing for a test on Cluster 4.

1. Explain how you would add two signed numbers and how you would subtract two signed numbers. Use examples to illustrate each.

2. Explain how you would multiply two signed numbers and how you would divide two signed numbers. Use examples to illustrate each.

3. What would you suggest to your classmates to avoid confusing addition and subtraction procedures with multiplication and division procedures?

4. **a.** Without actually performing the calculation, determine the *sign* of the product $(-0.1)(+3.4)(6.87)(-0.5)(+4.01)(3.9)$. Explain.

   b. Determine the *sign* of the product $(-0.2)(-6.5)(+9.42)(-0.8)(1.73)(-6.72)$. Explain.

   c. Determine the *sign* of the product $(-1)(-5.37)(-3.45)$. Explain.

d. Determine the *sign* of the product
$(-4.3)(+7.89)(-69.8)(-12.5)(+4.01)(-3.9)(-78.03)$. Explain.

e. What rule do the results you obtained in parts a–d suggest?

5. a. If -2 is raised to the power 4, what is the sign of the result? What is the sign if -2 is raised to the sixth power?

b. Raise -2 to the eighth power, tenth power, and twelfth power.

c. What general rule involving signs is suggested by the preceding results?

6. a. Raise -2 to the third power, fifth power, seventh power, and ninth power.

b. What general rule involving signs is suggested by the preceding results?

7. The following table contains the daily midnight temperatures in Batavia, NY, for a week in January.

DAY	SUN.	MON.	TUES.	WED.	THURS.	FRI.	SAT.
TEMPERATURE (°F)	−7°F	11°F	11°F	11°F	−7°F	−7°F	11°F

a. Determine the average daily temperature for the week.

b. Did you use the most efficient way to do the calculation? Explain.

8. a. What is the value of 3^2, 3^0, and 3^{-2}?

b. What is the value of any nonzero number raised to the zero power?

c. What is the meaning of any nonzero number raised to a negative power? Give two examples using -2 as the exponent.

9. a. The diameter of raindrops in drizzle near sea level is reported to be approximately 30×10^{-2} millimeter. Write this number in standard decimal notation and in scientific notation.

b. What is the measurement of these raindrops in centimeters (10 mm = 1 cm)? Write your answer in standard notation and in scientific notation.

c. From your experience with scientific notation, explain how to convert numbers from standard form to scientific form and vice versa.

How Can I Practice?

Calculate by hand or using your calculator.

1. $15 + (-39)$

2. $-43 + (-28)$

3. $-0.52 + 0.84$

4. $-7.8 + 2.9$

5. $-32 + (-45) + 68$

6. $-46 - 63$

7. $53 - (-64)$

8. $8.9 - (-12.3)$

9. $-75 - 47$

10. $-34 - (-19)$

11. $-4.9 - (-2.4) + (-5.6) + 3.2$

12. $16 - (-28) - 82 + (-57)$

13. $-1.7 + (-0.56) + 0.92 - (-2.8)$

14. $\frac{2}{3} + \left(-\frac{3}{5}\right)$

15. $-\frac{3}{7} + \left(\frac{4}{5}\right) + \left(-\frac{2}{7}\right)$

16. $-\frac{5}{9} - \left(-\frac{8}{9}\right)$

17. $2 - \left(\frac{3}{2}\right)$

18. $1 - \left(\frac{3}{4}\right) + \left(-\frac{3}{4}\right)$

19. $0 - \left(-\frac{7}{10}\right) + \left(-\frac{1}{2}\right) - \frac{1}{5}$

20. $\frac{-48}{12}$

21. $\frac{63}{-9}$

22. $\frac{121}{-11}$

23. $\frac{-84}{-21}$

24. $-125 \div -25$

25. $-2.4 \div 6$

26. $\frac{24}{-6}$

27. $-\frac{24}{6}$

28. $-0.8(12)$

29. $4(-0.06)$

30. $-9(-11)$

31. $-4(-0.6)(-5)(-0.01)$

32. $0.5(-7)(-2)(-3)$

33. $-4(-4)$

34. $(-4)(-4)$

35. $4(-4)$

36. You have $85.30 in your bank account. You write checks for $23.70 and $35.63. You then deposit $325.33. Later, you withdraw $130.00. What is your final balance?

37. You lose 4 pounds, then gain 3 pounds back. Later you lose another 5 pounds. If your aim is to lose 12 pounds, how many more pounds do you need to lose?

38. You and your friend Patrick go on vacation. When you are at sea level, Patrick decides to go scuba diving, but you prefer to do some mountain climbing. You decide to separate part of the day and do your individual activities. While you climb 2567 feet, Patrick is 49 feet underwater. What is the vertical distance between you and Patrick?

39. a. The temperature is 2°C in the morning but drops to −5°C by night. What is the change in temperature?

b. The temperature is −12°C in the morning, but is expected to drop 7° during the day. What will be the evening temperature?

c. The temperature is −17°C this morning, but is expected to rise 9° by noon. What will be the noon temperature?

d. The temperature is −8°C in the morning but drops to −17°C by night. What is the change in temperature?

e. The temperature is −14°C in the morning and −6°C at noon. What is the change in temperature?

40. a. $-6 \div 2 \cdot 3$ **b.** $(-3)^2 + (-7) \div 2$

c. $(-2.5)^3 + (-9) \div (-3)$ d. $(-14 - 4) \div 3 \cdot 2$

e. $(-4)^2 - [-8 \div (2 + 6)]$ f. $-4.4 \div (-0.2)^2 + 1.8$

g. $(-3)^3 \div 9 - 6$

41. a. At sea level, fog droplets on the average measure 20×10^{-3} millimeter in diameter. How many inches is this (1 mm = 0.03037 in.)?

       b. At sea level, the average diameter of raindrops is approximately 1 millimeter. How many inches is this?

       c. How many times larger are raindrops than fog droplets?

CLUSTER 5

Geometric Shapes
Topics: *Measurement, Perimeter, Area, Empirical Verification of Formulas*

Problem Solving with Geometry

You encounter problems involving geometric shapes every day. For example, you may wish to determine the amount of paint needed to redecorate a room, calculate the length of fencing needed for a new kennel for your dog, or design the shapes of pieces to make a quilt. These are all situations that involve measuring objects and calculating perimeters and areas. In the following problems, you will measure basic geometric figures and develop the formulas for calculating the perimeters and areas of these figures.

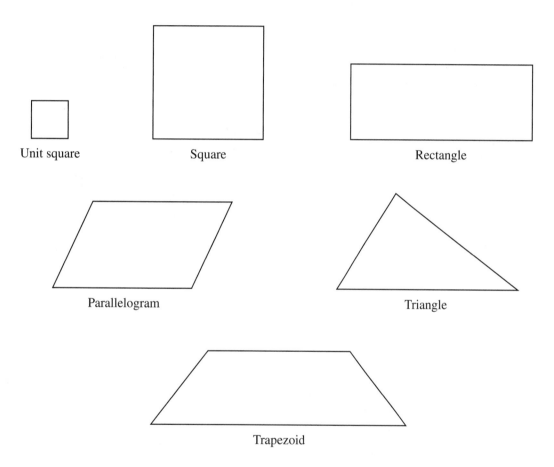

Unit square   Square   Rectangle

Parallelogram   Triangle

Trapezoid

1. Measure each side of each of the preceding geometric figures. Use a metric ruler and use centimeters for the unit of measure. Label each side with its measure. Note that 1 centimeter is abbreviated 1 cm.

2. Each basic geometric figure has its own special features. Study each geometric figure preceding Problem 1 and note what features it has. Put a check in each box of the following table where a feature applies.

	ALL SIDES EQUAL	OPPOSITE SIDES EQUAL	RIGHT ANGLE	BOTH PAIRS OF OPPOSITE SIDES PARALLEL	ONE PAIR OF OPPOSITE SIDES PARALLEL
TRAPEZOID					
PARALLELOGRAM					
RECTANGLE					
SQUARE					
TRIANGLE					

> The **perimeter** of a geometric figure is the measure of the distance around the figure. Perimeter is measured in linear units such as meters, feet, or miles.

3. Use your measurements from Problem 1 to determine the perimeter of each of the geometric figures. Be sure to include the unit of measurement.

	PERIMETER
TRAPEZOID	
PARALLELOGRAM	
RECTANGLE	
SQUARE	
TRIANGLE	

Sometimes you can use a formula as a shortcut to measuring the perimeter of a figure.

4. Draw a square and represent each side with the letter s. Write a formula for the perimeter P in terms of s.

5. Draw a rectangle. Let the longest side of the rectangle be represented by l (length), and let the shorter side be represented by w (width). Write a formula for the perimeter P of the rectangle in terms of l and w.

6. Draw a triangle and label its sides *a*, *b* and *c*. Write a formula for the perimeter of the triangle.

The formulas for the perimeters of a square, rectangle, and triangle are listed on the inside back cover of this text. Compare your answers to Problems 4, 5, and 6 with those formulas.

7. Use the appropriate geometric formula to answer each of the following.

   a. The side of a square, *s*, measures 5 centimeters. Determine the perimeter of the square.

   b. The length *l* and width *w* of a rectangle are 6 inches and 4 inches, respectively. Determine the perimeter of the rectangle.

   c. The lengths of the sides of a triangle are 3 feet, 4 feet, and 5 feet. What is the perimeter of the triangle?

The **area** of a geometric figure is the measure of the space enclosed by the sides of the figure. Area is measured in square units such as square meters, square feet, or square miles.

A unit square is one whose side length is 1 unit of measure. In this activity, you will use centimeters as your unit of measure, so the unit square will have a side length of 1 centimeter and an area of 1 square centimeter.

8. Draw copies of the unit square to fill the space enclosed by the sides of the larger square as seen on page 135. To determine the area of the larger square, simply count the number of whole and partial unit squares you drew. The total number of unit squares you drew is the area of the larger square in square centimeters. Record the area of the square. Be sure to use the correct units.

9. Your drawing in Problem 8 should suggest to you a way to obtain the area of a square from the measure of one of its sides, *s*. There is a shortcut formula for finding the area of a square. Determine a shortcut formula for the area *A* of a square in terms of *s*.

10. Draw copies of the unit square to fill the space enclosed by the sides of the rectangle on page 135. Determine the area of the rectangle by counting the number of unit squares you drew. Record the area of the rectangle. Be sure to use the correct units.

11. There is a shortcut formula for finding the area of a rectangle. Use your drawing from Problem 10 to determine the formula for the area A of a rectangle in terms of the measures of its length l and its width w.

12. Estimate the number of unit squares and fractions of unit squares you can draw in the space enclosed by the sides of the parallelogram on page 135. Use this estimate to approximate the area of the parallelogram. Record your result using the correct units.

13. a. In the accompanying figure, a shaded rectangle is placed over the original parallelogram. The length of the rectangle is the base b of the parallelogram, and the width of the rectangle is h, the height of the parallelogram. Measure h and b and then calculate the area of the shaded rectangle.

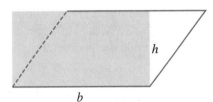

b. Compare your result for the area of the shaded rectangle with your estimate for the area of the parallelogram from Problem 12. Explain why the area of the shaded rectangle is the same as the area of the original parallelogram.

14. Write a formula for the area A of the parallelogram in terms of b and h.

15. Estimate the number of unit squares and fractions of unit squares you can draw in the space enclosed by the sides of the triangle on page 135. Use this estimate to approximate the area of the triangle. Record your result using the correct units.

16. a. In the accompanying figure, a shaded parallelogram is placed over the original triangle. Measure h, and calculate the area of the shaded parallelogram.

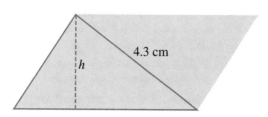

   b. How is the area of the original triangle related to the area of the shaded parallelogram?

17. a. Write a formula for the area A of a triangle in terms of the base b and the height h.

   b. Use the formula from part a to calculate the area of the triangle, and compare your answer with your estimate from Problem 15.

18. a. Estimate the number of unit squares and fractions of unit squares you can draw in the space enclosed by the sides of the trapezoid on page 135. Use this estimate to approximate the area of the trapezoid. Record your result with the correct units.

   b. The shortcut formula for the **area of a trapezoid** is

$$A = \frac{b + B}{2} \cdot h,$$

   where b and B are measures of the two parallel sides, or bases, and h is the distance between the bases. You can think of the formula as the average of the bases times the height. Use this formula to calculate the area of the trapezoid. Compare your result with your estimate from part a.

The shortcut formulas for the area of a square, rectangle, parallelogram, triangle, and trapezoid are listed on the inside back cover of this text. Check the formulas you obtained in this lab activity with the formulas printed there.

19. Use the appropriate geometric formula to answer each of the following.

   a. A square measures 1 foot 6 inches on each side. Calculate the area of the square in square feet by converting first to feet. Then convert the side measure to inches and calculate the area in square inches.

   b. Use unit conversions to show that the two areas obtained in part a are equivalent.

   c. The base of a triangle measures 5 inches and its height measures 3 inches. Calculate the area of the triangle.

   d. The bases of a trapezoid measure 2.2 meters and 4.6 meters and its height measures 1.8 meters. Calculate the area of the trapezoid.

   e. The length of a rectangle is 7.5 inches and the width is 2.3 inches. What is the area of the rectangle?

EXERCISES

Use a metric ruler to measure the dimensions of the following geometric figures, and use the applicable shortcut formula to calculate the perimeter and area of each figure.

1.

2.

3.

4.

5.

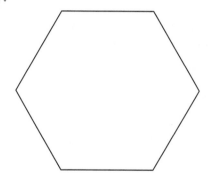

6.

7.

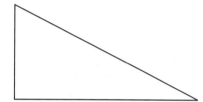

ACTIVITY 1.21

Home Improvements
Topic: *Using Geometry Formulas in Context*

Every summer you do some home improvement. This year you have $1000 budgeted for this purpose and the list looks like this: Replace the kitchen floor, add a wallpaper border to the third bedroom, seed the lawn, and paint the walls in the family room.

To stay within your budget, you need to determine the cost of each of these projects. You expect to do this work yourself.

1. The kitchen floor is divided into two parts. The first section is 12 by 14 feet, and the second section is a breakfast area, triangular in shape, which extends 5 feet out from the 12-foot side along a window. The cost of linoleum is $21 per square yard plus 6% sales tax. Linoleum is sold in 12-foot widths.

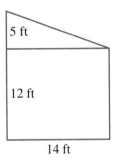

a. How long a piece of linoleum will you need to purchase if you want only one seam, where the breakfast area meets the main kitchen area, as shown? Remember that the linoleum is 12 feet wide.

b. How many square feet of linoleum must you purchase? How many square yards is that $(9 \text{ sq ft} = 1 \text{ sq yd})$?

c. How much will the linoleum cost?

d. How many square feet of waste will you have?

2. The third bedroom is rectangular in shape and has dimensions of $8\frac{1}{2}$ by 13 feet. On each 13-foot side, there is a window that measures 3 feet 8 inches wide and 4 feet high. The door is located on an $8\frac{1}{2}$-foot side and measures 3 feet wide from casing to casing. You are planning to put up the border around the room about halfway up the wall.

  a. How many feet of wallpaper border will you need to purchase?

  b. The border comes in rolls 5 yards in length. How many rolls will you need to purchase?

  c. The wallpaper border costs $10.56 per roll plus 6% sales tax. Determine the cost of the border.

  d. How much waste will you have (in feet)?

3. Your lawn needs to be seeded.

  a. Using the following scale, determine the size of your lot and your home. Determine how many square feet of lawn you need to seed.

  Scale: $\frac{1}{4}$ in. = 16 ft

b. Grass seed comes in bags of 5, 10, or 25 pounds. The 5-pound bag covers 2000 square feet of lawn and costs $12 a bag. The 10-pound bag covers twice as much lawn as the 5-pound bag and costs $23 a bag. The 25-pound bag costs $55, but since the label was ripped, you will need to determine proportionally how many square feet of lawn will be covered by the 25-pound bag.

c. Which bags of grass seed will you purchase?

d. What will be the total cost of your purchase? Remember to include the 6% sales tax.

4. Your family room needs to be painted. It has a cathedral ceiling. The front and back walls look like the diagram below. The two side walls measure 14 feet long and 12 feet high. The walls will need two coats of paint. The ceiling does not need paint.

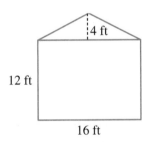

a. How many square feet will you be painting? (Remember, it will need two coats.)

b. Each gallon covers approximately 400 square feet. How many gallons of paint will you need to purchase?

c. Paint costs $19.81 per gallon plus 6% sales tax. What is the cost of the paint you need for the family room?

5. What is the total cost for all your home improvement projects?

6. Can you afford to do all the projects?

E X E R C I S E S

1. Your driveway is rectangular in shape and measures 15 feet wide and 25 feet long. Calculate the area of your driveway.

2. A flower bed is in the corner of your yard and is in the shape of a right triangle. The sides of the bed measure 6 feet 8 inches and 8 feet 4 inches. Calculate the area of the flower bed.

3. The front wall of your storage shed is in the shape of a trapezoid. The bottom measures 12 feet, the top measures 10 feet, and the height is 6 feet. Calculate the area of the front wall of your shed.

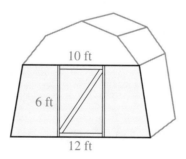

4. You live in a small, one-bedroom apartment. The bedroom is 10 by 12 feet, the living room is 12 by 14 feet, the kitchen is 8 by 6 feet, and the bathroom is 5 by 9 feet. Calculate the total floor space (area) of your apartment.

5. A stop sign is in the shape of a regular octagon. A regular octagon can be created using eight triangles of equal area. One triangle that makes up a stop sign has a base of 16 inches and a height of 18 inches. Calculate the area of the stop sign.

Objects in the shape of circles of varying sizes are found in abundance in everyday life. Coins, dart boards, and ripples made by a raindrop in a pond are just a few examples. The size of a circle is customarily described by the length of a line segment that passes through the center and starts and ends on the circle. You may recall that this line segment is called the **diameter** of the circle. Recall also that the **radius** of a circle is the line segment starting at the center of the circle and ending on the circle. The length of the radius is one-half the length of the diameter.

In this laboratory activity, you will estimate the perimeter (more commonly known as the **circumference**) and area of circles. You will then develop formulas to calculate the circumference and area of a circle, given its radius or diameter.

1. Find four objects that are circular in shape and list them in column 1 of the accompanying table.

   a. Use a tape measure or a piece of string and ruler to measure (in centimeters) the circumference C of each object. Record these numbers in column 2 of the table.

   b. The diameter is more difficult to measure accurately than is the circumference. Devise a method for measuring the length of the diameter and explain why your method will actually measure diameter.

   Now use your method to measure the diameters of your chosen objects and record the measurements in column 3 of the table.

   c. Use the information from parts a and b to complete columns 4 and 5 in the table.

COLUMN 1 OBJECT USED	COLUMN 2 CIRCUMFERENCE C, FOUND BY MEASURING	COLUMN 3 DIAMETER d, FOUND BY MEASURING	COLUMN 4 $\frac{C}{d}$, EXPRESSED AS A FRACTION	COLUMN 5 $\frac{C}{d}$, EXPRESSED AS A DECIMAL
1.				
2.				
3.				
4.				

   d. Are the four values in column 5 approximately the same? Should they be?

   e. Calculate the average of the four decimal values in column 5, and compare your average with the averages obtained by your classmates. What do you conclude?

In every circle, the ratio of the circumference to the diameter is the same. This ratio is represented by π, the Greek letter "pi." Locate the $\boxed{\pi}$ key on your calculator and enter it to determine the numerical value of π. _____ The exact value of π is a non-repeating decimal. You may recall two familiar approximations of π: the two-decimal place approximation, 3.14 and the fraction approximation, $\frac{22}{7}$. It is most accurate to use the $\boxed{\pi}$ key when performing any calculation with π, and to round your final answer to the specified decimal place.

> In every circle, the ratio of the circumference to the diameter is always the same. This ratio of circumference to diameter is represented by the Greek letter π.
>
> $$\frac{C}{d} = \pi$$
>
> The formula for the circumference of the circle is given by
>
> $$C = \pi d \qquad \text{or since } d = 2r, \qquad C = 2\pi r.$$

NOTE: These formulas are also found on the inside back cover of the book.

For example, the circumference of a circle with radius 2 centimeters is given by

$$C = 2\pi r = 2\pi(2 \text{ cm}) = 4\pi \text{ cm} \approx 12.57 \text{ cm}.$$

Recall: $\approx$ means "is approximately equal to."

2. Calculate the circumference of a circle with diameter 11 feet.

3. The area of a circle is defined as the space inside the circle and is measured in square units.

   a. For each of the four circles in Problem 1, estimate how many 1-by-1-centimeter unit squares and fractions of unit squares can fit in each circle by drawing these squares on the circles. Place these estimates in column 2 of the accompanying table.

   b. Calculate the radius of each circle from the diameter measurement in Problem 1b and record the result in column 3.

   c. Compute πr^2 for each circle, and place your answers in column 4.

COLUMN 1 OBJECT USED	COLUMN 2 ESTIMATED AREA	COLUMN 3 RADIUS, $r = \frac{d}{2}$	COLUMN 4 πr^2
1.			
2.			
3.			
4.			

d. Compare your answers in columns 2 and 4. What do you notice?

The **area** A of any circle of radius r is $A = \pi r^2$.

For example, the area of a circle with radius 3 centimeters is determined as follows:

$$A = \pi r^2 = \pi (3 \text{ cm})^2 = 9\pi \text{ cm}^2 \approx 28.27 \text{ cm}^2$$

4. Calculate the area of a circle with radius 4.6 feet.

5. You can demonstrate why the formula $A = \pi r^2$ for the area of a circle is reasonable.

    a. Cut a circle into eight equal pieces (called sectors). Rearrange the sectors into an approximate parallelogram (see accompanying figure).

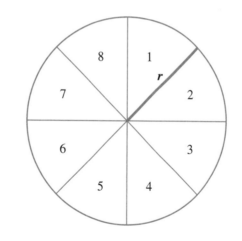

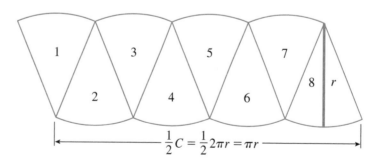

b. What measurement on the circle approximately represents the height of the parallelogram?

c. What measurement on the circle approximately represents the base of the parallelogram?

d. Recall that the area of a parallelogram is given by the product of its base and its height. Use your answers to parts b and c to determine the area of the parallelogram. What is the resulting expression?

A New Pool

Topic: Circumference and Area Formulas in Context

You are the proud owner of a new circular swimming pool of diameter 25 feet and are eager to dive in. However, you quickly discover that having a new pool requires making many decisions about other purchases.

1. Your first concern is a solar cover. You do some research and find that circular covers come in the following sizes: 400, 500, and 600 square feet. Which size is best for your needs? Explain.

2. You decide to build a circular concrete patio 6 feet wide all around your pool. What is the area covered by your patio? Explain.

3. State law requires that all pools be enclosed by a fence to prevent accidents. You decide to fence all around your pool and patio. How many feet of fencing do you need? Explain.

4. Your pool is uniformly 8 feet deep. How many cubic feet of water do you need to fill the pool? (Hint: Multiplying the area of the pool by its depth produces the volume of the pool, measured in cubic feet of water.) Explain.

EXERCISES

1. You have put on weight, and the radius of your waist has increased by an inch (assume that your waist is approximately circular). How much has your waist measurement increased?

2. A Norman window has the shape of a rectangle with a semicircular top. If the rectangle is 4 feet wide and 5 feet tall, what is the area of the window?

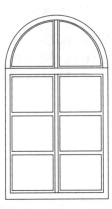

3. You want to know how much space is available between a basketball and the rim of the basket. One way to find out is to measure the circumference of each and use the circumference formula to determine the corresponding diameters. The distance you want to determine is the difference between the diameter of the rim and the diameter of the ball. Try it!

Carpenter's Square
Topics: *Right Triangles,*
Pythagorean Theorem,
Measurement, Angles,
Square Roots

In this activity you will experimentally verify an important right-triangle geometric formula used by surveyors, architects, and builders to check that two lines are perpendicular, that is, intersect at a 90° angle. You will need a protractor and ruler.

Recall that in a right triangle two sides (called *legs*) are perpendicular to each other. The side opposite the right angle is called the *hypotenuse* and is often denoted by the letter c; the legs are usually denoted by the letters, *a* and *b*.

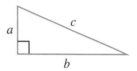

1. Use a protractor to construct three right triangles, one with legs of length 1 in. and 5 in., a second with legs of length 3 inches and 4 inches, and a third with legs of length 2 inches each.

2. Use a ruler to measure (in inches) the hypotenuse of each triangle, and record the lengths in the following table.

	a	b	c	a^2	b^2	c^2
TRIANGLE 1	1	5				
TRIANGLE 2	3	4				
TRIANGLE 3	2	2				

3. There does not appear to be a relationship between *a, b,* and *c,* but investigate further. Square *a, b,* and *c* and enter the values in the appropriate columns in the table in Problem 2.

4. What is the relationship between a^2, b^2, and c^2? Explain.

The Pythagorean theorem gives the relationship between the legs of a right triangle, *a* and *b*, and the hypotenuse, *c*. In words, the theorem states that the sum of the squares of the leg lengths is equal to the square of the hypotenuse length.

Symbolically, the theorem is written as: $c^2 = a^2 + b^2$ or $c = \sqrt{a^2 + b^2}$

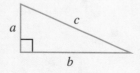

For example, if a right triangle has legs $a = 4$ centimeters and $b = 7$ centimeters, you can calculate the length of the hypotenuse as follows:

$$c = \sqrt{a^2 + b^2} = \sqrt{(4 \text{ cm})^2 + (7 \text{ cm})^2} = \sqrt{16 \text{ cm}^2 + 49 \text{ cm}^2}$$
$$= \sqrt{65 \text{ cm}^2} = \sqrt{65} \text{ cm} \approx 8.06 \text{ cm}$$

5. A right triangle has legs measuring 5 centimeters and 16 centimeters. Use the Pythagorean theorem to calculate the length of the hypotenuse. Check your answer by constructing the triangle and measuring its hypotenuse.

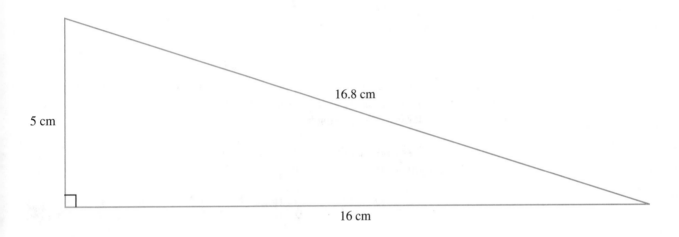

5 cm

16.8 cm

16 cm

6. In a right triangle, the sum of the measures of the two nonright angles is 90°. Verify this fact by using your protractor to obtain and then sum the measures of the nonright angles in each of the three right triangles in Problem 1.

	ONE ANGLE	OTHER ANGLE	SUM
TRIANGLE 1			
TRIANGLE 2			
TRIANGLE 3			

7. The experiment in Problem 6 also demonstrates that the sum of the three angles in a right triangle is 180°. This fact generalizes to any triangle. Verify this fact experimentally on each of the following three triangles:

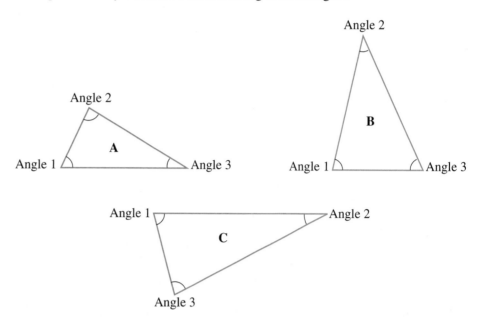

	ANGLE 1	ANGLE 2	ANGLE 3	SUM
TRIANGLE A				
TRIANGLE B				
TRIANGLE C				

The sum of the measures of the angles of a triangle is 180 degrees.

ACTIVITY 1.25

Septic Tank

Topics: *Pythagorean Theorem, Square Roots*

You are putting in a septic tank for your house. The building code in your area states that there must be at least 100 feet between the well and the septic tank. To make sure you meet code, you must find the distance between the septic tank and the well. Your house is located in the direct line of measurement from the well to the septic tank (see figure) preventing you from measuring this distance.

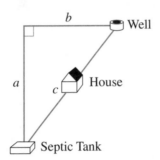

In this diagram, the distance between the well and the septic tank, represented by c, is the hypotenuse of a right triangle whose legs are a and b. Therefore, the Pythagorean theorem is applicable.

1. If $a = 65$ feet and $b = 85$ feet, do you meet code? Explain.

2. If $a = 55$ feet and $b = 75$ feet, do you meet code? Explain.

3. In calculating the square root of $a^2 + b^2$, is your answer the same as $a + b$? Explain.

4. If the distance between the septic tank and the well is exactly 100 feet and b is 80 feet, what is a?

Use the context of this activity to answer the following questions.

1. If $a = 60$ feet and $b = 80$ feet, do you meet code? Explain.

2. If $a = 70$ feet and $b = 40$ feet, do you meet code? Explain.

3. If $c = 130$ feet and $a = 50$ feet, determine b.

4. If $c = 260$ feet and $b = 240$ feet, determine a.

ACTIVITY 1.26

Basketballs, Soup Cans, Boxes, and Ice Cream Cones

Topic: *Volume of Three-Dimensional Objects*

Three-dimensional objects appear everywhere. One important measure associated with these figures is their *volume*—the amount of "space" inside the objects, measured in unit cubes. The following table contains formulas for finding volumes.

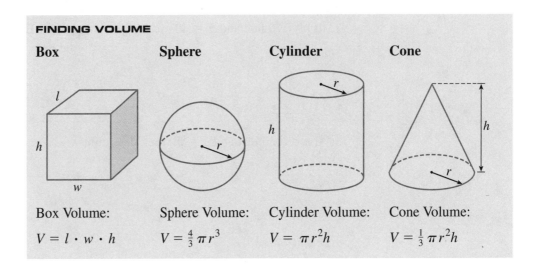

FINDING VOLUME

Box	Sphere	Cylinder	Cone
Box Volume:	Sphere Volume:	Cylinder Volume:	Cone Volume:
$V = l \cdot w \cdot h$	$V = \frac{4}{3}\pi r^3$	$V = \pi r^2 h$	$V = \frac{1}{3}\pi r^2 h$

For example, the volume of a standard basketball with radius 11 centimeters is given by

$$V = \tfrac{4}{3}\pi r^3 = \tfrac{4}{3}\pi(11 \text{ cm})^3 = \tfrac{5324}{3}\pi \text{ cm}^3 \approx 5575.28 \text{ cm}^3.$$

1. Measure the dimensions of a typical soup can and use these measurements to determine its volume in cubic centimeters.

2. You buy a can of soda and notice that the label reads 12 ounces. You decide to check on the accuracy of the label. You measure the height of the can and find that it is approximately 4.5 inches, and the can's diameter is 2.5 inches. You check a math table and find that 1 oz = 1.8 in^3. Is the label accurate? Explain.

3. Which one of the following containers encloses the greatest volume? Explain.

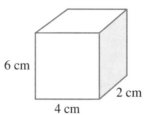

6 cm

2 cm

4 cm

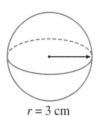

$r = 3$ cm

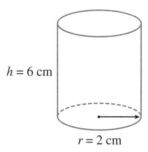

$h = 6$ cm

$r = 2$ cm

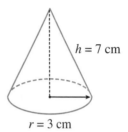

$h = 7$ cm

$r = 3$ cm

EXERCISES

1. You are assistant manager of the Tastee Ice Cream Shop in your hometown. Your manager has purchased a new cone size—the super deluxe. You have been asked to determine a price for soft ice cream cones made using the new cones.

a. Calculate the total volume of ice cream that fills the super deluxe cone in the accompanying diagram.

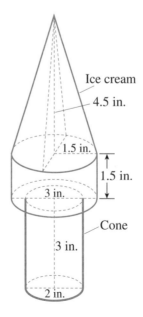

Ice cream

4.5 in.

1.5 in.

1.5 in.

3 in.

Cone

3 in.

2 in.

b. The average cost per cubic inch of ice cream is 3 cents. If each cone costs 2 cents, calculate the cost of the ice cream and the cone.

c. The price charged for this ice cream cone is set at double the cost of the ice cream and the cone. Calculate this price.

EXPLORING NUMERACY

If you could put a rope firmly around the equator and then loosen the rope by 1 inch, do you think you could slip a piece of paper through the space between the rope and the globe? A penny? If you put a rope around a basketball and then loosen the rope by 1 inch, how thick an object do you think you can slip through the space between the rope and the basketball? Explain.

Another New Pool
Topics: *Measurement, Scale Drawing, Volume*

A friend of yours is so excited by your new pool (see Activity 1.23) that she decides to buy a new in-ground swimming pool with the dimensions given by the following scale drawing.

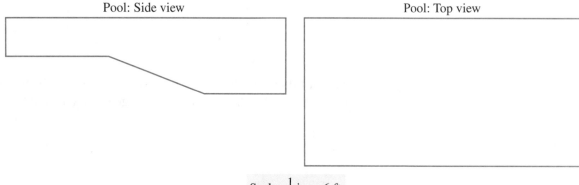

Pool: Side view   Pool: Top view

Scale: $\frac{1}{2}$ in. = 6 ft

1. Measure the scale drawing and use the measurements to determine the dimensions (in feet) of your friend's new pool.

2. Calculate the area of the top surface of water in her pool.

3. **a.** Calculate the area of the side of the pool (shown in the figure).

   b. Calculate the total volume of the pool. (*Hint:* What is the distance between the two sides?)

4. If the pool is to be filled with water to 6 inches from the top, calculate the amount of water, in cubic feet, needed to fill the pool.

5. Determine the number of gallons of water needed to fill the pool. (There are 7.48 gallons in 1 cubic foot of water.)

6. A garden hose can fill the pool at the rate of 4.5 gallons per minute. How many minutes will it take to fill the pool with the garden hose? How many hours? How many days?

7. A company that fills pools charges 3¢ per gallon for water delivered in a big tanker truck. How much will this company charge to fill the pool?

8. Because of the soil conditions in your area, your friend needs to know the weight of the water in the pool. Water weighs 62.4 pounds per cubic foot. Calculate the weight of the water in the pool to the nearest pound. Compare it with the weight of an average car.

9. The pool liner is guaranteed for five years. A new pool liner sells for $6.19 per square yard. How much will a new liner for your friend's pool cost? (Note that the liner covers all surfaces, the sides and bottom.)

10. To prevent accidents, state law requires that all pools be enclosed by a fence. How many feet of fencing are needed if a fence will be placed around the pool 4 feet from each side?

What Have I Learned?

In the Cluster 5 activities, you solved problems involving geometry. Review what you have learned by responding to the following questions.

1. Approximate the area of this irregular shape. Carefully explain your approach.

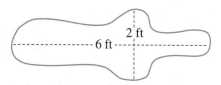

2. Measurement seldom produces the exact answer produced by the geometric formulas. Why do you think this is true?

3. You have a friend who has always had problems distinguishing perimeter from area. In language that your friend can understand, explain the difference. Expand your explanation to include the concept of volume. Include units in your explanation.

4. Why do you think so many shapes in the natural world are described by the formulas you investigated in these activities?

5. Geometry is one of the oldest areas of mathematics, with many important results dating back to the Greeks over 2000 years ago. Investigate some of this history by researching such names as Euclid or Archimedes. They and others may be found in history of mathematics books written by P. Beckmann, E. T. Bell, C. Boyer, F. Cajori, V. Katz, M. Kline, and D. Struick. The Internet is another good source of information. Write a summary of what you learn.

How Can I Practice?

1. Calculate the area of each of the following figures.

   a.

   $h = 4$ ft

   $b = 10$ ft

   b.

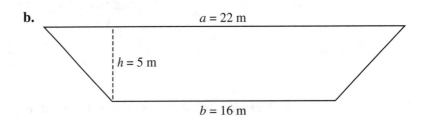

   $a = 22$ m

   $h = 5$ m

   $b = 16$ m

   c.

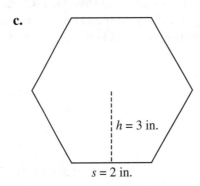

   $h = 3$ in.

   $s = 2$ in.

2. Calculate the area of each of the following figures.

   a. Rectangle topped by a semicircle

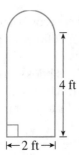

   4 ft

   $\leftarrow$ 2 ft $\rightarrow$

b. Rectangle topped by a right triangle

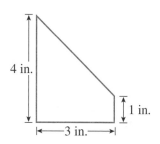

3. Determine the perimeter of each of the following.

   a. A parallelogram with a pair of parallel sides of 14 inches and 5 inches.

   b.

   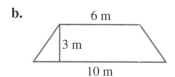

   c.

   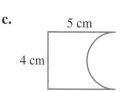

4. Determine the area of the *shaded region* in each of the following.

   a.

   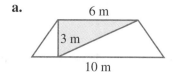

   b.

   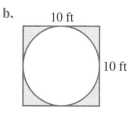

5. Determine the volume of each of the following.

   a.

   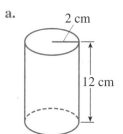

   b.

   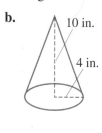

6. Consider the right triangle with dimensions as shown.

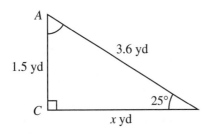

a. Determine the length x of the unknown side, rounded to the nearest thousandth.

b. Determine the measure of angle A.

c. Determine the area of the triangle.

7. You are remodeling your house. In one room, you want to put in a Norman window, which is a rectangle with a semicircle on top. The width of the window is 30 inches, and the total height of the window is 60 inches.

a. Determine the total area of the window, rounded to the nearest hundredth. Label the result appropriately.

b. Determine the total outside perimeter of the window, rounded to the nearest hundredth. Label the result appropriately.

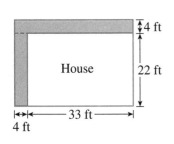

45 in.

30 in.

8. In addition to a Norman window, you also intend to put in a 4-foot-wide concrete walkway along two sides of your house, as shown.

a. Determine the area covered by the walkway only.

b. If the walkway is to be 4 inches thick, then determine the volume of concrete needed to build it, rounded to the nearest hundredth. Label the result appropriately.

House

4 ft

22 ft

33 ft

4 ft

1. Which number is greater, 1.0001 or 1.001?

2. Multiply: $32.09 \cdot 0.0006$

3. Evaluate 27^0.

4. Evaluate $|-9|$.

5. Determine the length of the hypotenuse of a right triangle if each of the legs has a length of 6 meters.

6. Evaluate $|9|$.

7. Write 203,000,000 in scientific notation.

8. How many quarts are there in 8.3 liters?

9. Write 2.76×10^{-6} in standard form.

10. How many ounces are there in a box of crackers that weighs 283 grams?

11. Determine the area of a U.S. one-dollar bill.

12. Determine the circumference of a quarter.

13. Determine the area of a triangle with base 14 inches and height 18 inches.

14. Twenty-five percent of the students in your history class scored between 70 and 80 on the last exam. In a class of 32 students, how many students does this include?

15. You have decided to accept typing jobs to earn some extra money. Your first job is to type a 420-page thesis. If you can type 12 pages per hour, how long will it take you to finish the job?

16. If you make 10 mistakes in typing 8 pages, how many mistakes will you make typing a 420-page thesis?

Exercise numbers appearing in color are answered in the Selected Answers section of this book.

17. How many pieces of string 1.6 yards long can be cut from a length of string 9 yards long?

18. Reduce $\frac{16}{36}$ to lowest terms.

19. Change $\frac{16}{7}$ to a mixed number.

20. Change $4\frac{6}{7}$ to an improper fraction.

21. Add: $\frac{2}{9} + \frac{3}{5}$

22. Subtract: $4\frac{2}{5} - 2\frac{7}{10}$

23. Multiply: $\frac{4}{21} \cdot \frac{14}{9}$

24. Divide: $1\frac{3}{4} \div \frac{7}{12}$

25. You mix $2\frac{1}{3}$ cups of flour, $\frac{3}{4}$ cup of sugar, $\frac{3}{4}$ cup of mashed bananas, $\frac{1}{2}$ cup of walnuts, and $\frac{1}{2}$ cup of milk to make banana bread. How many cups of mixture do you have?

26. Your stock starts the day at $30\frac{1}{2}$ points and goes down $\frac{7}{8}$ of a point during the day. What is its value at the end of the day?

27. You have just opened a container of orange juice that has 64 fluid ounces. You have three people in your household who drink one 8-ounce serving of orange juice per day. In how many days will you need a new container?

28. You want to serve $\frac{1}{4}$-pound hamburgers at your barbecue. There will be seven adults and three children at the party, and you estimate that each adult will eat two hamburgers and each child will eat one. How much hamburger meat should you buy?

29. Your friend tells you that her height is 1.67 meters and her weight is 58 kilograms. Convert her height and weight to feet and pounds, respectively.

30. Convert the following units:

10 mi = _____ ft

3 qt = _____ oz

5 pt = _____ oz

6 gal = _____ oz

3 lb = _____ oz

31. The following table lists the minimum distance of each listed planet from Earth, in millions of miles. Convert each distance into scientific notation.

PLANET	DISTANCE (IN MILLIONS OF MILES)	DISTANCE (IN SCIENTIFIC NOTATION)
Mercury	50	
Venus	25	
Mars	35	
Jupiter	368	
Saturn	745	
Uranus	1606	
Neptune	2674	
Pluto	2658	

32. A recipe you obtained on the Internet indicates that you need to melt 650 grams of chocolate in $\frac{1}{5}$ liter of milk to prepare icing for a cake. How many pounds of chocolate and how many ounces of milk do you need?

33. You ask your friend if you may borrow his eraser. He replies that he has only a tiny piece of eraser—2.5×10^{-5} kilometers long. Does he really have such a tiny piece that he cannot share it with you?

34. $\left(3\frac{2}{3} + 4\frac{1}{2}\right) \div 2 =$

35. $18 \div 6 \cdot 3 =$

36. $18 \div (6 \cdot 3) =$

37. $\left(3\frac{3}{4} - 2\frac{1}{3}\right)^2 + 7\frac{1}{2} =$

38. $\left(5\frac{1}{2}\right)^2 - 8\frac{1}{3} \cdot 2 =$

39. $6^2 \div 3 \cdot 2 + 6 \div \left(-3 \cdot 2\right)^2 =$

40. $6^2 \div 3 \cdot -2 + 6 \div 3 \cdot 2^2 =$

41. $-4 \cdot 9 + -9 \cdot -8 =$

42. $-6/3 - 3 + 5^0 - 14 \cdot -2 =$

CHAPTER 1

Gateway Review

1. Write the number 2.0202 in words.

2. Write the number fourteen and three thousandths in standard form.

3. What is the place value of the 4 in the number 3.06704?

4. Add: $3.02 + 0.5 + 7 + 0.004$

5. Subtract 9.04 from 21.2.

6. Multiply: $6.003 \cdot 0.05$

7. Divide 0.0063 by 0.9.

8. Round 2.045 to the nearest hundredth.

9. Change 4.5 to a percent.

10. Change 0.3% to a decimal.

11. Change 7.3 to a fraction.

12. Change $\frac{3}{5}$ to a percent.

13. Write the numbers in order from largest to smallest:
    1.001   1.1   1.01 $1\frac{1}{8}$

14. Evaluate 3^3.

15. Evaluate 6^0.

16. Evaluate 4^{-2}.

17. Evaluate $\sqrt{36}$.

18. Evaluate $\sqrt{\frac{4}{100}}$.

19. The square root of 18 falls between what two whole numbers?

20. Evaluate $|-12|$.

Exercise numbers appearing in color are answered in the Selected Answers section of this book.

21. Write 0.0000543 in scientific notation.

22. Write 3.7×10^4 in standard form.

23. Determine the average of your exam scores: 25 out of 30, 85 out of 100, and 60 out of 70. Assume that each exam counts equally.

24. Change $\frac{9}{2}$ to a mixed number.

25. Change $5\frac{3}{4}$ to an improper fraction.

26. Reduce $\frac{15}{25}$ to lowest terms.

27. Add: $\frac{1}{6} + \frac{5}{8}$

28. Subtract: $5\frac{1}{4} - 3\frac{3}{4}$

29. Multiply: $\frac{2}{9} \cdot 3\frac{3}{8}$

30. Divide $\frac{6}{11}$ by $\frac{8}{22}$.

31. Multiply: $2\frac{5}{8} \cdot 2\frac{2}{7}$

32. Solve the proportion for x:

$\frac{x}{8} = \frac{9}{4}$

33. What is 20% of 80?

34. Twenty-five percent of what number is 50?

35. Thirty is what percent of 60?

36. Find the length of one side of a right triangle when the other side is 7 centimeters and the hypotenuse is 12 centimeters. Write your answer rounded to the nearest whole number.

37. Find the circumference of a circle whose radius is 3 inches.

38. Determine the perimeter of the following figure.

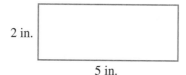

2 in.

5 in.

39. Find the area of a triangle with a base of 5 feet and a height of 7 feet.

40. Find the area of a circle whose radius is 1 inch. Round your answer to the nearest tenth.

41. Find the volume of a road cone with a radius of 10 centimeters and a height of 20 centimeters. Round your answer to the nearest whole number.

42. The number of ice-cream stands in your town decreased from 10 to 7 in one summer. What is the percent decrease in the number of stands?

43. A campus survey reveals that 30% of the students are in favor of the proposal for no smoking in the cafeteria. If 540 students responded to the survey, how many students are in favor of the proposal?

44. You read 15 pages of your psychology text in 90 minutes. At this rate, how many pages can you read in 4 hours?

45. The scale at a grocery store registered 0.62 kilogram. Convert this weight to pounds (1 kg = 2.2 lb). Round your answer to the nearest pound.

46. Your turtle walks at the rate of 2 centimeters per minute. What is the turtle's rate in inches per second (1 in. = 2.54 cm)?

47. Use the formula $C = \frac{5}{9}(F - 32)$ to determine the Celsius temperature when the Fahrenheit temperature is 50°.

48. $-5 + 4 =$

49. $2(-7) =$

50. $-3 - 4 - 6 =$

51. $-12 \div (-4) =$

52. $-5^2 =$

53. $(-1)(-1)(-1) =$

54. $-5 - (-7) =$

55. $-3 + 2 - (-3) - 4 - 9 =$

56. $3 - 10 + 7 =$

57. $\left(-\frac{1}{6}\right)\left(-\frac{3}{5}\right) =$

58. $7 \cdot 3 - \frac{4}{2} =$

59. $4(6 + 7 \cdot 2) =$

60. $3 \cdot 4^3 - \frac{6}{2} \cdot 3 =$

61. $5(7 - 3) =$

62. A deep-sea diver dives from the surface to 133 feet below the surface. If the diver swims down another 27 feet, determine his depth.

63. You lose $200 on each of three consecutive days in the stock market. Represent your total loss as a product of signed numbers, and write the result.

64. You have $85 in your checking account. You write a check for $53, make a deposit of $25, and then write another check for $120. How much is left in your account?

65. A poll is conducted asking 120 students about their preference for soft drinks. The results are given in the following table.

DRINK	Pepsi	Coke	7-Up	Dr. Pepper	Mountain Dew	Other	Total
FREQ.	37	33	12	10	9	19	120

a. Draw a bar graph to represent these data.

b. What percent of students prefer Dr. Pepper?

66. The mass of the hydrogen atom is about 0.00000000000000000000002 gram. If 1 gram is equal to 0.0022 pound, determine the mass of a hydrogen atom in pounds. Express the result in scientific notation.

Variable Sense

Arithmetic is the branch of mathematics that deals with counting, measuring, and calculating. Algebra is the branch that deals with variables and the relationships between and among variables. Variables and their relationships, expressed in table, graph, verbal, and symbolic forms, will be the central focus of this chapter.

CLUSTER 1

Interpreting and Constructing Tables and Graphs

ACTIVITY 2.1

Blood–Alcohol Levels
Topic: *Interpreting Relationships Numerically, Graphically, and Verbally*

Suppose you are asked to give a physical description of yourself. What categories might you include? You probably would include gender, race, and age. Can you think of any other categories? Each of these categories represents a distinct *variable*.

> A **variable**, usually represented by a letter, is a quantity or quality that may change in value from one particular instance to another.

The particular responses are the **values** of each variable. The first category, or variable, gender, will have just two possible values: male or female. The second variable, race, will have several possible values: Caucasian, African American, Hispanic, Asian, Native American, and so on. The third variable, age, will have a large range of possible values: from 17 years up to possibly 100 years of age. Note that the values of the variable age are numerical but the values of gender and race are not. The variables you will deal with in this text are numerical.

In many situations, we are not just interested in analyzing data pertaining to individual variables. Often we will look for **relationships** between two or more variables. One way that we can represent a relationship between two variables is by means of a **table.** Typically, one variable is designated the *input,* and the other is called the *output.*

> The **input** is the source of the values, and the **output** is the number that corresponds to or is matched with the input. The **input/output** designation may represent a cause-and-effect relationship, but this is not always the case.

This activity presents a variety of common input/output relationships.

In 1992, the U.S. Department of Transportation recommended that states adopt a 0.08 percent blood-alcohol concentration as the legal measure of drunk driving. The following table presents a numerical description of the relationship between the number of beers consumed in an hour by a 200-pound person (input) and his corresponding blood-alcohol concentration (output). You may symbolically represent the input variable, the number of beers consumed in an hour, by the letter n. Similarly, you may represent the output variable, blood-alcohol concentration, by B.

NUMBER OF BEERS IN AN HOUR, n	1	2	3	4	5	6	7	8	9	10
BLOOD-ALCOHOL CONCENTRATION, B	0.018	0.035	0.053	0.070	0.087	0.104	0.121	0.138	0.155	0.171

Based on body weight of 200 pounds

1. What is the blood-alcohol concentration for a 200-pound person who has consumed four beers in one hour? Nine beers in one hour?

2. What is the value of B when $n = 8$?

Notice that a table can reveal *numerical* patterns and relationships between the input and output variables. We can also describe these patterns and relationships *verbally* using such terms as *increases*, *decreases*, or *remains the same*.

3. According to the table, as the number of beers consumed in one-hour increases, what happens to the blood-alcohol concentration?

4. From the table, how many beers can a 200-pound person consume in one hour without exceeding the recommended legal measure of drunk driving?

A visual display (**graph**) of data on a grid is often helpful in detecting trends or other information not apparent in a table.

> The input is referenced on the horizontal axis, and the output is referenced on the vertical axis.

The following graph provides a visual description of the beer consumption and blood-alcohol level data from the table. Note that each of the ten input/output data pairs in the table corresponds to a plotted point on the graph. For example, drinking 7 beers in one hour is associated with a blood-alcohol concentration of 0.121. If you read across to 7 along the input (horizontal) axis and move up to 0.121 on the output (vertical) axis, you locate the point that represents the ordered pair of numbers (7, 0.121). Similarly, you would label the other points on the graph by ordered pairs of the form (*n*, *B*), where *n* is the input value and *B* is the output value.

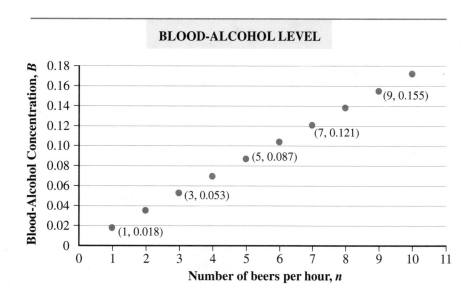

BLOOD-ALCOHOL LEVEL

5. **a.** From the graph, what would you approximate the blood-alcohol concentration to be when a person consumes two beers in an hour?

   Write this input/output correspondence using ordered pair notation.

   b. Estimate the number of beers that a 200-pound person needs to consume in one hour to have a blood-alcohol concentration of 0.104.

   Write this input/output correspondence using ordered pair notation.

   c. When *n* = 4, what is the approximate value of *B* from the graph?

   Write this input/output correspondence using ordered pair notation.

   d. As the number of beers consumed in one-hour increases, what is happening to the blood-alcohol concentration?

6. What are some advantages and disadvantages of using a graph when you are trying to communicate the relationship between the number of beers consumed in one hour and blood-alcohol concentration?

7. What advantages and disadvantages do you see in using a table?

We know that weight is a determining factor in a person's blood-alcohol concentration. The following table presents a numerical description of the relationship between the number of beers consumed in an hour by a 130-pound person (input) and her corresponding blood-alcohol concentration (output). There are other factors that may influence blood-alcohol concentration, such as the rate at which the individual's body processes alcohol, the amount of food eaten prior to drinking, and the concentration of alcohol in the drink, but these effects will not be considered here.

NUMBER OF BEERS, n	1	2	3	4	5	6	7	8	9	10
BLOOD-ALCOHOL CONCENTRATION, B	0.027	0.054	0.081	0.108	0.134	0.160	0.187	0.212	0.238	0.264

Based on body weight of 130 pounds

8. According to the table, how many beers can a 130-pound person consume in one hour without exceeding the recommended legal measure of driving while intoxicated?

9. How does lower body weight affect the blood-alcohol concentration for a given number of beers consumed in one hour?

E X E R C I S E S

1. Medicare is a government entitlement program that helps pay for the medical expenses of senior citizens. As the U.S. population ages and greater numbers of senior citizens join the Medicare rolls each year, the expense and quality of service becomes an increasing concern. The results of one Medicare study are presented in the following graph. Use the graph to answer the following questions.

ON THE UP AND UP

**As the U.S. Population Ages,
Medicare Expenditures Continue to Climb**

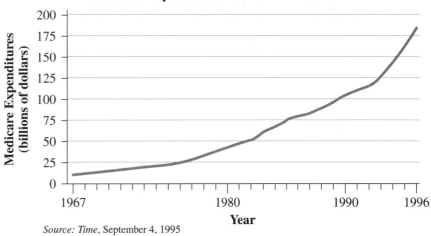

Source: *Time*, September 4, 1995

a. Identify the input variable and the output variable in this situation.

b. Use the graph to estimate the Medicare expenditures to complete the following table.

YEAR, y	MEDICARE EXPENDITURES, e (IN BILLIONS OF DOLLARS)
1967	
1972	
1977	
1984	
1989	
1995	

c. What letters in the table in part b are used to represent the input variable and the output variable?

d. Estimate the year in which the expenditures reached $100 billion.

e. Estimate the year in which the expenditures reached $25 billion.

f. During which 10-year period was the change in Medicare expenditures the least?

g. During which 10-year period was the change in Medicare expenditures the greatest?

h. For which time period does the graph indicate the most rapid increase in Medicare expenditures?

i. From the graph, predict the Medicare expenditures in the year 2000.

j. What assumptions are you making about the change in Medicare expenditures from 1996 to 2000?

2. When the input variable is measured in units of time, the relationship between input (time) and output indicates how the output *changes* over time. For example, the U.S. oil-refining industry has undergone many changes since the Middle East oil embargo of the 1970s. Among the factors that have had a major impact on the industry are changing crude oil prices, changing demand, increased imports, economics, and quality control. Since 1981, 120 U.S. oil refineries have been dismantled. Most were small and inefficient. The following graph shows the fluctuations in the number of U.S. refineries from 1974 to 1990.

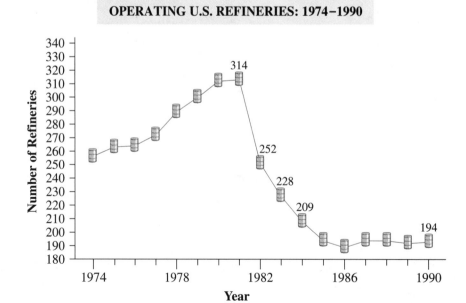

OPERATING U.S. REFINERIES: 1974–1990

Source: House Committee on Energy and Commerce

a. Identify the input variable.

b. On which axis does the input variable appear?

c. Identify the output variable.

d. On which axis does the output variable appear?

e. Make a table showing each year (input) and the number of refineries in operation that year (output).

YEAR	NUMBER OF OPERATING REFINERIES	YEAR	NUMBER OF OPERATING REFINERIES
1974		1983	
1975		1984	
1976		1985	
1977		1986	
1978		1987	
1979		1988	
1980		1989	
1981		1990	
1982			

f. Summarize the changes that the U.S. oil-refining industry has undergone since the 1970s. Include specific information. For example, indicate the intervals during which the number of U.S. oil refineries increased, decreased, and remained about the same. Also, indicate the years in which the number of refineries increased and decreased the most.

ACTIVITY 2.2

College Expenses
Topics: *Variables,*
Replacement Values,
Writing Verbal Rules,
Constructing Tables

You are considering taking some courses at your local community college on a part-time basis for the upcoming semester. For a student who carries fewer than 12 credits (the full-time minimum), the tuition is $143 for each credit hour taken. You have a limited budget, so you need to calculate the total cost based on the number of credit hours.

When you use a letter or symbol to represent a variable in the process of solving a problem, you must state what quantity (with units of measure) the variable is describing and what collection of numbers are meaningful replacement values for the variable.

1. a. Describe in words the input variable and the output variable in the college tuition situation.

b. Choose a letter or symbol to represent the input variable and a letter to represent the output variable.

c. Would a replacement value of 0 be reasonable for the input? Explain.

d. Would a replacement value of 15 be reasonable for the input? Explain.

e. What are the replacement values for the input?

f. What is the tuition bill if you register only for a three-credit-hour accounting course?

g. Use the replacement values you found in part e to complete the following table.

NUMBER OF HOURS									
TUITION									

h. Write a rule in words (called a verbal rule) to determine the total tuition bill for a student carrying fewer than 12 credit hours.

i. What is the change in the tuition bill as the number of credit hours increases from 2 to 3? Is the change in the tuition bill the same amount for each unit increase in credit hours?

Other possible college expenses include parking fines, library fines, and food costs. Students whose first class is later in the morning often do not find a parking space and park illegally. Campus police have no sympathy and readily write tickets. Last semester your first class began at 11:00 A.M. The following table shows your cumulative number of tickets as the semester progressed and the total in fines you owed the college at that time.

2. Discover the arithmetic relationship common to every input/output pair in the table. Describe in words how you can calculate the total fine, given your cumulative number of tickets. Then use your rule to complete the table.

NUMBER OF TICKETS LAST SEMESTER	2	3	5	8	10	12
TOTAL FINES	$25	$37.50	$62.50			

3. The college library fine for an overdue book is $0.25 per day. You leave a library book at a friend's house and totally forget about it.

a. Fill in the following table using the number of days the book is overdue as the input variable.

NUMBER OF DAYS OVERDUE	TOTAL FINE
2	
5	
8	
10	
14	

b. Write a verbal rule describing the arithmetic relationship between the input variable and the output variable.

4. At the beginning of the semester, you buy a meal ticket worth $110. The daily lunch special costs $3 in the college cafeteria.

a. Fill in the following table using the number of lunch specials you purchase as the input variable.

NUMBER OF LUNCH SPECIALS PURCHASED	REMAINING BALANCE ON YOUR MEAL TICKET
0	
10	
20	
30	
35	

b. Write a verbal rule describing the arithmetic relationship between the input variable and the output variable, remaining balance.

c. What are possible replacement values for the input variable? Explain.

E X E R C I S E S

1. Suppose the variable n represents the number of notebooks purchased by a student in the college bookstore for the semester.

   a. Can n be replaced by a negative value, such as -2? Explain.

   b. Can n be replaced by the number 0? Explain.

   c. What is a possible collection of replacement values for the variable n in this situation?

2. Determine the arithmetic relationship between the input and output variables in each of the following tables. Write a rule in words for obtaining the output from its corresponding input. Then complete the tables.

a.

INPUT	OUTPUT
2	4
4	8
6	12
8	16
10	
20	
25	

b.

INPUT	OUTPUT
2	4
3	9
4	16
5	
6	
7	
8	

c.

INPUT	OUTPUT
−2	−4
0	−2
2	0
4	2
6	
8	
10	

3. Complete the following tables using the verbal rule in column 2.

a.

INPUT	OUTPUT IS 10 MORE THAN THE INPUT
1	
3	
5	
7	
9	
12	
14	

b.

INPUT	OUTPUT IS 3 TIMES THE INPUT, PLUS 2
−5	
−4	
−3	
−2	
−1	
0	
1	
2	

4. Suppose the variable n represents the number of Beanie Babies a student owns.

   a. Can n be replaced by a negative value, such as −2? Explain.

   b. Can n be replaced by the number 0? Explain.

   c. What is a possible collection of replacement values for the variable n in this situation?

5. Suppose the variable t represents the average daily Fahrenheit temperature in Oswego, a city in upstate New York, during the month of February in any given year.

   a. Can t be replaced with a temperature of −3°F? Explain.

   −3°F.

   b. Can t be replaced with a temperature of −70°F? Explain.

c. Can *t* be replaced with a temperature of 98°F? Explain.

d. What is a possible collection of replacement values for the variable *t* that would make this situation realistic?

6. You are considering taking a part-time job at McDonald's. The job pays $6.25 per hour.

a. What would your gross pay be if you work 22 hours one week?

b. Write a rule in words (verbal rule) for determining your gross weekly pay based on the number of hours worked.

c. Complete the following table using the rule in part b.

NUMBER OF HOURS WORKED (INPUT)	WEEKLY PAY (OUTPUT)
9	
12	
15	
20	
22	
28	
30	

d. What are realistic replacement values for the input variable, the number of hours worked?

ACTIVITY 2.3

Earth's Temperature

Topics: *Constructing Graphs, Scaling, Plotting Points, Labeling Points*

Living on Earth's surface, we experience a relatively narrow range of temperatures. You may know what $-20°F$ (or $-28.9°C$) feels like on a bitterly cold winter day. Or you may have sweated through $100°F$ (or $37.8°C$) during summer heat waves. If you were to travel below Earth's surface and above Earth's atmosphere, you would discover a wider range of temperatures. The following graph displays a relationship between the altitude and temperature. Note that the altitude is measured from Earth's surface. That is, Earth's surface is at altitude 0.

TEMPERATURE VERSUS ALTITUDE

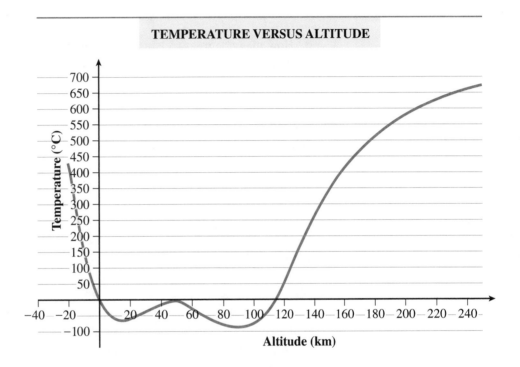

This graph uses a **rectangular coordinate system**, also known as a Cartesian coordinate system. It contains both a **horizontal axis** and a **vertical axis**.

1. a. How is the horizontal axis labeled?

   b. What is the practical significance of the positive values of this quantity?

   c. What is the practical significance of the negative values of this quantity?

   d. How many kilometers are represented between the tick marks on the horizontal axis?

2. a. How is the vertical axis labeled?

b. What is the practical significance of the positive values of this quantity?

c. What is the practical significance of the negative values of this quantity?

d. How many degrees Celsius are represented between the tick marks on the vertical axis?

3. Consult the graph to determine at which elevations or depths the temperature is high enough to cause water to boil?

- The distance between adjacent tick marks on an axis is determined by the particular replacement values for the variable represented on the axis. This is called **scaling.**

- The axes are often labeled and scaled differently. The way axes are labeled and scaled depends on the context of the problem. Note that the tick marks to the left of 0 on the horizontal axis are negative; similarly, the tick marks below 0 on the vertical axis are negative.

- The vertical scale is often different from the horizontal scale.

- On each axis, equal distance between adjacent pairs of tick marks must be maintained.

In the rectangular (Cartesian) coordinate system, the horizontal axis (commonly called the x-axis) and the vertical axis (commonly called the y-axis) are number lines that intersect at their respective 0 values at a point called the **origin**. The two perpendicular coordinate axes divide the plane into four **quadrants**. The quadrants are labeled counterclockwise, using Roman numerals, with Quadrant I being the upper-right quadrant.

Each point in the plane is identified by an **ordered pair** of numbers (x, y) that can be thought of as the point's "address" relative to the origin. The ordered pair representing the origin is $(0, 0)$. The first number, x, of an ordered pair (x, y) is called the **horizontal coordinate** because it represents the horizontal distance (to the right if x is positive, to the left if x is negative) of the point from the origin. Similarly, the second number, y, is called the **vertical coordinate** because it represents the vertical distance (up if y is positive, down if y is negative) of the point from the origin.

In contextual situations involving relationships between two variables, the ordered pair notation is interpreted as (input value, corresponding output value).

4. The following graph displays eight points selected from the Temperature graph on page 189.

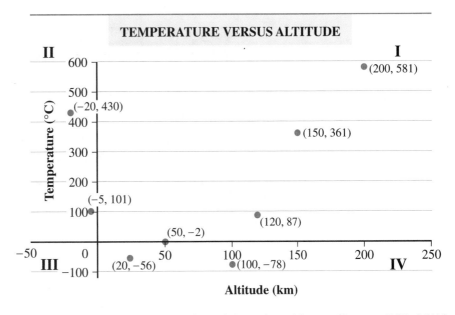

a. What is the practical meaning of the point with coordinates (150, 361)?

b. What is the practical meaning of the point with coordinates (100, −78)?

c. What is the practical meaning of the point with coordinates (−20, 430)?

5. In which quadrant are the points (120, 87), (150, 361), and (200, 581) located?

6. In which quadrant are the points (−5, 101) and (−20, 430) located?

7. In which quadrant are the points (20, −56) and (100, −78) located?

8. Are there any points located in Quadrant III? What is the significance of your answer?

9. Determine the sign (positive or negative) of the *x*- and *y*-coordinates of a point in each quadrant. For example, any point located in quadrant I has a positive *x*-coordinate and a positive *y*-coordinate.

QUADRANT	SIGN (+ OR −) OF x-COORDINATE	SIGN (+ OR −) OF y-COORDINATE
I		
II		
III		
IV		

10. In the following coordinate system, the horizontal axis is labeled the *x* axis, and the vertical axis is labeled the *y* axis. Determine the coordinates (*x*, *y*) of points *A* to *N*. For example, the coordinates of *A* are (30, 300).

A _____   *B* _____   *C* _____

D _____   *E* _____   *F* _____

G _____   *H* _____   *I* _____

J _____   *K* _____   *L* _____

M _____   *N* _____

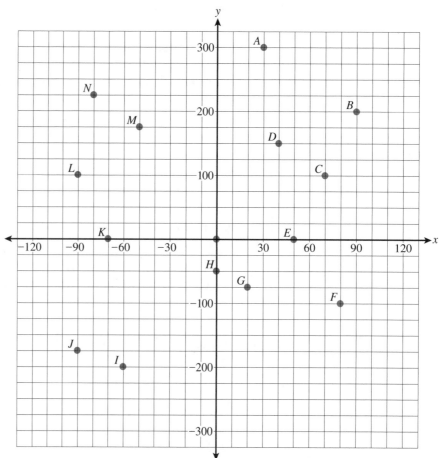

11. Which points (*A* to *N*) from Problem 10 are located:

   a. in the first quadrant?   **b.** in the second quadrant?

   c. in the third quadrant?   **d.** in the fourth quadrant?

12. a. Which points are on the horizontal axis? What are their coordinates?

   b. What is the *y*-value of any point located on the *x*-axis?

   c. Which points are on the vertical axis? What are their coordinates?

   d. What is the *x*-value of any point located on the *y*-axis?

13. You are working for the National Weather Service and are asked to do a study of the average daily temperatures for Anchorage, Alaska. You calculate the mean of the average daily temperatures for each month. You decide to place the information on a graph in which the date is the input variable and the temperature is the output variable. You also decide that January 1950 will correspond to the date 0 on the input scale. Determine the quadrant in which you would plot the points that correspond to the following data.

   a. The average daily temperature for January 1936 was $-15°$F.

   b. The average daily temperature for July 1963 was 63°F.

   c. The average daily temperature for July 1910 was 71°F.

   d. The average daily temperature for January 1982 was $-21°$F.

14. Measurements in wells and mines have shown that the temperatures within the earth generally increase with depth. The following table shows average temperatures for several depths below sea level.

EARTH'S INTERNAL TEMPERATURE

DEPTH (KM) BELOW SEA LEVEL	0	25	50	75	100	150	200
TEMPERATURE (°C)	20	600	1000	1250	1400	1700	1800

a. Represent the data from the table graphically. Place depth (input) along the horizontal axis and temperature (output) along the vertical axis.

b. Each tick mark on the horizontal axis will represent how many units?

c. Each tick mark on the vertical axis will represent how many units?

d. Explain your reasons for the particular scales that you used.

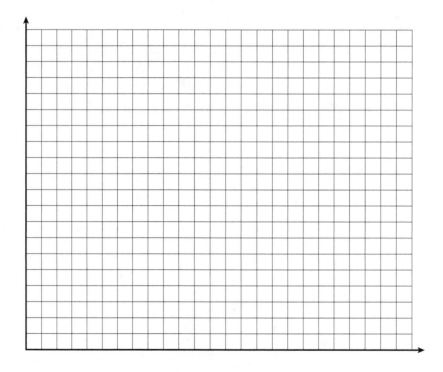

e. Which representation (table or graph) presents the information and trends in this data more clearly? Explain your choice.

EXERCISES

Points in a coordinate plane lie in one of the four quadrants or on one of the axes. In Exercises 1–18 place the letter corresponding to the phrase that best describes the location of the given point.

a. The point lies in the first quadrant.　　**b.** The point lies in the second quadrant.

c. The point lies in the third quadrant.　　**d.** The point lies in the fourth quadrant.

e. The point lies at the origin.　　　　　　**f.** The point lies on the positive x-axis.

g. The point lies on the negative x-axis.　**h.** The point lies on the positive y-axis.

i. The point lies on the negative y-axis.

1. $(2, -5)$ _____　　2. $(-3, -1)$ _____　　3. $(-4, 0)$ _____

4. $(0, 0)$ _____　　5. $(0, 3)$ _____　　6. $(-2, 4)$ _____

7. $(8, 6)$ _____　　8. $(0, -6)$ _____　　9. $(5, 0)$ _____

10. $(4, -6)$ _____　　11. $(2, 0)$ _____　　12. $(0, 6)$ _____

13. $(-7, -7)$ _____　　14. $(12, 5)$ _____　　15. $(0, -2)$ _____

16. $(12, 0)$ _____　　17. $(-10, 2)$ _____　　18. $(-13, 0)$ _____

19. The following table presents the average recommended weights for given heights for 25–29 year old medium-framed women (wearing 1-inch heels and 3 pounds of clothing). Consider height to be the input variable and weight to be the output variable. As ordered pairs, height and weight take on the form (h, w). The horizontal (input) axis will be designated the h-axis, and the vertical (output) axis will be designated the w-axis. Since all values of the data are positive, the points will lie in Quadrant I only.

h, HEIGHT (in.)	58	60	62	64	66	68	70	72
w, WEIGHT (lb)	115	119	125	131	137	143	149	155

Source: The World Almanac and Book of Facts, 1996.

Plot the ordered pairs in the height-weight table on the following grid. Note that there is equal spacing between tick marks on each of the axes. The distance between tick marks on the horizontal axis represents 2 inches. The distance between tick marks on the vertical axis represents 5 pounds.

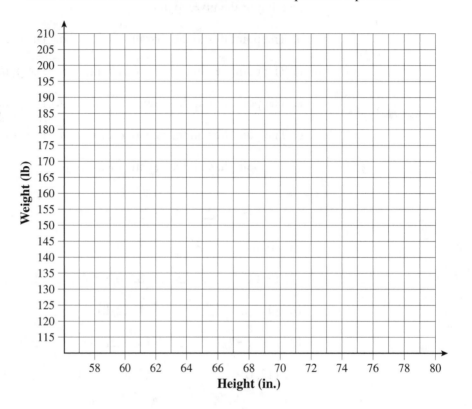

20. Graph the information from the following table, which you created in Activity 2.2, Problem 2 for the number of parking tickets and total fines. Choose a scale for each axis that will enable you to plot all six points. Label the horizontal axis to represent the cumulative number of tickets received and the vertical axis to represent the total in fines owed to the college at that time.

NUMBER OF TICKETS	2	3	5	8	10	12
TOTAL FINES	$25	$37.50	$62.50	$100	$125	$150

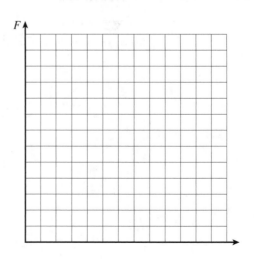

21. In Activity 2.2, Exercise 6, you created the following table showing the relationship between the number of hours worked at McDonald's part-time (input) and the weekly pay based on a rate of $6.25 per hour (output).

NUMBER OF HOURS WORKED (INPUT)	WEEKLY PAY (OUTPUT)
9	$56.25
12	$75.00
15	$93.75
20	$125.00
22	$137.50
28	$175.00
30	$187.50

a. Graph the information found in the table. Make sure you label each axis.

b. The distance between tick marks on the horizontal axis will represent how many units?

c. The distance between tick marks on the vertical axis will represent how many units?

d. Explain your reasons for the particular scales that you used.

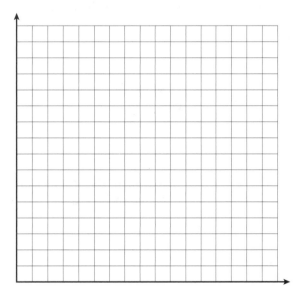

e. Which representation (table or graph) presents the information and trends more clearly? Explain your choice.

LAB ACTIVITY 2.4

How Many Cups Are in That Stack?

Topics: *Measurement, Relationships Defined by Tables, Converting Tables to Graphs*

In this lab, you will stack and measure the heights of disposable Styrofoam or plastic cups placed inside one another. With a centimeter ruler, measure and record the height (to the nearest tenth of a centimeter) of stacks containing increasing numbers of cups.

NUMBER OF CUPS	1	2	3	4	5	6	7	8
HEIGHT OF STACK (cm)								

1. Graph the data with the inputs (number of cups) on the horizontal axis and the outputs (height of stack) on the vertical axis.

2. Use the information you have gathered to *estimate* the height (to the nearest tenth of a centimeter) of 10 stacked cups.

3. Now construct a stack of 10 cups. Measure the stack and compare it with your estimate. Was your estimate reasonable? If not, why not?

4. Estimate the height of 16 cups. Is a stack of 16 cups twice as tall as a stack of 8 cups? Justify your answer.

5. Estimate the height of 20 cups. Is a stack of 20 cups twice as tall as a stack of 10 cups? Justify your answer.

6. Describe the formula or method you used to estimate the height of a stack of cups.

7. If your shelf has clearance of 40 centimeters, how many cups can you stack to fit on the shelf? Explain how you obtained your answer.

Body Parts

Topics: *Measurement, Relationships Defined by Tables—Converting Tables to Graphs*

Variables arise in many common measurements. Your height is one measurement that has probably been recorded from the day you were born. In this project, you are asked to pair up and measure five lengths associated with your body: height (h); arm span (a), the distance between the tips of your two middle fingers with arms outstretched; wrist circumference (w); the length of the foot (f); and neck circumference (n). For consistency, let's agree to measure in inches.

1. Gather the data for your entire class, and record in the following table.

STUDENT	HEIGHT (h)	ARM SPAN (a)	WRIST (w)	FOOT (f)	NECK (n)

2. What are some relationships you can identify, based on eyeballing the data? For example, how do the heights relate to the arm spans?

3. Graph one relationship, such as height (h) versus wrist (w). Make sure you label your axes and indicate the scales.

What Have I Learned?

1. Describe several ways that relationships between variables may be represented. What are the advantages and disadvantages of each?

2. Obtain a graph from your local newspaper, from a magazine, or from a textbook in your major field. Identify the input and output variables. The input variable is referenced on which axis? Describe any trends in the graph.

3. You are going to graph some data from a table in which the input values begin at 0 and stop at 150 and the output values range from 0 to 2000. Assume that your grid is a square with 16 tick marks across and down.

   a. The distance between tick marks on the horizontal axis will represent how many units?

   b. The distance between tick marks on the vertical axis will represent how many units?

   c. Will you use all four quadrants? Explain.

How Can I Practice?

1. You are a scuba diver and plan a dive in the St. Lawrence River. The water depth in the diving area does not exceed 150 feet. Let x represent your depth in feet *below* the surface of the water.

   a. What are the possible replacement values to represent your depth from the surface?

   b. Would a replacement value of 0 be reasonable? Explain.

   c. Would a replacement value of -200 be reasonable? Explain.

   d. Would a replacement value of 12 be reasonable? Explain.

2. You bought a company in 1968 and have tracked the company's profits and losses from its beginning in 1940 to the present. You decide to graph the information where the number of years since 1968 is the input variable and profit or loss for the year is the output variable. Note that the year 1968 corresponds to 0 on the horizontal axis. Determine the quadrant or axis on which you would plot the points that correspond to the following data. If your answer is on an axis, indicate between which quadrants the point is located.

   a. The loss in 1942 was $1500.

   b. The profit in 1998 was $6000.

   c. The loss in 1980 was $1000.

   d. In 1985, there was no profit or loss.

   e. The profit in 1940 was $500.

   f. The loss in 1968 was $800.

3. Fish need oxygen to live, just as you do. The amount of dissolved oxygen (D.O.) in water is measured in parts per million (ppm). Trout need a minimum of 6 ppm to live.

The data in the table show the relationship between the temperature of the water and the amount of dissolved oxygen present.

TEMP. (°C)	11	16	21	26	31
D.O. (in ppm)	10.2	8.6	7.7	7.0	6.4

a. Represent the data in the table graphically. Place temperature (input) along the horizontal axis and dissolved oxygen (output) along the vertical axis.

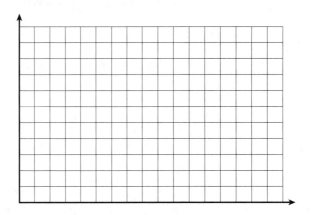

b. What general trend do you notice in the data?

c. In which of the 5° temperature intervals given in the table does the dissolved oxygen content change the most?

d. Which representation (table or graph) presents the information and trends more clearly?

4. When you were born, your uncle invested $1000 for you in your local bank. Your bank has compounded the interest continuously at a rate of 6%. The following graph shows how your investment grows.

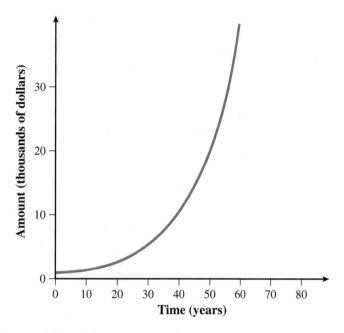

Time (years)

a. Which variable is the input variable?

b. How much money did you have when you were 10 years old?

−$2000.

c. Estimate in what year your original investment will have doubled.

d. If your college bill is estimated to be $2600 in the first year of college, will you have enough to pay the bill with these funds? (Assume that you attend when you are eighteen.) Explain.

e. Assume that you expect to be married when you are 25 years old. You figure that you will need about $5000 for your share of the wedding and honeymoon expenses. Assume also that you left the money in the bank and did not use it for your education. Will you have enough money to pay your share of the wedding and honeymoon expenses? Explain.

Solving Equations Numerically, Graphically, and Algebraically

ACTIVITY 2.6

The Write Way to Learn Algebra

Topics: *Translating English Phrases into Symbolic (Algebraic) Expressions, Equivalent Expressions, Terminology, Verbal Rules into Symbolic Form*

To communicate in a foreign language, you must learn the language's alphabet, grammar, and vocabulary. You must also learn to translate between your native language and the foreign language you are learning. The same is true for algebra, which, with its symbols and grammar, is the language of mathematics. To become confident and comfortable with algebra, you must practice speaking and writing it and translating between it and the natural language you are using (English, in this book).

Of course, you have been doing this already in this book. For example, in geometry, the verbal statement "The perimeter of a rectangle is equal to two times its length plus two times its width" is translated into algebra (symbols) as $P = 2l + 2w$, where P represents the perimeter, l the length, and w the width. $P = 2l + 2w$ is an example of a symbolic statement in which $2l + 2w$ on the right-hand side of the equation is called an **expression.**

The algebraic statement $y = 3x + 2$ translates to "The output is equal to three times the input plus two" (see Activity 2.2, Exercise 3b). In this example, $3x + 2$ is the expression.

> An **algebraic expression** is a shorthand code for a sequence of operations to be performed on the input value to produce a corresponding output value.

In this activity, you will first practice your skills with expressions and then use them in equations.

1. Let x represent the input variable. Translate each of the following phrases into an algebraic expression.

   a. The product of 3 and the input

   b. Four times the input

   c. The input increased by 5

   d. The input squared

   e. Ten less than twice the input

In the algebraic expression $3 \cdot x$ from Problem 1a, the number 3 is called the **coefficient** of the variable x. The multiplication symbol (seen here as a raised dot) is not necessary when multiplying a variable by a number. Because the arithmetic operation between a variable and its coefficient is always understood to be multiplication, $3 \cdot x$ is written as $3x$.

2. a. Complete the following table.

x	x · 3	3x
−4		
−1		
0		
3		

b. Do the expressions $x \cdot 3$ and $3x$ produce the same output value when given the same input value?

c. In part b, the input value times 3 gives the same result as 3 times the input value. What property of multiplication does this demonstrate?

d. Use a grapher to sketch a graph of $y_1 = x \cdot 3$ and $y_2 = 3x$ on the same coordinate axes. How do the graphs compare? Are you surprised?

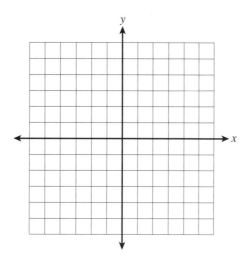

The expressions $x \cdot 3$ and $3x$ are called **equivalent expressions**, and you can write $x \cdot 3 = 3x$. It is common practice to write the coefficient of the variable first.

Algebraic expressions are said to be **equivalent** if identical inputs always produce identical outputs.

3. a. Why is it true that $1x = x$? That is, why are $1x$ and x equivalent? Use a numerical replacement for x to illustrate.

b. Why is it true that $-1x = -x$? That is, why are $-1x$ and $-x$ equivalent? Use a numerical replacement for x to illustrate.

4. a. Use a grapher to sketch a graph of $y_1 = x + 5$ and $y_2 = 5 + x$ on the same coordinate axes. How are the graphs related?

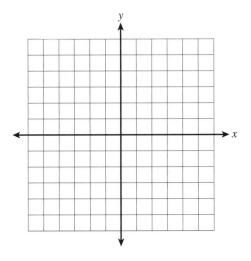

b. How does the graph show that the expressions $x + 5$ and $5 + x$ are equivalent? What property of addition does this demonstrate?

5. a. Is the expression $x - 3$ equivalent to $3 - x$? Complete the following table to help justify your answer.

x	x - 3	3 - x
−5		
−3		
0		
1		
3		

b. What correspondence do you observe between the output values in columns 2 and 3? How is the expression $3 - x$ related to the expression $x - 3$?

c. Use a grapher to sketch a graph of $y_1 = x - 3$ and $y_2 = 3 - x$. How do the graphs compare?

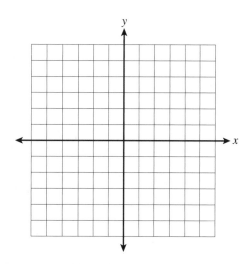

d. Is the operation of subtraction commutative?

6. In the expression $3x$, the 3 and x are called **factors**. In the expression $x + 5$, the x and 5 are called **terms**. Explain the difference between a factor and a term.

7. a. Is the algebraic expression $2x$ equivalent to the expression x^2? Explain.

b. In the expression $2x$, the number 2 is called the _____.

c. In the expression x^2, the number 2 is called the _____.

8. The output is twice the input, plus 5. Let y represent the output variable and x represent the input variable.

a. Translate the given **verbal rule** describing the relationship between output and input into a **symbolic rule.**

b. Complete the following table using the symbolic rule in part a.

x	-3	1.27	15.4
y			

c. Sketch a graph of the input-output rule.

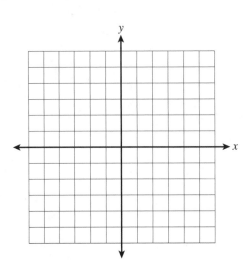

9. For each of the following, identify the input and output variables, represent each variable by a letter, translate the verbal rule (words) into a symbolic rule, and complete the given table.

a. The sales tax is 8% of the selling price.

Input variable _____   Letter representation _____

Output variable _____   Letter representation _____

Symbolic rule _____

SELLING PRICE	95	320	10,240
SALES TAX			

b. In a certain company, the profit (in millions of dollars) is equal to the revenue (in millions of dollars) decreased by expenses of $10 million.

Input variable _____   Letter representation _____

Output variable _____   Letter representation _____

Symbolic rule _____

REVENUE (MILLIONS OF $)	25	75	185
PROFIT (MILLIONS OF $)			

c. The number of inches of water that accumulates in the soil is the number of inches of snow fallen divided by 12.

Input variable _____

Letter representation _____

Output variable _____

Letter representation _____

Symbolic rule _____

INCHES OF SNOW	10	24	35
INCHES OF WATER			

SUMMARY

1. An **algebraic expression** is a shorthand code for a sequence of operations to be performed on the input value to produce a corresponding output value.

2. **Variable:** a quantity, usually represented by a letter or symbol, that changes in value.

3. **Constant:** a quantity that does not change in value (e.g., the number 2).

4. **Numerical coefficient:** a number that multiplies a variable or expression.

5. **Factors:** numbers, variables and/or expressions that are multiplied together to form a product.

6. **Terms:** parts of an expression that are separated by plus or minus signs.

7. **Equation:** a symbolic statement that includes the symbol $=$ to indicate that two quantities have the same value.

EXERCISES

Let x represent the input variable. Translate each of the phrases in Exercises 1–6 into an algebraic expression.

1. The input decreased by 7

2. Five times the difference between the input and 6

3. Seven increased by the quotient of the input and 5

4. Twelve more than one-half of the square of the input

5. Twenty less than the product of the input and -2

6. The sum of three-eighths of the input and five

7. Determine the coefficient of x in the following expressions:

    a. $7x$ **b.** x **c.** $-x$ **d.** $\dfrac{3}{4}x$ **e.** $\dfrac{2x}{3}$ **f.** $\dfrac{x}{5}$

8. Identify the factors in each of the following algebraic expressions.

    a. $5x$ **b.** $\frac{x}{2}$

    c. $-2x^3$ **d.** $2(x + 3)$

9. Identify the terms in each of the following expressions.

    a. $x + 10$ **b.** $2x - 3$

    c. $x^2 + 2x + 5$ **d.** $3x$

10. The algebraic expression $8x^2 - 10x + 9y - z + 7$ is an example of a polynomial.

    a. How many terms are in this polynomial?

    b. What is the coefficient of the first term?

    c. What is the coefficient of the fourth term?

    d. What is the constant term?

    e. What are the factors in the first term?

11. a. Translate each of the following into an algebraic expression. Let x represent the input variable.

    i. The input divided by 2   **ii.** Two divided by the input

b. Are the expressions in parts a(i) and a(ii) equivalent? Complete the following table to help justify your answer.

x	INPUT DIVIDED BY 2	2 DIVIDED BY INPUT
−4		
−2		
2		
4		

c. Is the operation of division commutative?

d. Is 0 a possible replacement value for the input in part a(i)? If yes, what is the output value?

e. Is 0 a possible replacement value for the input in part a(ii)? Explain.

In Exercises 12–15, identify the input and output variables, represent each variable by a letter, translate the verbal rule into a symbolic rule, and complete the given table.

12. At a certain hotel, you accept the offer of breakfast for two for just $10 more than the room charge.

Input variable _____   Letter representation _____

Output variable _____   Letter representation _____

Symbolic rule _____

INPUT	75	110	155
OUTPUT			

13. The average amount of precipitation in Boston during the month of March is four times the average amount of precipitation in Phoenix.

Input variable _____   Letter representation _____

Output variable _____   Letter representation _____

Symbolic rule _____

INPUT	2	6	10
OUTPUT			

14. The daily cost of renting a compact car is $30 plus 10 cents per mile.

Input variable _____   Letter representation _____

Output variable _____   Letter representation _____

Symbolic rule _____

INPUT	85	150	240
OUTPUT			

15. The markdown of an item is 20% of the original cost of the item.

Input variable _____   Letter representation _____

Output variable _____   Letter representation _____

Symbolic rule _____

INPUT	30	55	110
OUTPUT			

Cub Scout Fund-Raiser

Topics: *Relationships Defined by Tables, Graphs, and Verbal and Symbolic Forms, Introduction to Solving Equations Numerically and Graphically*

Suppose you are the leader of a local Cub Scout pack that is running a fund-raiser at the neighborhood roller rink. The admission charge is $4.50 per person, $2.00 of which is used to pay the rink's rental fee and the remainder is donated to your pack.

1. What are the variables that naturally arise from the description of the Cub Scout fund-raiser?

2. On what does the total amount donated depend?

3. a. Which variable can best be designated as the input variable?

   b. What are its units of measurement?

4. a. Which variable can best be designated as the output variable?

   b. What are its units of measurement?

5. Create a table to represent the relationship between input and output.

INPUT	OUTPUT($)
1	
2	
3	
4	
5	
6	
7	
8	

6. Graph the relationship in Problem 5. Recall that the input axis is horizontal and the output axis is vertical. Label and scale each axis appropriately. You might want to extend your graph to include points that are not listed in the table.

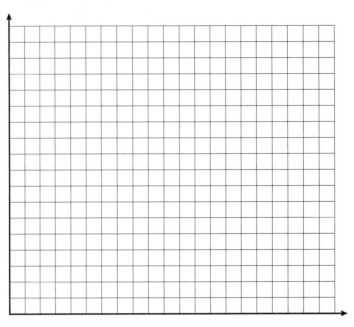

7. What difficulty would you encounter in using your table or graph to determine the amount of money raised if 84 tickets are sold?

8. For any input value, explain in your own words how to determine the corresponding output value. This description is called a *verbal rule*.

9. Let x represent the input, the number of admission tickets sold. Let y represent the output, the amount of money donated to the Cub Scout pack. Write a *symbolic rule* describing the input/output relationship given in Problem 8.

To determine the output value corresponding to any input value using a symbolic rule, replace the input symbol, commonly x, with its assigned value and perform the arithmetic operation(s) to obtain the output value.

This process is known as **evaluating** the expression for a given input value. Note that in this process, the variable to be determined (output) is isolated on one side of the equation.

10. Use the symbolic rule from Problem 9 to determine the amount of money raised if 84 tickets are sold. Stated another way, determine y when $x = 84$.

11. Complete the following table by using your calculator to evaluate the appropriate expression using the rule from Problem 9.

x, NUMBER OF TICKETS SOLD	y, AMOUNT RAISED ($)
26	
94	
278	

12. If you have a graphing calculator with table capability, use the table feature to check your results in Problem 11.

13. Suppose the goal is to raise a total of at least $200 for the Cub Scout pack. Is 200 an input value or an output value? (Note its units.)

You want to determine how many tickets must be sold to achieve a goal of $200. You can write this problem symbolically by replacing y with 200 in the rule $y = 2.50x$ to produce the **equation**

$$200 = 2.50x.$$

The input variable, x, is not isolated on one side of the equation, and you must solve the equation $200 = 2.50x$ to determine the value of x. There are three different methods for solving such equations: numerical, graphical, and algebraic. Although the algebraic methods of solution are fundamental mathematical skills, the numerical and graphical methods are often quite useful, but frequently overlooked. Problems 14 and 15 will illustrate the numerical and graphical solution techniques. The algebraic method is explored in the next activity.

14. Use a table of values to estimate how many tickets must be sold to raise at least $200. Begin by determining a value for x such that the expression $2.50x$ on the right-hand side of the equation is close to 200. For example, the value $x = 100$ produces a value of 250, so you might start with a value such as $x = 70$ or $x = 75$. Then increase or decrease your guess for x until you obtain an output of 200. This approach, which consists of a guess, check, and repeat, is the **numerical method** for solving the equation $200 = 2.50x$ for x.

x, NUMBER OF TICKETS SOLD	y, AMOUNT RAISED

15. a. Graph the symbolic rule $y = 2.50x$. Be sure to scale your axes so that the vertical (output) axis includes the value 200. Your graph should resemble the one below.

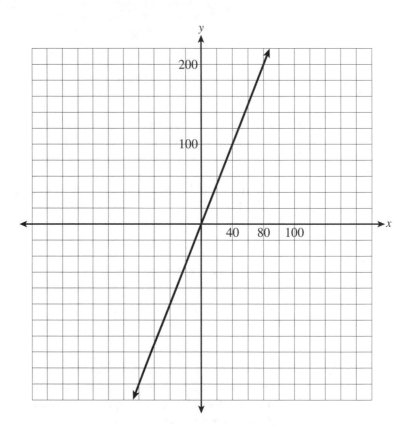

   b. Use the graph from part a to estimate how many tickets must be sold to raise at least $200. Remember, you need to determine x when $y = 200$. This approach is the **graphical method** for solving the equation $200 = 2.50x$ for x.

E X E R C I S E S

1. Your favorite gas station lists the price of regular unleaded gasoline as $1.239 per gallon.

   a. Let x be the input variable representing the number of gallons purchased. Let C be the output variable representing the total cost of the fuel. Write an equation relating x and C.

b. Use the equation from part a to complete the following table.

x, NUMBER OF GALLONS PURCHASED	5	10	15	20
C, TOTAL COST OF THE PURCHASE				

c. You have only $10 with you. How many gallons can you purchase? Estimate your answer using a numerical approach.

d. Use your grapher to graph the equation determined in part a. Use the graph to estimate the number of gallons you can purchase with $10.

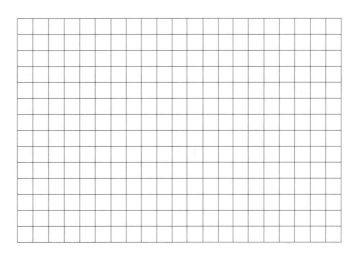

2. You are president of the band booster club at your local college. You have arranged for the college's jazz band to perform at a local bookstore for three hours. In exchange for the performance, the Booster Club will receive $\frac{3}{4}$ of the store's gross receipts during that three-hour period.

a. Let x be the input variable representing the gross receipts of the bookstore during the performance. Let y be the output variable representing the share of the gross receipts that the bookstore will donate to the band boosters. Write an equation relating x and y.

b. Use the equation from part a to complete the following table.

x, TOTAL GROSS RECEIPTS	250	500	750	1000
y, BOOSTERS' SHARE				

c. If the bookstore presents you with a check for $650, what were the gross receipts during the performance? Estimate your answer using a numerical approach.

d. Graph the equation in part a. Use a graphing approach to estimate the gross receipts from part c. Be sure to scale your axes appropriately.

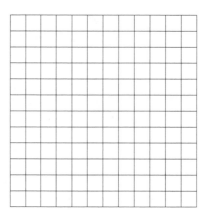

3. A tennis ball is dropped from the top of the Empire State Building in New York City. The following table gives the distance (output) that the ball is from ground level at a given time (input) after it is dropped.

TIME (sec)	1	2	3	4	5	6	7
DISTANCE (ft)	1398	1350	1270	1158	1014	838	630

The Empire State Building is 1414 feet tall. Use the information in the table to approximate the time when the ball would have fallen half the height of the building.

In Exercises 4 and 5, solve the resulting equation both numerically (using a table) and graphically.

4. If $y = x + 4$, determine x when $y = 7$.

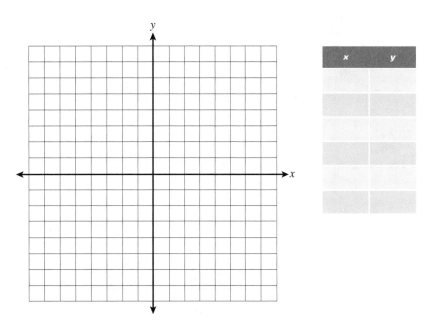

5. If $y = 3x$, determine x when $y = -12$.

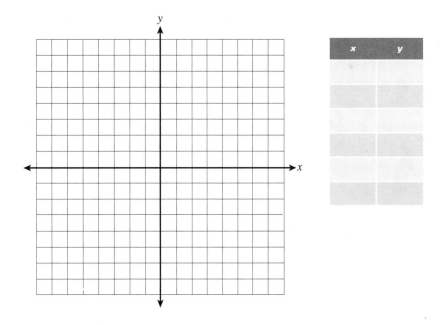

ACTIVITY 2.8

Let's Go Shopping

Topic: *Evaluating and Solving Equations of the Form y = ax and y = x + b Algebraically*

The sales tax collected on taxable items in Nassau County, New York, is 8.5%. The tax you must pay depends on the price of the item you are purchasing.

1. What is the sales tax on a dress shirt that costs $20?

2. **a.** Because you are interested in determining the sales tax given the price of an item, which variable is the input?

   b. What are its units of measurement?

3. **a.** Which variable is the output?

   b. What are its units of measurement?

4. Determine the sales tax you must pay on the following items.

ITEM	PRICE (INPUT, x)	SALES TAX ($) (OUTPUT, y)
Calculator	12.00	
Shirt	25.00	
Microwave Oven	200.00	
Car	15,000.00	

5. **a.** Express the arithmetic relationship between these variables using a verbal rule.

   b. Express the arithmetic relationship between these variables using a symbolic rule, with x representing the input (price) and y representing the output (sales tax).

   c. Describe the arithmetic operation you used to obtain an output, y, from an input value, x.

   Start with x (input) → _____ to obtain y (output).

6. Now determine the price of a fax machine for which you paid a sales tax of $21.25. In this situation, you know the sales tax (output), and you want to determine the price (input). You can accomplish this by "reversing the direction" in Problem 5 and by replacing the operation with its "opposite" (**inverse**).

   Start with $y = 21.25$ (output) → _____ to obtain x (input).

   Use this reverse process to determine the price of the fax machine.

7. Use the reverse process you developed in Problem 6 to determine the price of a color scanner for which you paid a sales tax of $61.20.

The reverse process used in Problems 6 and 7 can be done in a more formal and systematic way using the symbolic rule $y = 0.085x$. For example, in Problem 7, the sales tax for a color scanner was $61.20. Substitute 61.20 for y in $y = 0.085x$ and proceed as follows:

$$61.20 = 0.085x \qquad \text{Equation to be solved}$$

$$\frac{61.20}{0.085} = \frac{0.085x}{0.085} \qquad \text{Divide by 0.085 or,}$$
$$\qquad\qquad\qquad \text{equivalently, multiply by } \tfrac{1}{0.085}.$$

$$720 = x$$

This approach is the **algebraic method** of solving the equation $61.20 = 0.085x$. Note that this process is completed when the variable, x, has been isolated on one side of the equation. The value obtained for x (here, 720) is called the **solution** of the equation.

8. Suppose you have a $15 coupon that can be used on any purchase over $100 at your favorite clothing store.

a. Determine the discount price of a sports jacket having a retail price of $116.

b. Identify the input variable and the output variable.

c. Let x represent the input and y represent the output. Write a symbolic rule that describes the relationship between the input and output. Include, as part of your rule, an algebraic statement describing the requirement that the discount be applied only to purchases totaling more than $100.

d. Complete the following table using the symbolic rule determined in part c.

x, RETAIL PRICE	y, DISCOUNT PRICE
80	
135	
184	
205	

e. Beginning with the input value x, write the arithmetic operation you used to obtain the output y.

Start with x (input) $\rightarrow$ _____ to obtain y (output).

9. Suppose you are asked to determine the retail price if the discount price of a suit is $187.

a. Is 187 an input value or an output value?

b. Reverse the direction in Problem 8e by replacing the operation with its "opposite" (inverse).

Start with $y = 187$ (output) $\rightarrow$ _____ to obtain x (input).

Use this reverse process to determine the retail price of the suit.

Solve the resulting equation for x using an algebraic approach. Substitute 187 for y in the symbolic rule $y = x - 15$ as follows.

$$187 = x - 15 \quad \text{Equation to be solved}$$
$$\underline{+15 \qquad +15} \quad \text{Add 15}$$
$$202 = x$$

In this activity, you solved two equations for x using an algebraic approach. In each situation, your strategy was to isolate the input variable, x, on one side of the equation by using a "reverse" process. To reverse or undo a multiplication, you divided. To reverse or undo a subtraction, you added.

10. Use an algebraic approach to determine the input x for the given output value. That is, set up and solve the appropriate equation.

a. $y = 7.5x$

Determine x when $y = 90$.

b. $z = -5x$

Determine x when $z = -115$.

c. $y = \frac{x}{4}$

Determine x when $y = 2$.

d. $p = \frac{2}{3}x$

Determine x when $p = -18$.

11. Use an algebraic approach to determine the input x for the given output value.

a. $y = x - 10$

Determine x when $y = -13$.

b. $y = 13 + x$

Determine x when $y = 7$.

c. $p = x + 4.5$

Determine x when $p = -10$.

d. $x - \frac{1}{3} = s$

Determine x when $s = 8$.

To reverse or undo multiplication of the input by a coefficient, divide each side of the equation by that coefficient.

To reverse or undo division of the input by a coefficient, multiply each side of the equation by that coefficient.

To reverse or undo addition of a value to the input, subtract that value from each side of the equation.

To reverse or undo subtraction of a value from the input, add that value to each side of the equation.

A solution of an equation is a replacement value for the variable that results in equal values for both sides of the equation.

E X E R C I S E S

1. You've decided to enroll in a local college as a part-time student (fewer than 12 credit hours). Full-time college work does not fit into your present financial or personal situation. The cost per credit hour at your college is $128.

   a. What is the cost of a 3-credit-hour literature course?

   b. Write a symbolic rule to determine the total tuition for a given number of credit hours. Let n represent the number of credit hours taken (input) and y represent the total tuition paid (output).

   c. Complete the following table.

CREDIT HOURS (INPUT)	TUITION PAID ($) (OUTPUT)
1	
2	
3	
4	

d. Suppose you have enrolled in a psychology course and a computer course, each of which is 3 credit hours. Use the symbolic rule in part b to determine the total tuition paid.

e. Use your grapher to verify your answer for part d both numerically (using a table) and graphically.

CREDIT HOURS	TUITION PAID

f. Use the symbolic rule from part b to determine algebraically the number of credit hours carried by a student with the following tuition bill.

   i. $384   ii. $640

2. Have you ever played a game based on the popular TV game show *The Price Is Right*? The idea is to guess the price of an item. You win the item by coming closest to the correct price without going over. If your opponent goes first, a good strategy is to overbid her regularly by a small amount, say, $15. Then your opponent can win only if the item's price falls in that $15 region between her bid and yours.

You can model this strategy by defining two variables: the input x will represent your opponent's bid, and the output y will represent your bid.

a. Write a symbolic rule to represent this input/output relationship.

b. Determine your bid if your opponent's bid is $475.

c. Complete the following table.

OPPONENT'S BID	YOUR BID
$390	
$585	
$1095	

d. Use the symbolic rule determined in part a to calculate your opponent's bid if you have just bid $605. Stated another way, determine x if $y = 605$.

For part a of Exercises 3–8, evaluate the equation by calculating the output when given the input. In part b, use an algebraic approach to solve the equation for the input when given the output.

3. $3.5x = y$

   a. Determine y when $x = 15$.   **b.** Determine x when $y = 144$.

4. $z = -12x$

   a. Determine z when $x = -7$.   **b.** Determine x when $z = 108$.

5. $y = 15.3x$

   a. Determine y when $x = -13$.   **b.** Determine x when $y = 351.9$.

6. $y = x + 5$

   a. Determine y when $x = -11$.   **b.** Determine x when $y = 17$.

7. $y = x + 5.5$

   a. Determine y when $x = -3.7$.   **b.** Determine x when $y = 13.7$.

8. $z = x - 11$

   a. Determine z when $x = -5$.   **b.** Determine x when $z = -4$.

ACTIVITY 2.9

Are They the Same?

Topics: *Translating English Phrases into Algebraic Expressions, Grouping Symbols, Equivalent Expressions*

A major road-construction project in your neighborhood is forcing you to take a 3-mile detour each way when you leave and return home. To compute the round-trip mileage for each of your routine trips, you will have to double the usual one-way mileage and add the extra 3 miles due to the detour.

Does it matter in which order you perform these operations? Try it both ways and determine if there is a difference.

USUAL MILEAGE (INPUT)	RULE 1: TO OBTAIN THE OUTPUT, DOUBLE THE INPUT, THEN ADD 3.	RULE 2: TO OBTAIN THE OUTPUT, ADD 3 TO THE INPUT, THEN DOUBLE.
8		
15		
24		

1. Do rules 1 and 2 generate the same output values?

2. From which sequence of operations do you obtain the correct round-trip mileage?

3. For an input n, write the output y for each rule symbolically.

   Rule 1:

   Rule 2:

4. Using the symbolic rules you wrote in Problem 3, write the calculator keystroke sequence needed to determine each output y for input $n = 15$. Then perform these keystrokes and compare your results with the outputs in the table.

5. Sketch a graph of rules 1 and 2 on the same axes and compare the graphs. What do you observe?

6. Use a table or graph to determine if the expression $2n + 3$ is equivalent to the expression $(n + 3) \cdot 2$.

7. For the following rules, complete the tables and answer the accompanying questions.

Rule 3: To obtain the output, multiply the input by 3, then subtract 2 from the product.

Rule 4: To obtain the output, subtract 2 from the input, then multiply the difference by 3.

INPUT	OUTPUT
-1	
0	
2	
5	
10	

INPUT	OUTPUT
-1	
0	
2	
5	
10	

Notice that each pair of rules contains the same two arithmetic operations, but they are performed in the opposite order.

8. What do you notice about the outputs generated by rules 3 and 4?

9. For each rule, write the output y symbolically in terms of the input n.

   a. Rule 3:   **b.** Rule 4:

10. Write the calculator keystroke sequence needed to evaluate each output y for input $n = 5$. Perform the keystrokes and compare your results with the outputs in the preceding tables.

11. Is the expression $3n - 2$ equivalent to the expression $(n - 2) \cdot 3$? Explain.

12. For the following rules, complete the tables and answer the accompanying questions.

Rule 5: To obtain the output, multiply the input by 2, then add 10 to the product.

Rule 6: To obtain the output, add 5 to the input, then multiply the sum by 2.

INPUT	OUTPUT
-1	
0	
2	
5	
10	

INPUT	OUTPUT
-1	
0	
2	
5	
10	

13. What do you notice about the outputs generated by rules 5 and 6?

14. For each rule, write the output y symbolically in terms of the input n.

   a. Rule 5:   **b.** Rule 6:

15. Write the calculator keystroke sequence needed to evaluate each output y for input $n = 5$. Then perform the keystrokes and compare your results with the outputs in the tables.

16. Sketch a graph of rules 5 and 6 on the same axes and compare the graphs. What do you observe?

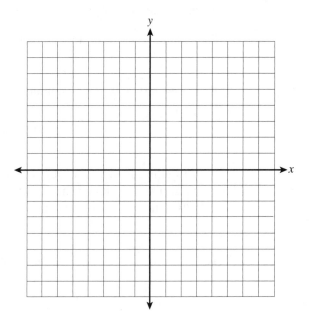

17. Is the expression $2n + 10$ equivalent to the expression $2 \cdot (n + 5)$?

EXERCISES

1. Below each symbolic rule, write in words how the output is obtained. Make sure you include the operations indicated by each rule as well as the order in which each operation is performed. Then complete the tables.

a. $y_1 = x^2$

b. $y_2 = -x^2$

c. $y_3 = (-x)^2$

Change the sign of the input, then square.

x	y_1
-2	
-1	
0	
2	
3	

x	y_2
-2	
-1	
0	
2	
3	

x	y_3
-2	
-1	
0	
2	
3	

2. Which, if any, of the expressions x^2, $-x^2$, and $(-x)^2$ are equivalent? (That is, which expressions generate identical outputs?) Explain.

 3. Sketch the graphs of the rules in Exercise 1. Compare the graphs and state what you observe.

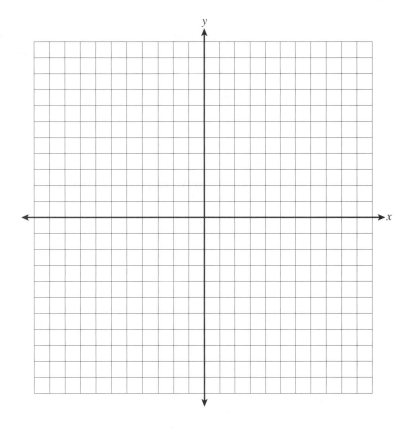

In each of the following exercises, the output is determined by an expression that involves two or more arithmetic operations. You will need to pay close attention to the appropriate order in which to perform these operations.

4.

x	$y_1 = 2x + 1$	$y_2 = 2(x + 1)$
-1		
0		
2		
5		

Are the expressions $2x + 1$ and $2(x + 1)$ equivalent? Why or why not?

5.

x	$y_3 = x + 1 \cdot 5$	$y_4 = (x + 1) \cdot 5$
-1		
0		
2		
5		

Are the expressions $x + 1 \cdot 5$ and $(x + 1) \cdot 5$ equivalent? Why or why not?

6.

x	$y_5 = 1 + x^2$	$y_6 = (1 + x)^2$
-1		
0		
2		
5		

Are the expressions $1 + x^2$ and $(1 + x)^2$ equivalent? Why or why not?

7.

x	$y_7 = 2x^2 - 4$	$y_8 = 2(x^2 - 2)$
-3		
-1		
0		
1		
3		

Are the expressions $2x^2 - 4$ and $2(x^2 - 2)$ equivalent? Why or why not?

8.

x	$y_9 = 3x^2 + 1$	$y_{10} = (3 \cdot x)^2 + 1$
-1		
0		
2		
5		

Are the expressions $3x^2 + 1$ and $(3 \cdot x)^2 + 1$ equivalent? Why or why not?

9.

x	$y_{11} = \sqrt{x + 4}$	$y_{12} = \sqrt{x} + 2$
-4		
-2		
0		
2		
4		

Are the expressions $\sqrt{x + 4}$ and $\sqrt{x} + 2$ equivalent? Why or why not?

ACTIVITY 2.10

**Weather Fahrenheit
or Celsius**

Topic: *Evaluating and
Solving Equations of the
Form y = ax + b*

There are two distinct temperature scales that are widely used in the United States. The National Weather Service reports the daily temperature in degrees Fahrenheit. The scientific community, as well as most of Europe, reports temperature in degrees Celsius. The Celsius and Fahrenheit temperature readings are related by the following *algebraic* rule:

$$F = 1.8C + 32.$$

In this representation, C takes the role of the *input* (its units are degrees Celsius, °C) and F becomes the *output* (its units are degrees Fahrenheit, °F).

Note that you may choose to call the input x and the output y, in which case the rule becomes

$$y = 1.8x + 32.$$

Suppose you are planning a trip to Paris and read in a travel brochure that the average temperature at that time of year is 25°C. To determine what type of clothing to pack, you can use the symbolic rule to convert 25°C to the more familiar degrees Fahrenheit.

1. Is the given temperature, 25°C, a replacement value for input or output?

2. You evaluate $F = 1.8C + 32$ using a sequence of two operations. Using the order of operations indicated by the symbolic rule, begin with the input value (C) and write the *sequence* of operations you will use to obtain the output (F).

    Start with C → _____ → _____ to obtain F.

3. Use the sequence of operations from Problem 2 to determine the Fahrenheit reading for each Celsius temperature in the following table.

INPUT °C	−10	0	15	25
OUTPUT °F				

4. Suppose you know the temperature in degrees Fahrenheit (output) and you want to convert this value to degrees Celsius. For example, convert room temperature, which is about 68°F, to degrees Celsius. You can accomplish this by "reversing the direction" in Problem 2 and by replacing each operation with its "opposite" (inverse).

    a. Start with F → _____ → _____ to obtain C.

    b. Use the sequence of operations from part a to determine the Celsius equivalent to 68°F.

   c. Use the sequence of operations from part a to determine the Celsius equiv-
      alent to normal body temperature, 98.6°F.

You have been using a "reverse" process to solve equations such as $68 = 1.8C + 32$
and $98.6 = 1.8C + 32$. Often you will want to use a more systematic procedure to
solve these types of equations.

For example, in Problem 4b, you used the two-step reverse process:

   Start with $F = 68 \rightarrow$ <u>subtract 32</u> $\rightarrow$ <u>divide by 1.8</u> $\rightarrow$ gives $C = 20$.

Now consider the corresponding algebraic procedure. In the algebraic rule
$F = 1.8C + 32$, replace F with 68:

$$68 = 1.8C + 32 \qquad \text{Equation to be solved}$$
$$\underline{-32 \qquad\qquad\quad -32} \qquad \text{As in Problem 4b, first subtract 32,}$$
$$36 = 1.8C$$

$$\frac{36}{1.8} = \frac{1.8C}{1.8} \qquad \text{then divide by 1.8, or,}$$
$$\qquad\qquad\qquad \text{equivalently, multiply by } \tfrac{1}{1.8}$$
$$20 = C \qquad \text{to obtain the Celsius reading.}$$

Note that the algebraic procedure emphasizes that an equation can be thought of as
a scale whose arms are in balance. The equals sign can be thought of as the balancing
point.

As you perform the appropriate inverse operation to solve the equation for C, each
step in the process must not upset the balance. If 32 is subtracted from one side, then
32 must be subtracted from the other side. Similarly, if one side is divided by 1.8,
then the other side must be divided by 1.8. The sequence of steps results in isolat-
ing the variable C on one side of the equation.

   5. Determine the Celsius reading corresponding to the temperature at which water
      boils, 212°F. Use an algebraic approach.

6. Substitute the given y-value and use an algebraic (symbolic) approach to solve the resulting equation in each of the following.

a. If $y = 3x - 4$ and $y = 10$, determine x.

b. If $y = \frac{2}{3}x + 6$ and $y = 0$, determine x.

c. If $-2x + 15 = y$ and $y = -3$, determine x.

SUMMARY

To solve an equation of the form $y = ax + b$ for the variable x:

1. Reverse, or undo, the addition of b by adding the opposite of b to each side of the equation.

2. Reverse, or undo, the multiplication of the variable by the coefficient a by dividing each side of the equation by the coefficient a.

EXERCISES

Complete the following tables using algebraic methods. Indicate the equation that results when you replace x or y with its assigned value. Then solve the equation, showing each step.

1. $y = 2x + 3$

x	y
4	
	19

Here $x = 4$, so $y = 2 \cdot 4 + 3$.

Here $y = 19$, so $19 = 2x + 3$.

2. $y = 4x - 11$

x	y
6	
	53

3. $y = -5x + 80$

x	y
12	
	35

4. $y = \frac{2}{3}x + 21$

x	y
9	
	47

5. $y = \frac{x}{5} - 2$

x	y
24	
	18

6. $y = 6 + 3x$

x	y
0	
	24

7. $y = -2x - 12$

x	y
0	
	0

8. $y = 30 - 2x$

x	y
12	
	24

9. $y = 21 - \frac{3}{4}x$

x	y
16	
	0

10. $y = 2.25x - 18$

x	y
0	
	0

Use an algebraic approach (reverse process) to solve each of the following equations for the input x.

11. $18 = 2x + 8$

12. $-33 = -5x - 3$

13. $-26 - 3x = -14$

14. $24 + 8x = -38$

15. $3x - 14 = 14$

16. $\frac{2}{3}x - 27 = 66$

17. $-71 = -87 - x$

18. $-\frac{x}{4} + 15 = 39$

19. $1.25x - 22 = 0$

20. $12 - \frac{1}{5}x = 9$

21. $\frac{9}{8}x - 1 = 0$

22. $0.75x - 2.5 = 4$

23. Let P represent the perimeter of an isosceles triangle that has two equal sides of length a and a third side of length b. The formula for the perimeter is $P = 2a + b$. Determine the length of the third side of an isosceles triangle having a perimeter of 15 meters and equal sides measuring 4 meters each.

24. The recommended weight for an adult male is given by the formula $w = \frac{11}{2}h - 220$, where w represents the recommended weight of the person in pounds and h represents the height of the person in inches. Determine the height of an adult male whose recommended weight is 165 pounds.

25 **a.** The temperature on a bank thermometer shows 45°C. What is the temperature in degrees Fahrenheit?

b. A patient has a temperature of 103°F. What is the patient's temperature in degrees Celsius?

26. Housing prices in your neighborhood have been increasing steadily since you bought your home in 1990. The relationship between the market value of your home and the length of time you have been living there can be expressed algebraically by the rule

$$V = 130{,}000 + 3500x$$

where x is the length of time (in years) in your home and V is the market value (in dollars).

a. Complete the following table.

YEAR	x	MARKET VALUE
1990		
1995		
2000		

b. Determine the value of your home in 1994.

c. In which year will the value of your home reach $186,000?

Tables of Tables

Topics: *Tables, Graphing, Symbolic Rules, Evaluating and Solving Equations of the Form y = ax + b*

Suppose you are a waiter at a restaurant that has square tables that seat only one person on a side. To seat a group of more than four, you must slide additional tables over to form a long, rectangular table. Draw pictures of a 1-table, 2-table, 3-table, and 4-table configuration, and count the number of guests that can be seated with each.

1. Construct a table of values whose input variable is the number of square tables used and whose output variable is the total number of guests that can be served.

NUMBER OF SQUARE TABLES	TOTAL NUMBER OF GUESTS SERVED

2. Graph the data obtained in Problem 1. Use a straightedge to connect the points and extend the graph.

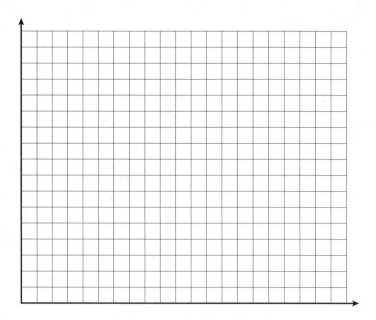

3. How many guests can be served with a 5-table configuration?

4. Will 10 tables seat twice as many as 5 tables? Why or why not?

5. How many guests can be served with a 10-table configuration?

6. Determine a symbolic rule that describes the relationship between the input and the output. Let x represent the input variable and y represent the output variable.

7. Use the symbolic rule that you derived in Problem 6 to write and solve an equation to determine the number of tables needed to serve the following groups of people.

   a. A party of 10

   b. A party of 18

   c. A party of 26

E X E R C I S E S

1. You've decided to enroll in college and take a few courses. Full-time college work does not fit into your present financial or personal situation. The bursar's bill includes a one-time $25 fee for placement testing in addition to the $143-per-credit-hour tuition charges.

   a. Complete the following table.

CREDIT HOURS (INPUT)	TOTAL BILL (OUTPUT)
1	
2	
3	
4	

   b. What is the total cost if you register for a 5-hour biology course?

   c. Write a symbolic rule to express the relationship between number of credit hours and total bill paid. Use x to represent the input and y to represent the output.

   d. What would you predict the cost to be if you wanted to enroll in a psychology course and a computer course? Assume that each course is a 3-credit-hour course.

   e. Use the symbolic rule that you derived in part c to determine the number of credit hours carried by a student whose tuition bill is:

   i. $454 _____   **ii.** $883 _____

   iii. $1598 _____

2. The history club on campus is planning a trip to the County Historical Museum. The admission per student is $2, and the total cost of the bus rental is $78.

   a. Complete the following table.

NUMBER OF STUDENTS	COST OF TRIP
10	
15	
20	
25	

   b. What is the cost to the club if 40 students sign up?

   c. Write a symbolic rule to express the relationship between number of students signed up and total cost of the trip. Use x to represent the input and y to represent the output.

   d. Use the algebraic rule that you derived in part c to determine the number of students signed up if the trip cost is:

      i. $108 _____

      ii. $140 _____

      iii. $152 _____

      iv. $512 _____

ACTIVITY 2.12

Depth of Snow Deposit

Topics: *Evaluating and Solving Equations of the Form y = ax + b, Formulas*

During the winter you often notice snow fences, especially along highways bordered by fields. The association between the depth of the winter snow deposit in the field and stored soil water content from the spring snow melt is illustrated in the following graph.

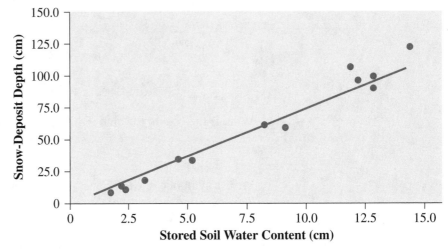

Source: Gray and Hale, *Handbook of Snow*, Pergamon Press, 1981, p. 88.

The data that led to this graph were collected in Akron, Colorado, between 1964 and 1968. Each plotted point comes from recording the depth of the winter's snow and determining stored soil water content in the spring.

The line shown on the graph is known as a **regression line**. In this example, it is represented by the equation

$$d = 7.5x - 2.5,$$

where *d* is the snow depth (output) and *x* is the stored soil water content (input). Both quantities are measured in centimeters. The equation and the line are said to model the relationship between the variables. When the line is determined by a mathematical technique known as the "method of least squares," it is considered to be the line that best describes the collected data.

1. Complete the following table using the mathematical model $d = 7.5x - 2.5$.

STORED SOIL WATER	SNOW DEPTH
5	
8.5	
13	

2. As the stored soil water content increases, does the snow deposit depth increase or decrease? Explain.

It would probably be more useful to a farmer to ask the opposite question; that is, what is the stored water content in the soil based on the winter's snow deposit depth?

3. a. If the winter's snow deposit depth is 35 centimeters, what will be the stored soil water content?

   b. If the winter's snow deposit depth is 110 centimeters, what will be the stored soil water content?

4. Solve the equation $d = 7.5x - 2.5$ for x.

5. Complete the following table using the equation obtained in Problem 4.

SNOW DEPTH (cm)	STORED SOIL WATER (cm)
35	
80	
110	
120	

6. Sketch a graph of the data in Problem 5. Use a straightedge to connect the points and extend the graph.

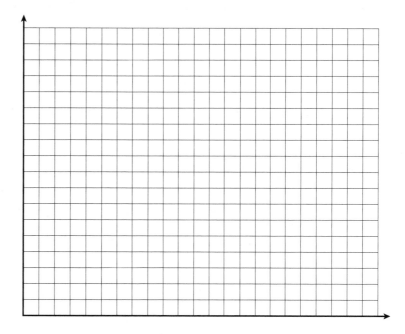

7. As the snow deposit depth increases, does the stored soil water content increase or decrease? Explain

8. Solve the formula $P = 2a + b$ for a.

9. **a.** Solve the formula $F = 1.8C + 32$ for C.

   b. Use the new formula to complete the following table.

°F	0	32	95
°C			

EXERCISES

1. The windchill produced by a 30 mph wind can be approximated by the formula $w = 1.6t - 49$, where w is the windchill and t is the temperature in degrees Fahrenheit.

   a. On a cold day in Buffalo, the wind was blowing at 30 mph. If the windchill was reported to be $-10°$ F, then what was the temperature on that day?

   b. Solve the formula $w = 1.6t - 49$ for t.

   c. Use the new formula and repeat part a.

2. A worker's weekly earnings are given by the formula $E = s + \frac{3}{2}rn$, where E represents the weekly earnings, s the weekly salary, r the hourly rate, and n the number of overtime hours worked.

   a. Suppose, as a worker, your weekly salary is $300 and the hourly rate is $6.80. If you work 7.5 hours of overtime, determine your earnings for the week.

   b. If your earnings for a given week are $402, determine the number of overtime hours you worked.

3. Solve each of the following formulas for the given variable.

   a. $y = mx + b$ for x.　　　　　　　b. $P = 2l + 2w$ for w.

4. Archaeologists and forensic scientists use the length of human bones to estimate the height of individuals. A person's height h, in centimeters, is determined from the length of the femur f (the bone from the knee to the hip socket), in centimeters, using the following formulas:

$$\text{Male: } h = 69.089 + 2.238f$$

$$\text{Female: } h = 61.412 + 2.317f$$

   a. A partial skeleton of a man is found. The femur measures 50 centimeters. How tall was the man?

   b. What is the length of the femur for a female who is 150 centimeters tall?

   c. Solve the formula $h = 61.412 + 2.317f$ for f.

   d. Use the new formula from part c to solve part b.

The basal energy rate is the daily amount of energy (measured in calories) needed by the body at rest to maintain the basic life processes of respiration, cell metabolism, circulation, glandular activity, and maintenance of body temperature. As you may suspect, the basal energy rate differs for individuals, depending on their gender, age, height, and weight. The formula for the basal energy rate for males is

$$B = 655.096 + 9.563W + 1.85H - 4.676A$$

where B is the basal energy rate (in calories), W is the weight (in kilograms),

H is the height (in centimeters), and A is the age (in years).

This algebraic statement is an example of a **formula** describing the relationship between the input variables W, H, and A, and the output variable, B. The following problems demonstrate the use of formulas in a variety of situations.

1. Suppose you are a nurse in a hospital charged with the care of comatose patients. Determine the minimum daily caloric intake (basal energy rate) for a 27-year-old male who is 5 feet 10 inches and weighs 185 pounds. Be careful! You need to make two conversions in measurement before you substitute values into the formula.

The perimeter of a two-dimensional geometric figure is the measure of the distance around it. The area of a geometric figure is the size of the two-dimensional region it encloses. The following table gives the perimeter and area formulas for several common geometric figures. Note that the perimeter of a circle is called its circumference and is denoted by C.

FIGURE	NAME	PERIMETER, P	AREA, A
□	Square	$P = 4s$	$A = s^2$
▭	Rectangle	$P = 2l + 2w$	$A = lw$
△	Triangle	$P = a + b + c$	$A = \frac{1}{2}bh$
○	Circle	$C = 2\pi r$	$A = \pi r^2$

2. a. Determine the perimeter and area of a soccer field. Be sure to indicate which formula and unit of measurement you are using.

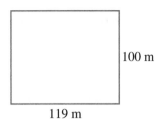

100 m

119 m

b. The top of a circular barbecue grill has diameter 18 inches. Determine the area of the grill top to the nearest square inch.

18 in.

Meteorology and business are other areas in which formulas are very useful.

3. Windchill is a term commonly used during the winter months to describe the effect that wind has on the perceived outdoor temperature. The higher the wind speed, the lower the windchill factor.

a. The windchill produced by a 30 mile-per-hour wind can be approximated by the formula $w = 1.6t - 49$, where w is the windchill and t is the temperature in degrees Fahrenheit. Complete the following table.

t	−15	−5	5
w			

b. Conversely, given the windchill, w, produced by a 30-mph wind, the Fahrenheit temperature, t, can be approximated using the formula

$$t = \frac{w + 49}{1.6}.$$

Use the formula to complete the following table.

w	−36	−24	−10
t			

c. On a cold day in New York City, the wind was blowing at 30 mph. If the windchill was reported to be −5°F, then what was the temperature on that day? Did you use the formula from part a or part b? Explain your choice.

4. Two commonly used formulas from the field of business and finance express the relationship among four quantitative variables:

Simple Interest   **Compound Interest**

$$A = P(1 + r \cdot t) \qquad A = P(1 + r)^t$$

Where:

P is the principal, the amount of the original deposit (in dollars)

r is the rate, the decimal equivalent of the annual interest earned

t is the time during which the money has been in the account (in years)

A is the amount of the current balance in the account (in dollars).

The formulas presume that no additional deposits or withdrawals have been made. Suppose that you make an initial deposit of $1000 in an account paying 10% interest per year.

a. What is the balance at the end of 1 year in a simple-interest account?

b. What is the balance at the end of 1 year if compounded annually?

c. What is the balance at the end of 2 years in a simple-interest account?

d. What is the balance at the end of 2 years if compounded annually?

e. What is the balance at the end of 10 years in a simple-interest account?

f. What is the balance at the end of 10 years if compounded annually?

g. Explain the difference in how interest is accrued in an account bearing simple interest and an account in which the interest is compounded annually.

5. You are the manager of an appliance store. To make a profit, you must sell a product for more than it cost you. The retail price is the price at which the store sells the product.

a. The *retail price* is the sum of what the product costs (*wholesale*) and the *markup*. Write a formula that describes the relationship between retail price p, wholesale cost c, and markup m.

b. A buyer for your store purchases a shipment of CD players for $150.75 each. Use the formula in part a to determine the retail price if the markup is $75.

c. The retail price of a 32-inch television set is $595. Use your formula from part a to determine the markup if the wholesale cost is $375.

Suppose you want to determine the markup for several different items and you know the retail price and cost of each item. In such a situation, you could substitute the values for p and c into the formula $p = c + m$ and then solve the resulting equation for m (see Problem 5c). However, it would be more efficient if you first solve the equation $p = c + m$ for m, substitute the values for p and c into the new formula, and then evaluate.

> To **solve** the equation $p = c + m$ for m means to isolate the variable m on one side of the equation, with all other expressions on the "opposite" side.

6 a. Solve the equation $p = c + m$ for m.

b. Use the formula you determined in part a to solve problem 5c.

7. Carpet-Land advertises a sale in which all carpet prices have been reduced. You would like to buy some carpeting for a rectangular room that measures $10\frac{1}{2}$ by $13\frac{1}{4}$ feet.

a. Use the appropriate geometric formula to determine the total amount of carpeting needed. What are the units of measurement?

b. The length of a rectangular picture frame is 14 inches, and its area is 238 square inches. Use the formula in part a to determine the width of the frame.

c. Solve the formula $A = lw$ for w. Then use this new formula to repeat part b.

8. a. The area A of a triangle is given by the formula $A = \frac{1}{2}bh$. Solve the formula for h.

b. Use your formula from part a to determine the height of a triangle that has a base of length 6 meters and an area of 30 square meters.

9. Solve each of the following equations for the given variable.

a. $I = prt$ for t **b.** $V = \pi r^2 h$ for h

c. $P = 2a + b$ for b **d.** $y = x - 7$ for x

EXERCISES

1. If the wind speed is 15 mph, the windchill can be approximated by the formula $w = 1.4t - 32$, where t is the temperature in degrees Fahrenheit. Complete the following table.

t	-15	-5	5
w			

2. If the wind speed is 15 mph, the temperature can be approximated by the formula

$$t = \frac{w + 32}{1.4}$$

where w is the windchill. Use this formula to *approximate* the Fahrenheit temperature t, and complete the following table.

w	−36	−24	−10
t			

3. You called last night and complained to your mother about the cold. You told her it was −10°F with a wind of 30 mph. She told you to be thankful because in her town it was −20°F with a wind of 15 mph! Was your mother's response justified? Explain. Recall that the windchill produced by a 30-mile-per-hour wind can be approximated by $w = 1.6t − 49$.

4. An alternative formula used for determining windchill is

$$W = 91.4 − (91.4 − T) \cdot \left(0.478 + 0.301\sqrt{V} − 0.02V\right)$$

where W is the windchill temperature (in degrees Fahrenheit)
       T is the air temperature (in degrees Fahrenheit)
       V is the wind velocity (in miles per hour).

   a. Determine the windchill temperature if the air temperature measures 10°F and the wind is blowing at 25 mph.

   b. Use your calculator to determine the windchill temperature if the air temperature measures −10°F and the wind is blowing at 40 mph.

5. The basal energy rate for a female is given by the formula

$$B = 66.473 + 13.752W + 5.003H − 6.755A$$

where B is the basal energy rate (in calories), W is the weight (in kilograms), H is the height (in centimeters), A is the age (in years).

   A female patient who is 70 years old, weighs 120 pounds, and stands 5 feet 8 inches tall, is prescribed a total daily caloric intake of 1000 calories. Determine if this patient is being properly fed.

6. The profit that a business makes is the difference between the revenue (the money it takes in) and the costs.

   a. Write a formula that describes the relationship between the profit p, revenue r, and costs c.

   b. It cost a publishing company $85,400 to produce a textbook. The revenue from the sale of the textbook was $315,000. Determine the profit.

   c. The sales from another textbook amounted to $877,000, and the company earned a profit of $465,000. Use your formula from part a to determine the cost of producing the book.

7. The distance traveled is the product of the rate (speed) at which you are traveling and the amount of time you will be traveling at that rate.

   a. Write a symbolic rule that describes the relationship between the distance d, rate r, and time t.

   b. A gray whale can swim 20 hours a day at an average speed of approximately 3.5 mph. How far does the whale swim in a day?

   c. A Boeing 747 flew 1950 miles at an average speed of 575 mph. Use the formula from part a to determine the flying time.

   d. Solve the formula in part a for the variable t. Then use this new formula to rework part c.

8. Solve each of the formulas for the given variable.

   a. $p = c + m$ for c

   b. $d = rt$ for r

   c. $E = IR$ for I

   d. $A = \frac{1}{2}bh$ for b

   e. $y = mx + b$ for b

   f. $B = \dfrac{lwt}{12}$ for t

   g. $C = 2\pi r$ for r

   h. $m = g - vt^2$ for g

Refer to the geometric formulas in this activity to help answer the following questions.

9. Determine, to the nearest hundredth, the circumference and area of each circle having the given radius. Include appropriate units.

   a. 3 m

   b. 5 ft

   c. 6 cm

   d. 7 yd

   e. 6.8 mm

   f. 7.25 km

10. The diameter of a bicycle wheel is 120 centimeters. What is the circumference of the wheel?

11. A salad bar is designed in the shape of a semicircle. If the radius of the salad bar is 16 feet, what is its area?

12. A pipe has a diameter of 6 inches. A welder must connect a wire clamp around the outside of the pipe. Determine the length of the wire needed to wrap once around the pipe.

13. The inside of a rectangular container has a length of 40 inches and a width of 16 inches. If you fill the container to a depth of 8 inches, what is the volume of the filled portion of the container? (Recall the formula $V = lwh$.)

14. Your sister won a lottery and gave your 2-year-old daughter $10,000 toward her college education. If you invest the money in a long-term certificate of deposit (CD) at a rate of 9% compounded annually, how much will be in the account when your daughter graduates from high school at age 18?

What Have I Learned?

1. Describe the difference between terms and factors.

2. Use examples to describe how solving the equation $4x - 5 = 11$ for x is similar to solving the equation $4x - 5 = y$ for x.

3. In the formula $d = rt$, assume that the rate r is 60 mph. The formula then becomes the equation $d = 60t$.

   a. Which variable is the input variable?

   b. Which is the output variable?

   c. Which variable from the original formula is now a constant?

   d. Discuss the similarities and differences in the equations $d = 60t$ and $y = 60x$.

4. Describe in words what operation(s) must be performed to isolate the variable x in each of the following.

   a. $10 = x - 16$ b. $-8 = \frac{1}{2}x$

   c. $-2x + 4 = -6$ d. $ax - b = c$

5. The area A of a triangle is given by the formula $A = \frac{1}{2}bh$, where b represents the base and h represents the height. The formula can be rewritten as $b = \frac{2A}{h}$ or $h = \frac{2A}{b}$. Which formula would you use to determine the base b of a triangle, given its area A and height h? Explain.

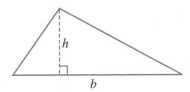

6. **a.** You are designing a cylindrical container as new packaging for a popular brand of coffee. The current package is a cylinder with a diameter of 4 inches and a height of 5.5 inches. The volume of a cylinder is given by the formula

$$V = \pi r^2 h,$$

where V is the volume (in cubic inches); r is the radius (in inches)
   h is the height (in inches).

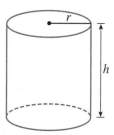

How much coffee does the current container hold? Round your answer to the nearest tenth of a cubic inch.

You have been asked to alter the dimensions of the container so that the new package will contain less coffee. To save money the company plans to sell the new package for the same price as before.

You will do this in either of two ways:

1. By increasing the diameter and decreasing the height by $\frac{1}{2}$ inch each, (resulting in a slightly wider and shorter can), or,

2. By decreasing the diameter and increasing the height by $\frac{1}{2}$ inch each (resulting in a slightly narrower and taller can).

b. Determine which new design, if either, will result in a package that holds less coffee than the current one.

c. By what percent will you have decreased the volume?

How Can I Practice?

1. Consider the expression $5x^3 + 4x^2 - x - 3$.

   a. How many terms are there?

   b. What is the coefficient of the first term?

   c. What is the coefficient of the third term?

   d. If there is a constant term, what is its value?

   e. What are the factors of the second term?

2. Let x represent the input variable. Translate each of the phrases into an algebraic expression.

   a. Input increased by 10

   b. The input subtracted from 10

   c. Twelve divided by the input

   d. Eight less than the product of the input and -4

   e. The quotient of the input and 4, increased by 3

   f. One-half of the input squared, decreased by 2

3. a. Write a symbolic rule that represents the relationship in which the output variable y is 35 less than the input variable x.

   b. What is the output corresponding to an input value of 52?

   c. What is the input corresponding to an output value of 123?

4. a. Write a symbolic rule that represents the relationship in which the output variable t is 10 more than 2.5 times the input variable r.

b. What is the output corresponding to an input of 8?

c. What is the input corresponding to an output of -65?

5. Solve each of the following equations for x using an algebraic approach.

   a. $x + 5 = 2$ **b.** $2x = -20$

   c. $x - 3.5 = 12$ **d.** $-x = 9$

   e. $13 = x + 15$ f. $4x - 7 = 9$

   g. $10 = -2x + 3$ h. $\frac{3}{5}x - 6 = 1$

6. The cost of printing a brochure to advertise your lawn-care business is a flat fee of $10 plus $0.03 per copy. Let C represent the total cost of printing and x represent the number of copies.

   a. Write in symbolic form an equation that will relate C and x.

   b. Organize the data into a table of values. Begin with 1000 copies, increase by increments of 1000, and end with 5000 copies.

NUMBER OF COPIES, x	TOTAL COST, C

c. Graph the data obtained in part b. Use a straightedge to connect the points and extend the graph. Scale the axes appropriately so that you can plot the ordered pair corresponding to 10,000 copies.

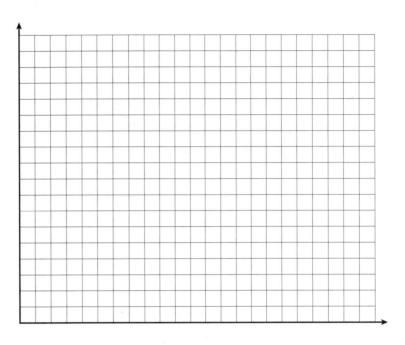

d. What is the total cost of printing 8000 copies?

e. You have $300 to spend on advertising. How many copies can you have printed for that amount?

7. Your car needs a few new parts to pass inspection. The labor cost is $32 an hour, and the parts cost a total of $148. Whether you can afford these repairs depends on how long it will take the mechanic to install the parts.

a. Write a verbal rule that will enable you to determine a total cost for repairs.

b. Write the symbolic form of this rule, letting x be the input variable and y be the output variable. What does x represent? (Include its units.) What does y represent? (Include its units.)

c. Use the symbolic rule in part b to create a table of values. Label the output and the input variables.

HOURS	TOTAL COST ($)
1	
2	
3	
4	
5	

d. How much will it cost if the mechanic works 5 hours?

e. You have $250 available in your budget for car repair. Determine if you have enough money if the mechanic says that it will take him $3\frac{1}{2}$ hours to install the parts.

f. You decide that you can spend an additional $100. How long can you afford to have the mechanic work?

g. Solve the symbolic rule from part b for the input variable x. Why would it ever be to your advantage to do this?

8. a. The formula used to convert a temperature in degrees Fahrenheit to a temperature in degrees Celsius is $C = \frac{5}{9}(F - 32)$. Use this formula to determine the Celsius temperature when the Fahrenheit temperature is 59°.

b. Solve the formula in part a for F.

c. Use your result from part b to determine the Fahrenheit temperature corresponding to a Celsius temperature of 15°.

9. Solve each of the following equations for the given variable.

a. $d = rt$ for r

b. $P = a + b + c$ for b

c. $A = P + Prt$ for r

d. $y = 4x - 5$ for x

e. $w = \frac{4}{7}h + 3$ for h

Mathematical Modeling Involving Algebraic Expressions

ACTIVITY 2.14

Do It Two Ways

Topics: *Distributive Property, Properties of Exponents*

You earn $6 per hour at your job and are paid every other week. You work 25 hours the first week and 15 hours the second week. What is your gross salary?

Determine two different methods for computing the gross salary.

These two methods for computing your gross salary demonstrate the **distributive property** of multiplication over addition.

DISTRIBUTIVE PROPERTY

The distributive property is expressed algebraically as

$$a \cdot (b + c) = a \cdot b + a \cdot c$$

factored form   expanded form

Note that in factored form, you add first, then multiply. In expanded, form you calculate the individual products first, then add the results.

The distributive property can also be interpreted geometrically. Consider the following diagram:

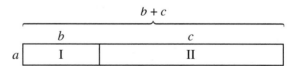

1. **a.** Write an expression for the area of rectangle I.

   b. Write an expression for the area of rectangle II.

2. The area of the rectangle having width a and total top length $b + c$ is given by the expression _____.

3. Use the geometric interpretation of $a \cdot b$, $a \cdot c$ and $a(b + c)$ as areas, from Problems 1 and 2, to explain why $a \cdot (b + c)$ equals $a \cdot b + a \cdot c$. Explain, using the results from Problems 1 and 2.

The distributive property is frequently used to transform one algebraic expression into an equivalent expression.

For example,

$$2 \cdot (x + 5)$$

can be transformed into

$$2 \cdot x + 2 \cdot 5,$$

which can be written as

$$2x + 10$$

The *factored form*, $2 \cdot (x + 5)$, indicates that you start with the input x, add 5, and then multiply by 2. The *expanded form*, $2x + 10$, indicates that you start with the input x, multiply by 2, and then add 10.

4. a. Complete the following table to demonstrate that the expression $2(x + 5)$ is equivalent to the expression $2x + 10$:

INPUT	OUTPUT 1	OUTPUT 2
x	$y_1 = 2(x + 5)$	$y_2 = 2 \cdot x + 10$
1		
2		
4		
10		

b. Explain how the table demonstrates the equivalence of the two expressions.

 c. Use your grapher to sketch the graphs of both output equations. How are the graphs related?

You can visualize the process of writing an expression such as $4(3x - 5)$ in expanded form using the distributive property in two ways.

First, you can make a table that looks very similar to a rectangular area problem. The factor 4 is placed on the left of the table. The terms of the expression $3x - 5$ are placed along the top. You multiply each term along the top by 4 and then add the resulting products.

	$3x$	-5
4	$12x$	-20

Therefore, $4(3x - 5) = 12x - 20$.

Second, you can draw arrows to emphasize that each term within the parentheses is multiplied by the factor 4:

$$4(3x - 5) = 4(3x) - 4(5) = 12x - 20$$

5. Use the distributive property to write each of the following expressions in expanded form.

a. $5(x + 6)$

b. $-10x(y + 11)$

c. $-(2x - 7)$ Note: What is the number preceding the parentheses?

The distributive property can be extended to sums of more than two terms within the parentheses. For example, it can be used to multiply $5(3x - 2y + 6)$.

Using Table Approach:

	$3x$	$-2y$	6
5	$15x$	$-10y$	30

$5(3x - 2y + 6) = 15x - 10y + 30$

Using Arrows Approach:

$5(3x - 2y + 6) = 5(3x) - 5(2y) + 5(6) = 15x - 10y + 30$

6. Write out the procedure used to multiply $5(3x - 2y + 6)$ verbally.

7. Use the distributive property to write the expression $-3(2x^2 - 4x - 5)$ in expanded form.

Suppose you are asked to write the expression $x^2(x^3 + 4x - 3)$ in expanded form. What is the first product, x^2 times x^3? Recall that in the expression x^2, the base x is used as a factor two times while in the expression x^3, the base x is used as a factor three times.

$$x^2 \cdot x^3 = \underbrace{x \cdot x \cdot x \cdot x \cdot x}_{\substack{\text{base } x \text{ is used} \\ \text{as a factor 5 times}}} = x^5$$

This demonstrates that $x^2 \cdot x^3$ is equivalent to x^5.

8. a. Complete the following table.

x	$y_1 = x^2 \cdot x^3$	$y_2 = x^5$
4		
3		
1		
-2		

b. How does the table demonstrate that $x^2 \cdot x^3$ is equivalent to x^5?

c. Use your graphing calculator to sketch the graphs of $y_1 = x^2 \cdot x^3$ and $y_2 = x^5$. How do the graphs compare?

Perform the given multiplication. What pattern do you observe?

9. a. $x^2 \cdot x^4$ **b.** $y^4 \cdot y^6$

c. $a^2 \cdot a^2 \cdot a^3$ **d.** $x^5 \cdot x^3 \cdot x$

The results of Problems 8 and 9 lead to Property 1 of exponents.

PROPERTY 1 OF EXPONENTS

If m and n represent real numbers, then $b^m \cdot b^n = b^{m+n}$

10. Does $x^3 z^4 = xz^7$? Explain.

Problem 10 illustrates that when the bases are different, the exponential factors cannot be combined. For example, $x^3 y^2$ cannot be simplified. However, a product such as $(x^3 y^2)(x^2 y^5)$ can be rewritten as $x^3 x^2 y^2 y^5$ which can be simplified to $x^5 y^7$.

11. Multiply the following.

a. $(-2x^3)(3x^4)$ **b.** $(3a^2)(-4a^3)$

c. $(a^2 b^2)(a^4 b^5)(a^3)$

SUMMARY

To multiply a series of factors:

1. Multiply the numerical coefficients.

2. Simplify the product of the variable factors that have the same base by using Property 1 of exponents. That is, if m and n represent real numbers, then $b^m \cdot b^n = b^{m+n}$.

---

12. Use the distributive property and Property 1 of exponents to write each of the following expressions in expanded form.

a. $x^2\left(x^3 + 4x - 3\right)$ **b.** $-2x\left(x^2 - 3x + 10\right)$

c. $3a^3\left(a^3 + a^2 - 3a + 5\right)$

13. The volume of a solid is a measure of the amount of space it encloses. Many common solids have formulas that are used to determine their volume. For example, the volume V of a rectangular box is the product of its width w times its length l times its height h.

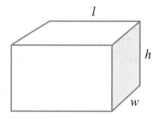

a. Write a formula for the volume V of a rectangular box that has width w, length l, and height h.

b. What is the unit of measurement for volume? Determine the volume of a box with length 5 feet, width 3 feet, and height 4 feet.

c. A rectangular box whose length, width, and height are of equal size is a **cube**. Write a formula for the volume of a cube, where x represents the length of one of its edges.

d. Determine the volume of a cube with edge 2 cm.

Suppose that the length of one of the edges of a cube is a^2. Then the volume of the cube can be written as $V = \left(a^2\right)^3$.

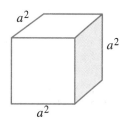

You can rewrite the expression $\left(a^2\right)^3$ more simply. First, note that the expression $\left(a^2\right)^3$ indicates that the base a^2 is used as a factor three times. Therefore,

$$\left(a^2\right)^3 \;=\; \underbrace{a^2 a^2 a^2}_{\substack{\text{base } a^2 \text{ used as} \\ \text{a factor 3 times}}} \;=\; \underbrace{a^{2+2+2}}_{\substack{\text{Property 1} \\ \text{of exponents}}} \;=\; a^6$$

The procedure just completed indicates how $\left(a^2\right)^3$ can be simplified without expanding. Do you see how? Problem 14 provides additional examples that you can use to confirm your observation or help you to discover the property.

14. Perform the given operation (raising a power to a power).

    a. $\left(t^3\right)^5$ **b.** $\left(y^2\right)^4$ **c.** $\left(x^6\right)^3$

The pattern demonstrated by Problem 14 leads to property 2 of exponents.

PROPERTY 2 OF EXPONENTS

If m and n represent real numbers, then $\left(x^m\right)^n = x^{m \cdot n}$.

---

15. Use the properties of exponents to simplify each of the following.

    a. $\left(3^2\right)^4$ **b.** $\left(y^{11}\right)^5$

    c. $2\left(a^5\right)^3$ **d.** $x\left(x^2\right)^3$

    e. $-3\left(t^2\right)^4$ **f.** $\left(5xy^2\right)\left(3x^4y^5\right)$

16. a. Choose two values of x and evaluate the expressions $(2x)^3$ and 2^3x^3 for each value. Is the expression $(2x)^3$ equivalent to 2^3x^3?

b. Choose two values of x and evaluate the expressions, $(2 + x)^3$ and $2^3 + x^3$ for each value. Is the expression $(2 + x)^3$ equivalent to $2^3 + x^3$?

Appendix

Refer to Appendix C for additional properties of exponents.

E X E R C I S E S

1. Use the distributive property to expand the algebraic expression $10(x - 8)$. Then evaluate the factored form and the expanded form for the input values 5, -3, and $\frac{1}{2}$. What did you discover about these two algebraic expressions? Explain.

Use the distributive property to expand each of the algebraic expressions in Exercises 2–11.

2. $6(4x - 5)$

3. $-7(t + 5.4)$

4. $2.5(4 - 2x)$

5. $3(2x^2 + 5x - 1)$

6. $-(3p - 17)$

7. $-(-2x - 3y)$

8. $-3(4x^2 - 3x + 7)$

9. $-(4x + 10y - z)$

10. $\frac{5}{6}(\frac{3}{4}x + \frac{2}{3})$

11. $-\frac{1}{2}(\frac{6}{7}x - \frac{2}{5})$

12. Show, using the properties of exponents, how to simplify each of the following.

    a. $x \cdot x \cdot x \cdot x \cdot x$

    b. $x^3 \cdot x^7 \cdot x$

    c. $(x^4)^3$

Simplify the expressions in Exercises 13–22 using the properties of exponents.

13. $t^3 \cdot 7^7$ **14.** $x \cdot x^5$

15. $-x^2 \cdot x^9 \cdot x$ **16.** $r^3 \cdot r^3 \cdot r^3$

17. $(y^2)^5$ **18.** $(-x^3)^6$

19. $(-2x^3)(-4x^2)$ **20.** $(3x)(-5y)(4x^3)$

21. $(a^2bc)(a^3b^3c^2)(b^4)$ **22.** $(-3s^2t)(s^4t^3)(2s^3t)$

23. Show, using the properties of exponents and the distributive property, how to simplify the expression $2x(x^3 - 3x^2 + 5x - 1)$.

Use the distributive property and the property of exponents to expand the algebraic expressions in Exercises 24–31.

24. $x(x + 3)$ **25.** $-x(7 - 3x)$

26. $x(5 - x)$ **27.** $-x(2x^2 + 6)$

28. $3x^3(2x - 1)$ **29.** $2x(x^2 + x + 9)$

30. $-5x^3(4x^4 - 3x^2 - 3)$ **31.** $2x^7(-5x^5 + x - 1)$

32. Simplify the expression $4(a + b + c)$ and the expression $4(abc)$. Are the results the same? Explain.

33. **a.** The radius of a cylindrical container is three times its height. Write a formula for the volume of the cylinder in terms of its height. Remember to indicate what each variable in your formula represents.

   b. A tuna fish company wants to make a tuna fish can where the radius of the container is equal to its height. Write a formula for the volume of this cylinder in terms of its height. Be sure to indicate what each variable in your formula represents. Write the formula in its simplest form.

   c. Another manufacturing company says that the sizing of the can is the most important factor in their profit. They prefer the height of the can to be 4 inches greater than the radius. Write a formula for the volume of this cylinder in terms of its radius. Be sure to indicate what each variable in your formula represents. Write the formula in its expanded form.

   d. In the cylinder shown here, the relationship between the height and radius is expressed in terms of x. Write a formula, in expanded form, for the volume of the cylinder.

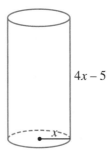

$4x - 5$

x

34. A certain telephone regularly retails for y dollars. The manager of the local discount store reduces the price of the phone by $5. If 12 of these telephones are sold at the reduced price, write an expression in expanded form (without parentheses) that represents the amount of revenue received by the store.

35. John invested $5000 in an account earning simple interest at the rate of 10%. Henry invested the same amount earning interest at the rate of 8% compounded annually. At the end of 12 years, who has more money in the bank and by how much?

36. When you rent an apartment, you are often required to give the landlord a security deposit, which is returned if you leave the apartment undamaged. In New York State the landlord is required to pay the tenant interest on the deposit at an annual rate of 5% compounded annually. The landlord, however, may invest the money at a higher or lower rate of interest. If the landlord invests a $1,000 deposit at an annual rate of 6% compounded annually, determine the net gain by the landlord at the end of 4 years.

While attending college, you live with a group to share expenses, including the phone bill. You check the list of long-distance calls for the month, and your calls are for 12, 7, 8, and 10 minutes in duration. The bill shows that the average cost per minute is $0.162.

1. One way to determine how much of the phone bill you need to pay is to multiply the number of minutes of each call by $0.162 and then add the results. Complete the following calculation.

$$0.162(12) + 0.162(7) + 0.162(8) + 0.162(10) =$$

2. Another way you can determine how much of the phone bill you owe is to add the number of minutes first and then multiply the sum by $0.162.

$$0.162(12 + 7 + 8 + 10) =$$

3. **a.** Compare your results in Problems 1 and 2. Are they equal? Do you think they should be?

   b. What is the property that is being demonstrated in Problems 1 and 2?

In Problem 3, you determined that the numerical expressions in Problems 1 and 2 are equivalent and recognized that the distributive property allows you to write the product as an equivalent sum of terms. For example,

$$\underbrace{0.162(12 + 7 + 8 + 10)}_{\text{product}} = \underbrace{0.162(12) + 0.162(7) + 0.162(8) + 0.162(10)}_{\text{sum}}$$

The distributive property also allows you to reverse the multiplication process and write the sum as a product. This means that to write a sum as a product, you would divide each term by the common factor and then place the common factor outside the parentheses that contain the sum of the remaining factors. For example,

$$\underbrace{0.162(12) + 0.162(7) + 0.162(8) + 0.162(10)}_{\text{sum}} = \underbrace{0.162(12 + 7 + 8 + 10)}_{\text{product}}$$

In this example, 0.162 is a factor of each term on the left and is called a **common factor.** The process of writing a sum equivalently as a product is called **factoring.** The process of dividing each term by a common factor is called **factoring out the common factor.**

4. a. One of your housemates made calls that were for 11, 9, 6, and 5 minutes in duration. Use the new numbers from your friend's calls to describe in words the procedure you would use to write the sum

$$0.162(11) + 0.162(9) + 0.162(6) + 0.162(5)$$

in equivalent factored form as the product

$$0.162(11 + 9 + 6 + 5).$$

b. Use the procedure from part a to write the sum $23(5) + 16(5) - 4(5)$ as a product. What is the common factor?

A common factor is called a **greatest common factor** if there are no additional factors common to the terms in the expression. In the expression $12x + 30$, the numbers 2, 3 and 6 are common factors but 6 is the greatest common factor. When an expression is written in factored form and none of its factors can themselves be factored any further, the expression is said to be in **completely factored form**. Therefore, $12x + 30$ is written in completely factored form as $6(2x + 5)$.

5. a. What is the greatest common factor of $8x + 20$? Rewrite the expression in completely factored form.

b. What are common factors in the sum $10x + 6x$? What is the greatest common factor? Rewrite the expression as an equivalent product.

SUMMARY

FACTORING OUT A COMMON FACTOR

1. Identity the common factors.
2. Divide the common factors out of each term. (Factor out the common factors.)
3. Place the sum of the remaining factors inside the parentheses, and place the common factors outside the parentheses as a coefficient.

---

6. To factor an expression means to write the expression as a product. Factor each of the following by factoring out the common factor.

   a. $2a + 6$ **b.** $5x + 3x - 7x$ **c.** $3x + 12$

   d. $4x^2 - 8x$ **e.** $6x + 18y - 24$

7. a. You do not know the cost per minute for your four long-distance calls of 12, 7, 8, and 10 minutes in duration, so you will represent the cost per minute by the letter x. Write four distinct algebraic expressions, each representing the cost of one of your calls.

    b. Your total share of the phone bill is the sum of these four expressions. Write this sum.

    c. What is the common factor?

    d. Factor out the common factor from each term by placing the common factor outside the parentheses and placing the remaining factors inside the parentheses using the distributive property.

    e. Rewrite the expression in part d by summing the values contained within the parentheses.

    f. Recall that terms are parts of an algebraic expression that are separated by plus or minus signs. How many terms are there in the expression in Problem 6b? In Problem 6e?

Because the number of terms in the expression $12x + 7x + 8x + 10x$ has been reduced from four terms to one term in Problem 6e, we say that the expression has been simplified to $37x$.

The terms in the expression $12x + 7x + 8x + 10x$ are called *like terms*. They differ only by their numerical coefficients. For example, $4x$ and $6x$ are like terms; x^2 and $-10x^2$ are like terms; $4x$ and $9y$ are not like terms; $2x^2$ and $2x^3$ are not like terms.

LIKE TERMS

Like terms are terms that contain identical variable factors, including exponents.

The distributive property provides a way to combine like terms. You will investigate this in more detail.

8. a. Your friend uses a different long-distance carrier. He has two different rates per minute (an evening rate and a day rate). The day rate is x cents per minute, and the evening rate is y cents per minute. His day calls are 4, 8, and 13 minutes long, and his evening calls are 25, 2, 9, and 14 minutes long. Write an expression in terms of x and y that represents the total cost of your friend's calls by forming a sum of the individual costs of the seven calls.

b. How many terms does the algebraic expression in part a contain?

c. Are there any like terms in the expression written in part a? If so, list the like terms.

d. Combine the like terms in part c to simplify the algebraic expression.

SUMMARY

COMBINING LIKE TERMS
Like terms can be combined by adding or subtracting their coefficients. This is a direct result of the distributive property.

For example, $15xy$ and $-8xy$ are like terms with coefficients 15 and -8, respectively. Thus, $15xy - 8xy = (15 - 8)xy = 7xy$.

9. Identify the like terms, if any, in each of the following expressions and combine them.

a. $3x - 5y + 2z - 2x$ **b.** $13s^2 + 6s - 4s^2$ **c.** $2x + 5y - 4x + 3y - x$

When there is no coefficient written immediately to the left of a set of parentheses, the number 1 is understood to be the coefficient of the expression in parentheses. For example, $45 - (x - 7)$ is $45 - 1(x - 7)$. Use the distributive property to multiply each term inside the parentheses by –1 and then combine like terms: $45 - x + 7 = 52 - x$.

10. Professor Sims brings calculators to class each day. There are 30 calculators in the bag she brings. She never knows how many students will be late for class on a given day. Her routine is to first remove 15 calculators and then to remove one additional calculator for each late arrival. If x represents the number of late arrivals on any given day, then the expression $30 - (15 + x)$ represents the number of calculators left in the bag after the late arrivals remove theirs.

a. Simplify the expression for Professor Sims so that she can more easily keep track of her calculators.

b. Suppose there are 4 late arrivals. Evaluate both the original expression and the simplified expression. Compare your two results.

11. Use the distributive property to simplify and combine like terms.

  a. $20 - (10 - x)$ **b.** $4x - (-2x + 3)$ **c.** $2x - 5y - (5x - 6y)$

  d. $4x^2 - x(x - 5)$ **e.** $2(x - 3) - 4(x + 7)$

E X E R C I S E S

In Exercises 1–8, factor out the greatest common factor, and write the result as a product.

1. $3x + 15$

2. $3xy - 7xy + xy$

3. $5x^2 - 10x$

4. $6x + 20xy - 10x$

5. $4 - 12x$

6. $x^4 + 3x^2y + x$

7. $4srt - 3s^2rt + 10srt^3$

8. $8p^3m + 4p^2 - 12p^4m$

9. a. How many terms are in the expression $2x^2 + 3x - x - 3$ as written?

  b. How many terms are in the simplified expression?

10. Are $3x^2$ and $3x$ like terms? Explain.

In Exercises 11–20, simplify the following expressions by combining like terms.

11. $5a + 2ab - 3b + 6ab$

12. $3x^2 - 6x + 7$

13. $100r - 13s^2 + 4r - 18s^3$

14. $3a + 7b - 5a - 10b$

15. $2x^3 - 2y^2 + 4x^2 + 9y^2$

16. $7ab - 3ab + ab - 10ab$

17. $xy^2 + 3x^2y - 2xy^2$

18. $9x - 7x + 3x^2 + 5x$

19. $2x - 2x^2 + 7 - 12$

20. $3mn^3 - 2m^2n + m^2n - 7mn^3 + 3$

21. You are purchasing insurance for several items in the prop room of the Two-Bit Players, a local theater group. In the inventory you record the following items and their estimated value: manual typewriters valued at $100 each, kerosene lanterns valued at $24 each, small tables valued at $38 each, and roll-top desks valued at $200 each.

a. Let t represent the number of manual typewriters, k represent the number of kerosene lanterns, s represent the number of small tables, and d represent the number of roll-top desks. Write an expression to represent the total value of the inventoried items.

b. The insurance agent disagrees with your estimates. He reduces the value of the typewriters by $20 each and increases the value of the desks by $75 each. Write an expression for the total value of the inventoried items that reflects the insurance agent's adjustments.

22. Show how to simplify the expression $2x - (4x - 8)$.

In Exercises 23–36, simplify the expressions using the distributive property and then combine like terms.

23. $30 - (x + 6)$

24. $18 - (x - 8)$

25. $2x - (25 - x)$

26. $27 - 6(4x + 3y)$

27. $12.5 - (3.5 - x)$

28. $3x + 2(x + 4)$

29. $x + 3(2x - 5)$

30. $4(x + 2) + 5(x - 1)$

31. $7(x - 3) - 2(x - 8)$ **32.** $11(0.5x + 1) - (0.5x + 6)$

33. $2x^2 - 3x(x + 3)$ **34.** $3x + 2x^2(x + 4) - 2x$

35. $x(2x - 2) - 3x(x + 4)$ **36.** $3x^2(x^2 + 5x) - 2x(x^2 - 2)$

37. You will be entering a craft fair with your latest metal wire lawn ornament. It is metal around the exterior and hollow in the middle, with the shape of a bird as illustrated. However, you can produce different sizes, and all will be in proportion, depending on the length of the legs, x.

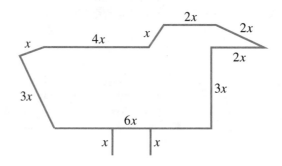

     a. Write an algebraic expression that would determine the amount of wire you would need to create the bird. Be sure to include his legs.

     b. Simplify the expression by combining like terms.

     c. For each lawn ornament, the amount of wire is determined from the expression in part b plus an extra 2 inches for the eye. If you have 400 inches of metal wire, write an expression showing how much wire you will have left after completing one ornament.

     d. Exactly how much wire is needed for an ornament whose legs measure 3 inches? How much wire will be left on the spool?

ACTIVITY 2.16

Math Magic
Topics: *Algebraic
Expressions, Simplifying
Expressions by Combining
Like Terms*

Algebraic expressions arise every time a sequence of arithmetic operations (instructions) is applied to a variable. For example, if your instructions are to double a variable quantity and then add 5, you would express this algebraically as $2x + 5$. If, however, you start with an algebraic expression, say $3x - 2$, you can "decode" the expression to determine the sequence of operations that is applied to x: *Multiply by 3, then subtract 2.*

Perhaps you have seen magicians on TV who astound their audiences by guessing numbers that a volunteer has secretly picked. Look at a few examples of "math magic" and see if you can algebraically decode the tricks.

Select a number, preferably an integer for ease of calculation. Don't tell anyone your number. Perform the following sequence of operations using the number you selected:

Add 1 to the number.
Triple the result.
Subtract 6.
Divide by 3.
Tell your instructor your result.

You might be very surprised that your instructor, a "math magician" of sorts, tells you the number that you originally selected. There is a hidden pattern in the sequence of operations that causes the number selected and the result of the sequence of operations to always have the same relationship.

1. The following table lists results from this trick using four selected numbers. What is the relationship between the number selected and the result?

NUMBER SELECTED	RESULT OF SEQUENCE OF OPERATIONS
0	−1
2	1
5	4
10	9

2. **a.** There are several ways to show how and why this trick works. You first generalize the situation by choosing a variable to represent the number selected. Using x as your variable, translate the first step into an algebraic expression, and simplify by removing parentheses (if necessary) and combining like terms.

b. Now, use the result from part a and translate the second step into an algebraic expression, simplifying when possible. Continue until you complete all the steps.

c. Use your result from part b to interpret verbally (in words) the relationship between the number selected and the result.

3. Rather than simplifying after each step, you can wait until after you have written a single algebraic expression using all the steps. The resulting expression represents the algebraic code for the number trick.

 a. Write all the steps for this number trick as a single algebraic expression.

 b. Simplify the expression in part a. How does this simplified expression compare with the result you obtained in Problem 2b?

4. Here is another trick you can try on a friend. The instructions are:

 Think of a number.
 Double it.
 Subtract 5.
 Multiply by 3.
 Add 9.
 Divide by 6.
 Add 1.
 Tell me your result.

 a. Try this trick with a friend several times. Record the result of your friend's calculations and the numbers he or she originally selected.

 b. Explain how you knew your friend's numbers.

 c. Show how your trick works no matter what number your partner chooses. Use the algebraic approach given in either Problem 2 or Problem 3.

5. You are a magician and you need a new number trick for your next show at the Magic Hat Club. Make up a trick like the ones you learned in Problems 1 and 4. Have a friend select a number and do the trick. Explain how you figured out the number. Show how your trick will work using any number.

6. Your magician friend had a show to do tonight but was suddenly called out of town because of an emergency. She asked you and some other magician friends to do her show for her. She wanted you to do the mind trick, and she left you the instructions in algebraic code. Translate the code into a sequence of arithmetic operations to be performed on a selected number. Try the trick on a friend before the show. Write down how you will figure out any chosen number.

Code: $[2(x - 1) + 8] \div 2 - 5$

7. Here are some more codes you can use. Try them out as you did in Problem 6. State what the trick is.

a. Code: $[(3n + 8) - n] \div 2 - 4$

b. Code: $\dfrac{(4x - 5) + x}{5} + 2$

> ### SUMMARY
>
> **SIMPLIFYING ALGEBRAIC EXPRESSIONS**
>
> Simplify the expression from the innermost parentheses outward using the order of operations. Use the distributive property when it applies, and remember to combine like terms.

8. Simplify the algebraic expressions.

  a. $[3 - 2(x - 3)] \div 2 + 9$

  b. $\{[3 - 2(x - 1)] - [-4(2x - 3) + 5] + 4\}$

  c. $\dfrac{2(x - 6) + 8}{2} + 9$

EXERCISES

1. You want to boast to a friend about the stock that you own without telling him how much money you originally invested in the stock. Let your original stock be valued at x dollars. You watch the market once a month for four months and record the following.

MONTH	1	2	3	4
STOCK VALUE	Increased $50	Doubled	Decreased $100	Tripled

  a. Use x as a variable to represent the value of your original investment, and write an algebraic expression to represent the value of your stock after the first month.

b. Use the result from part a to determine the value at the end of the second month, simplifying when possible. Continue until you determine the value of your stock at the end of the fourth month.

c. Do you have good news to tell your friend? Explain to him what has happened to the value of your stock over four months.

d. Instead of simplifying after each step, write a single algebraic expression that represents the four-month period.

e. Simplify the expression in part d. How does the simplified algebraic expression compare with the result in part b?

2. **a.** Write a single algebraic expression using the following sequence of operations beginning with a number represented by n.

Multiply by -2.
Add 4.
Divide by 2.
Subtract 5.
Multiply by 3.
Add 6.

b. Simplify the algebraic expression.

c. Your friend in Alaska tells you that if you replace n with the value 10, you will discover the average Fahrenheit temperature in Alaska for the month of December. Does that seem reasonable? Explain.

3. Show how you would simplify this expression.

$$2\{3 - [4(x - 7) - 3] + 2x\}$$

4. Simplify the following algebraic expressions.

   a. $5x + 2(4x + 9)$

   b. $2(x - y) + 3(2x + 3) - 3y + 4$

   c. $-(x - 4y) + 3(-3x + 2y) - 7x$

   d. $2 + 3[3x + 2(x - 3) - 2(x + 1) - 4x]$

   e. $6[3 + 2(x - 5)] - [2 - (x + 1)]$

   f. $2\{2xy + 3y[1 + 2(x - 3y) - 4x]\}$

   g. $\dfrac{5(x - 2) - 2x + 1}{3}$

ACTIVITY 2.17

More Room for Work

Topics: *Extending the Distributive Property, Multiplying Binomials*

The basement of your family home is only partially finished. Currently, there is a 9- by 12-foot room that is being used for storage. You want to convert this room to a home office but want more space than is now available. You decide to knock down two walls and enlarge the room.

1. What is the current area of the storage room?

2. If you extend the long side of the room by 4 feet and the short side by 2 ft, what will be the area of the new room?

Start with a diagram of the current room (9 by 12 feet), and extend the length by 4 feet and the width by 2 feet to obtain the following representation of the new room.

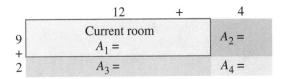

3. Calculate the area of each section, and record the results in the appropriate places on the diagram.

4 a. Determine the total area by summing the areas of the four sections.

   b. Compare your answer from part a to the area you calculated in Problem 2.

You are not sure how much larger a room you want. However, you plan to extend the length and the width each by the same number of feet, x.

Starting with a geometric representation of the current room (9 by 12 feet), you can extend the length and the width by x feet to obtain the following diagram of the new room.

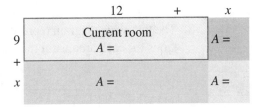

5. Determine the area of each section in your geometric model, and record your answer in the appropriate place. Note that the areas in several sections will be an expression in x.

6. Write an expression for the total area by summing the areas of the four sections. Combine like terms and write in descending order of powers of x.

Another way to represent the area of the new room is by multiplying the expression for the new length by the expression for the new width. The new length of the room is represented by $12 + x$, and the new width is represented by $9 + x$. If you write these two expressions as a product, you obtain $A = (12 + x)(9 + x)$.

> To perform the multiplication $(12 + x)(9 + x)$:
>
> **1.** Multiply each term of the second expression by every term of the first expression.
>
> **2.** Sum the results.
>
> **3.** Combine like terms.

Symbolically, we write

$$A = (12 + x)(9 + x)$$
$$= 12 \cdot 9 + 12 \cdot x + x \cdot 9 + x \cdot x$$
$$= 108 + 21x + x^2 \quad \text{or} \quad x^2 + 21x + 108$$

Compare the final expression with your result in Problem 6. They should be the same.

The expression $x^2 + 21x + 108$ represents the area of a 9-foot by 12-foot room whose length and width have each been expanded by x feet.

7. a. Use the expanded formula to determine the area of the remodeled room when $x = 3$ feet. What are the dimensions of the room? Remember to include the units.

   b. Determine the area of the remodeled room when $x = 4.2$ feet. What are the dimensions of the room? Include the units.

8. The following diagram represents the product $(x + 4)(x - 5)$. Calculate expressions for each area, sum the areas, and then combine like terms.

	x	$+$	4
x			
-5			

9. Multiply $(3x + 1)(x + 2)$ by first creating a geometric model (diagram), then calculating expressions for each area, summing the areas, and combining like terms.

	$3x$	1
x		
2		

In the enlarged room situation, you are multiplying a binomial (an expression containing exactly two terms) by a second binomial. In such cases, the multiplication process is often referred to as the FOIL method (**First** + **Outer** + **Inner** + **Last**), in which each letter refers to the position of the factors in the two binomials.

$$(12 + x)(9 + x)$$

The FOIL method can be viewed as an extension of the distributive property. Each term of one binomial is distributed over the second binomial. For the enlarged room situation,

$$(12 + x)(9 + x) \quad = \quad \underline{12(9 + x)} \quad + \quad \underline{x(9 + x)}$$

first row of geometric representation   plus   second row of geometric representation

$$= 108 + 12x + 9x + x^2$$
$$= 108 + 21x + x^2$$
$$= x^2 + 21x + 108$$

10. Use the extended distributive property (FOIL method) to multiply $(x + 4)(x - 5)$. Compare your answer with that in Problem 8.

11. Use the extended distributive property (FOIL method) to multiply $(3x + 1)(x + 2)$. Compare your answer with that in Problem 9.

E X E R C I S E S

1. Multiply each of the following pairs of binomial expressions. Remember to simplify by combining like terms.

   a. $(x + 1)(x + 7)$ b. $(w - 5)(w - 2)$

   c. $(x + 3)(x - 2)$ d. $(x - 6)(x + 3)$

   e. $(5 + 2c)(2 + c)$ f. $(x + 3)(x - 3)$

   g. $(3x + 2)(2x + 1)$ h. $(5x - 1)(2x + 7)$

   i. $(6w + 5)(2w - 1)$ j. $(4a - 5)(8a + 3)$

   k. $(x + 3y)(x - 2y)$ l. $(3x - 1)(x - 4)$

   m. $(2a - b)(a - 2b)$ n. $(4c + d)(4c - d)$

2. The distributive property may be extended to include factors containing any number of terms. Use the distributive property to multiply the following **polynomials** (an expression containing two or more terms). Recall that only like terms may be combined.

   a. $(x - 1)(3x - 2w + 5)$ b. $(x + 2)(x^2 - 3x + 5)$

   c. $(x - 4)(3x^2 + x - 2)$ d. $(a + b)(a^2 - 2b - 1)$

   e. $(x - 3)(x^2 + 3x + 9)$ f. $(x - 2y)(3x + xy - 2y)$

3. a. You have a circular patio that you wish to enlarge. The radius of the existing patio is 10 feet. If you extend the radius by x feet, express the area of the new patio in terms of x. Leave your answer in terms of π.

b. Use the FOIL method (extension of the distributive property) to express the area in expanded form.

c. You decide to increase the radius by 3 feet. Use the algebraic expression for area from part a to determine the area of the new patio.

d. Use your algebraic expression for area from part b to determine the area of the patio. Compare your results with your answer in part c. Are the two expressions equal? Explain.

4. **a.** You have an old rectangular birdhouse just the right size for wrens. The birdhouse has a length of 7 inches, a width of 4 inches, and a height of 8 inches. What is the volume of the birdhouse?

b. Using the old birdhouse as a model, you want to build a new birdhouse to accommodate larger birds. You will increase the length and width by the same amount, x, and leave the height unchanged. Express the volume of the birdhouse in factored form in terms of x.

c. Use the FOIL method (extension of the distributive property) to expand the expression you obtained in part b.

d. If you increase the length and width by 2 inches each, determine the volume of the new birdhouse. By how much have you increased the volume? By what percent have you increased the volume?

e. Instead of increasing both the length and width of the house, you choose to decrease the length and increase the width by the same amount, x. Express the volume of the birdhouse in factored form in terms of x.

f. Use the FOIL method to expand the expression you obtained in part e.

g. Use your result from part f to determine the volume of the birdhouse whose length has been decreased by 2 inches and whose width has been increased by 2 inches. Check your result by calculating the new length and width directly and then computing the volume.

What Have I Learned?

1. Determine numerically and algebraically which of the following three algebraic expressions are equivalent. Select any three input values and then complete the table. Describe your findings in a few sentences.

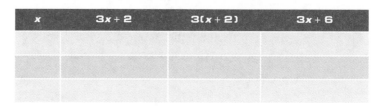

x	3x + 2	3(x + 2)	3x + 6

2. Explain the difference between the two expressions $-x^2$ and $(-x)^2$

3. Are $2x$ and $2x^2$ like terms? Why or why not?

4. Can the distributive property be used to simplify the expression $3(2xy)$? Explain.

5. What role does the negative sign to the left of the parentheses play in simplifying the expression $-(x - y)$?

6. Simplify the product $x^3 \cdot x^3$. Explain how you obtained your answer.

7. Is there a difference between $(2x)^2$ and $2x^2$? Explain.

8. Extend what you know about multiplying binomials to multiplying the following two trinomial expressions.

$$(x^2 + 3x - 1)(2x^2 - x + 3)$$

9. As you simplify the following expression, list each mathematical principle that you use.

$$(x - 3)(2x + 4) - 3(x + 7) - (3x - 2) + 2x^3x^5 - 7x^8 + (2x^2)^4 + (-x)^2$$

How Can I Practice?

1. Complete the following table. Then determine numerically and algebraically which of the following expressions are equivalent.

   a. $13 + 2(5x - 3)$ **b.** $10x + 10$ **c.** $10x + 7$

x	13 + 2(5x − 3)	10x + 10	10x + 7
1			
5			
10			

2. Select any three input values and then complete the following table. Then determine numerically and algebraically which of the following expressions are equivalent.

   a. $4x^2$ **b.** $-2x^2$ **c.** $(-2x)^2$

   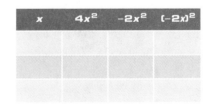

x	4x²	−2x²	(−2x)²

3. Use the properties of exponents to simplify the following.

   a. $3x^3 \cdot x$ **b.** $-x^2 \cdot x^5 \cdot x^7$ **c.** $(2x)(-6y^2)(x^3)$

   d. $8(2x^2)(3xy^4)$ **e.** $(p^4)^5 \cdot (p^3)^2$ **f.** $(3x^2y^3)(4xy^4)(x^7y)$

4. Use the distributive property to expand each of the following algebraic expressions.

   a. $6(x - 7)$ **b.** $3x(x + 5)$ **c.** $-(x - 1)$

d. $-2.4(x + 1.1)$ **e.** $4x(x^4 - 6x^2 - 1)$ **f.** $x^3(x^6 - x^3 + x)$

5. Factor out the greatest common factor(s) in each of the following.

   a. $5x - 30$ **b.** $6xy - 8xy^2$ **c.** $-6y - 36$

   d. $2xy^2 - 4xy + 10x$ **e.** $6xy^3 + 18x^2y - 12x^3y$ **f.** $a^3b^5 + 2a^2b^7$

6. Combine like terms to simplify the following expressions.

   a. $5x^3 + 5x^2 - x^3 - 3$ **b.** $xy^2 - x^2y + x^2y^2 + xy^2 + x^2y$

   c. $3ab - 7ab + 2ab - ab$

7. For each of the following algebraic expressions, list the specific operations indicated, in the order in which they are to be performed.

   a. $10 + 3(x - 5)$

   b. $(x + 5)^2 - 15$

   c. $(2x - 4)^3 + 12$

8. Simplify the following algebraic expressions.

   a. $4 - (x - 2)$ **b.** $4x - 3(4x - 7) + 4$

   c. $x(x - 3) + 2x(x + 3)$ **d.** $2[3 - 2(a - b) + 3a] - 2b$

   e. $3 - [2x + 5(x + 3) - 2] + 3x$ **f.** $\dfrac{7(x - 2) - (2x + 1)}{5}$

9. Use the distributive property to expand the following, and combine like terms where necessary.

   a. $(x + 2)(x - 3)$ **b.** $(a - b)(a + b)$ **c.** $(x + 3)(x + 3)$

   d. $(x - 2y)(x + 4y)$ **e.** $(2y + 1)(3y - 2)$ **f.** $(4x - 1)(3x - 1)$

   g. $(x + 3)(x^2 - x + 3)$ **h.** $(x - 1)(x^2 + 2x - 1)$

   i. $(a - b)(a^2 - 3ab + b^2)$

10. A certain stock is worth x dollars per share at the beginning of a recent month. By the end of the first week of that month, the stock's value has already doubled but then decreased by 3 points ($3 per share). Express, in symbolic form, the total value of your stock at the end of the week if you own 25 shares.

11. Another volatile stock began the last week of the year worth x dollars per share. The table shows the changes during that week. If you own 30 shares, express in symbolic form the total value of your stock at the end of the year.

DAY	1	2	3	4	5
CHANGE IN VALUE/SHARE	Doubled	Lost 10	Tripled	Gained 12	Lost half its value

12. **a.** You are drawing up plans to enlarge your square patio. You want to triple the length of one side and double the length of the other side. If x represents a side of your square patio, write an expression for the new area in terms of x.

    b. You discover from the plan that after doubling the one side of your patio, you must cut off 3 feet from that side to clear a bush. Express the area of your new patio in factored form in terms of x. Then, write your result in expanded form.

c. From the plan you also realize that if you add 5 feet to the side that you tripled, you could extend the patio to the walkway. What would be the new area of your patio with the changes in both sides incorporated into your plan? Express this area in factored form in terms of x and expand the result.

d. Your contractor comes by and says that it would be more economical if you would double both sides and then take away 3 feet from each side. What is the new area of your patio after making these adjustments? Write the expression in factored form in terms of x and expand. What is the shape of the new patio in this situation?

13. You are planning a trip with your best friend from college. You have only four days for your trip and plan to travel x hours each day.

The first day, you stop for sightseeing and lose 2 hours of travel time. The second day, you gain 1 hour because you do not stop for lunch. On the third day, you travel well into the night and double your planned travel time. And on the fourth day, you travel only a fourth of the time you planned because your friend is sick. You average 45 mph for the first two days and 48 mph for the last two days.

a. How many hours, in terms of x, did you travel the first two days?

b. How many hours, in terms of x, did you travel the last two days?

c. Express the total distance traveled over the four days as an algebraic expression in terms of x. Simplify the expression. Recall that

$$\text{Distance} = \text{average rate} \times \text{time}$$

d. Express the total distance that you would have traveled if you had traveled an equal number, x, of hours each day.

e. If 7 hours was your anticipated travel time for each day, how many miles did you actually go on your trip?

f. How many miles would you have gone if you traveled exactly 7 hours each day?

Problem Solving Involving Solution of Equations

ACTIVITY 2.18

Building Fences
Topic: *Solving Equations Using the Distributive Property*

As part of a community-service project, your fraternity is asked to put up a fence in the playground area at a local day-care facility. A local fencing company will donate 500 feet of fencing. The day-care center director specifies that the length of the rectangular enclosure be 20 feet more than the width. Your task is to determine the dimensions of the enclosed region.

1. a. What does the 500 represent with respect to the rectangular enclosure?

b. The length of the rectangle is described in terms of the width. If w represents the width, write an equation that expresses the length of the rectangle in terms of w.

c. One problem-solving strategy is to sketch a diagram that represents the situation. Sketch a rectangle, and label the sides with the expressions determined in part b.

The formula for the perimeter of a rectangle is

$$P = 2l + 2w,$$

where P represents the perimeter (distance around), l represents the length, and w represents the width. Substituting the information about the rectangular enclosure determined in Problem 1, you have

$$2l + 2w = P$$
$$2(w + 20) + 2w = 500.$$

To determine the dimensions of the enclosure, you must solve the equation $2(w + 20) + 2w = 500$ for w. The length can then be determined by substituting this value of w into the expression for l in Problem 1b.

To solve the equation $2(w + 20) + 2w = 500$ for w, the strategy is to isolate the variable w on one side of the equation with all the other numbers and variables on the other side. The following step-by-step process outlines a general procedure for solving similar equations.

$$2(w + 20) + 2w \ = \ 500 \quad \text{Use the distributive property to remove parenthesis.}$$

$$2w + 40 + 2w \ = \ 500 \quad \text{Combine like terms appearing on the same side of the equation.}$$

$$4w + 40 \ = \ 500 \quad \text{Add the opposite of the constant term to both sides.}$$

$$\underline{\quad -40 \qquad\qquad -40 \quad}$$

$$4w \ = \ 460$$

$$\frac{4w}{4} \ = \ \frac{460}{4} \quad \text{Divide both sides by the coefficient of the variable term.}$$

Therefore, $w = 115$ feet (width), and $w + 20 = 135$ feet (length).

2. The day-care center director would like the playground area to be larger than that provided by the 500 feet of fencing. The center staff also wants the length of the playground to be 30 feet more than the width. Additional donations help the center obtain a total of 620 feet of fencing.

   a. Write an equation to determine the width of the enlarged playground area.

   b. Determine the dimensions of the enlarged playground area.

   c. What is the area of the enlarged playground?

3. a. The day-care center staff actually obtains 660 feet of fencing but still wants to have the length 30 feet more than the width, as in Problem 2. Determine the dimensions if all 660 feet of fencing are used.

   b. How much more area would there be in the playground if 660 feet were used rather than 620 feet?

4. Suppose that the director requires that the length of the enclosure is to be 6 feet less than twice the width. Determine the dimensions of the rectangle with this new requirement, assuming that 660 feet of fencing will be used.

5. Determine the solution for each of the following equations.

   a. $2(x + 4) = 12$ **b.** $2x - 7(x + 3) - 10 = 4$

   c. $-7 = -3(x - 7) + 2x + 1$ **d.** $2\{3-[4(x + 9)-3x] + 4x\} = 12$

The same general strategy for solving equations can be used to solve formulas for a specified letter.

6. Solve each of the following formulas for the specified letter.

   a. $P = 2l + 2w$ for w

   b. $V(P + a) = k$ for P

   c. $w = 110 + \frac{11}{2}(h - 60)$ for h

E X E R C I S E S

1. Solve the equation $-2(3x + 8) - 19 = 31$. Give a step-by-step explanation of how to find the solution.

2. Determine the solution to each of the following equations.

   a. $3(x - 6) + 2x = 67$ b. $2(x + 7) - 3 = 121$

   c. $-3(x + 5) + 6x = 21$ d. $57 = -2(x - 6) + 12$

   e. $2x - (x + 15) = 63$ f. $-51 = 32 - 5(2x - 3)$

   g. $71 = -(x - 12) + 52$ h. $2[3x - 7(x - 1) + 3x] - 5x = -21$

   i. $\frac{1}{2}x - 2\left(\frac{3}{4}x + 6\right) = 0$ j. $\dfrac{x - 6}{3} + 4 = -10$

3. You are asked to grade some of the questions on a skills test. Here are five results you are asked to check. If an example is incorrect, find the error and show the correct solution.

a. $34 = 17 - (x + 5)$

$34 = 17 - x - 5$

$34 = 12 - x$

$22 = -x$

$x = -22$

b. $-47 = -6(x - 2) + 25$

$-47 = -6x + 12 + 25$

$-47 = -6x + 37$

$-6x = 84$

$x = 14$

c. $-93 = -(x - 5) - 13x$

$-93 = -x + 5 - 13x$

$-93 = -14x + 5$

$-14x = -98$

$x = 7$

d. $3(x + 1) + 9 = 22$

$3x + 4 + 9 = 22$

$3x + 13 = 22$

$3x = 9$

$x = 3$

e. $83 = -(x + 19) - 41$

$83 = -x - 19 - 41$

$83 = -x - 50$

$133 = -x$

$x = -133$

4. Explain how to solve for y in terms of x in the equation $2x + 4y = 7$.

5. Solve each formula for the specified letter.

a. $y = mx + b$ for x

b. $A = \dfrac{B + C}{2}$ for B

c. $A = 2\pi r^2 + 2\pi rh$ for h d. $F = \frac{9}{5}C + 32$ for C

e. $3x - 2y = 5$ for y f. $12 = -x + \frac{y}{3}$ for y

g. $A = P + Prt$ for P h. $z = \dfrac{x - m}{s}$ for x

6. You are able to get three summer jobs to help pay for college expenses. You work 20 hours per week in your job as a cashier and earn $6.50 per hour. The second and third jobs are both at a local hospital. You earn $8.50 per hour as a payroll clerk and $6.00 per hour as an aide. You always work 10 hours less per week as an aide than you do as the payroll clerk. Your weekly salary is determined by the number of hours that you work at each job.

   a. Determine the input and output variables.

   b. Explain how you would calculate the total amount earned each week.

   c. If x represents the number of hours that you work as a payroll clerk, how would you represent the number of hours that you work as an aide in terms of x?

   d. Write a symbolic rule that describes the total amount earned each week. Use x to represent the input variable and y to represent the output variable.

   e. If you work 12 hours as a payroll clerk, how much will you make in one week?

   f. What are realistic replacement values for x? Would 8 hours at your payroll job be a realistic replacement value? What about 50 hours?

g. When you don't work as an aide, what is your total weekly salary?

h. If you plan to earn a total of $505 in one week from all jobs, how many hours would you have to work at each job? Is the total number of hours worked realistic? Explain.

i. Solve the equation in part d for x in terms of y. When would it be useful to have the equation in this form?

7. In tennis, the length of a singles court is 3 feet less than 3 times its width.

   a. Let x represent the width of the court. Write an equation to express the length of the court in terms of x.

   b. Express the perimeter of the court in terms of x.

   c. If the perimeter of the singles court is 210 feet, write an equation that can be used to determine the width and length of the court.

   d. Solve the equation in part c. What is the length and width of the court?

8. A florist sells roses for $1.50 each and carnations for $0.85 each. Suppose you purchase a bouquet of one dozen flowers consisting of roses and carnations.

   a. Let x represent the number of roses purchased. Write an expression in terms of x that represents the number of carnations purchased.

   b. Write an expression that represents the cost of purchasing x roses.

302 CHAPTER 2 VARIABLE SENSE

c. Write an expression that represents the cost of purchasing the carnations.

d. What does the sum of the expressions in parts b and c represent?

e. Suppose you are willing to spend $14.75. Write an equation that can be used to determine the number of roses that can be included in a bouquet of one dozen flowers consisting of roses and carnations.

f. Solve the equation in part e. How many carnations should be included in the bouquet?

Refrigeration

Topic: *Solving Linear Equations Simultaneously*

The cost of a new Coolair refrigerator is $950. It depreciates (loses value) by $50 with each year of use. The algebraic form of the relationship between the number of years in use, x, and the refrigerator's resale value, v_1, is given by

$$v_1 = 950 - 50x.$$

The cost of a Freezo refrigerator is $1200, and it depreciates $100 per year. The algebraic form of the relationship between the number of years in use, x, and the resale value, v_2, is given by

$$v_2 = 1200 - 100x.$$

You want to determine when the two refrigerators will have equal value.

Numerical Method of Solution

INPUT TIME (IN YEARS)	OUTPUT 1 VALUE OF COOLAIR ($)	OUTPUT 2 VALUE OF FREEZO ($)
0	950	1200
1	900	1100
2	850	1000
3	800	900
4	750	800
5	700	700
6	650	600

Input v_1 and v_2 into your grapher as y_1 and y_2 respectively and use the table feature to duplicate the table. The sixth row in the table indicates that in five years the two refrigerators will have equal resale value.

Graphical Method of Solution

Input v_1 and v_2 into your grapher as y_1 and y_2 respectively. Graph them both in the same window and locate the x coordinate at their point of intersection. See the following screen. Use a window with xmin $= -3$, xmax $= 10$, ymin $= 0$, ymax $= 1300$.

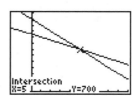

The point of intersection (5, 700) confirms that after 5 years, the two refrigerators have equal resale value, $700.

If the solution to the problem were not a whole number, $x = 5$, it might have been difficult (or at least time consuming) to determine the exact solution numerically or graphically. Solving an equation algebraically always produces an exact solution if one exists.

Algebraic Method of Solution

You are looking for the input value x for which the value of the Coolair, v_1, is equal to the value of the Freezo, v_2. Setting the expressions for v_1 and v_2 equal to each other yields the equation

$$950 - 50x = 1200 - 100x,$$

in which the input variable occurs on both sides of the equation.

To solve such an equation, you extend the notion of adding (or subtracting) the same quantity to both sides so that all terms containing the factor x appear on one side of the equation and all the remaining terms appear on the opposite side.

$950 - 50x = 1200 - 100x$	Eliminate the $-100x$ term from the right by adding $100x$ to both sides.
$+100x +100x$	Combine like terms $-50x$ and $100x$ on the left.
$950 + 50x = 1200$	
$-950 -950$	Subtract 950 from both sides.
$50x = 250$	
$\div 50 \quad \div 50$	Divide both sides by 50.
$x = 5$	Interpret by stating that in 5 years, the two refrigerators will have equal resale value.

1. The cost of an American car is \$13,600, and its value depreciates \$500 a year. The algebraic form of the relationship between the number of years in use, x, and the car's resale value, c_1, is given by

$$c_1 = 13,600 - 500x.$$

The cost of a Japanese car is \$16,000, and it depreciates \$800 a year. The algebraic form of the relationship between the number of years in use, x, and the car's resale value, c_2, is given by

$$c_2 = 16,000 - 800x.$$

a. Construct a table, as on page 303, to investigate the value of each car after 1 year, 5 years, 8 years, 10 years, and 12 years.

NUMBER OF YEARS	VALUE OF AMERICAN CAR	VALUE OF JAPANESE CAR

b. Write an equation that you can use to determine the year in which the two cars will have equal resale value.

c. Solve the equation you wrote in part b and interpret your results in a sentence.

d. What is the common resale value of the two cars at that time?

2. Check your result in Problem 1.

   a. Evaluate both c_1 and c_2 using your result.

 b. Graph both equations and find their point of intersection. Use y_1 for c_1 and y_2 for c_2.

SUMMARY

General strategy for solving equations for a variable, such as x.

1. Remove parentheses, if necessary, by applying the distributive property.
2. Combine like terms that appear on the same side of the equation.
3. Isolate the variable. That is, add and/or subtract terms appropriately and combine like terms on the same side of the equal sign so that the product of the coefficient and the variable is on one side and all other numbers and variables are on the other side.
4. Solve for the variable by dividing each side of the equation by the coefficient of the variable.
5. Check the result to be sure that the value of the variable produces a true statement.

You can practice the skill of solving equations in Problem 3.

3. Determine the solution for each of the following equations.

   a. $18x - 4 = 6x$ **b.** $2x + 14 = 8x + 2$

   c. $10 = 5(x - 1)$ **d.** $6(a + 4) - 5 = 2a - 5$

e. $12x - 9 = 4(6 + 3x)$ **f.** $2(8x - 4) + 10 = 22 - 4(5 - 4x)$

EXERCISES

1. Three friends worked together on a homework assignment that included two equations to solve. Although they are certain that they solved the equations correctly, they are puzzled by the results they obtained. Here they are.

   a. $3(x + 4) + 2x = 5(x - 12)$

   $$3x + 12 + 2x = 5x - 60$$

   $$5x + 12 = 5x - 60$$

   $$5x - 5x = -12 - 60$$

   $$0 = -72?$$ What does the result $0 = -72$ indicate about the solution to the original equation?

    Graph the equations $y_1 = 3(x + 4) + 2x$ and $y_2 = 5(x - 12)$. Do the graphs intersect?

   What does this indicate about the solution to the original equation?

   b. $2(x + 3) + 4x = 6(x + 1)$

   $$2x + 6 + 4x = 6x + 6$$

   $$6x + 6 = 6x + 6$$

   $$6x - 6x = 6 - 6$$

   $$0 = 0?$$ What does the result $0 = 0$ indicate about the solution to the original equation?

    Graph the equations $y_1 = 2(x + 3) + 4x$ and $y_2 = 6(x + 1)$. Do the graphs intersect?

   What does this indicate about the solution to the original equation?

2. Solve the following equations. Check your solutions numerically and/or graphically.

   a. $9x + 8 = 6x + 35$

   b. $7x - 13 = 5x + 11$

   c. $21 + 3(x - 4) = 4(x + 5)$

   d. $2(x - 3) = 8x + 48$

   e. $3 - 2(x - 4) = 5 - 2x$

   f. $18 + 2(4x - 3) = 8x + 12$

   g. $\frac{2}{3}x - 10 = \frac{1}{2}x + 5$

   h. $3.15 - 0.4x = 1.8x + 4.25$

3. You have hired an architect to design a home. She gives you the following information regarding the installation and operating costs for two types of heating systems: solar and electric.

TYPE OF SYSTEM	INSTALLATION COST	OPERATING COST PER YEAR
Solar	$25,600	$200
Electric	$5500	$1200

Although the installation cost for solar heating is much more than that for the electric system, the operating cost per year for the solar system is much lower. Therefore, it is reasonable that the total cost for the electric system will eventually "catch up" and surpass the total cost for the solar system.

   a. Write a symbolic rule to represent the relationship between the number of years in use, x, and the total cost, C, for a solar heating system.

   b. Write a symbolic rule to represent the relationship between the number of years in use, x, and the total cost, C, for an electric heating system.

c. Construct a table, as on page 303, to investigate the cost of each heating system after 1 year, 5 years, 10 years, 20 years, and 25 years.

TIME (IN YEARS)	COST OF SOLAR HEATING SYSTEM ($)	COST OF ELECTRIC HEATING SYSTEM ($)

d. Write an equation that you can use to determine the year in which the total cost for the electric heating system will "catch up" and surpass the total cost for the solar heating system.

e. Solve the equation you wrote in part d and interpret your results in a sentence.

f. What will the cumulative cost of each heating system be at that time?

Every day you are subjected to forces from all directions. Natural forces such as gravity keep you on the ground. Powerful engines create forces that propel our cars, trains, planes, and buses. The effects of many forces, whether natural or the result of human endeavor, can be described using algebraic rules. In this activity, you will work with algebraic rules in which the variable is involved in a squaring or square root process.

Heads Up

When you drop an object from any height, gravity will cause it to fall to Earth. The relationship between the time t (input) and the distance s (output) that the object has fallen is given by the well-known algebraic rule

$$s = 16t^2,$$

where t is measured in seconds and s is measured in feet.

1. To determine how far the object has fallen t seconds after being dropped, you need to perform a sequence of two operations. Use the order of operations indicated by the algebraic rule to write the sequence of operations you will use to obtain the output.

   Start with t (in sec) $\rightarrow$ _____ $\rightarrow$ _____ to obtain s (in ft).

2. How far has the object fallen 1.5 seconds after being dropped?

3. How far has the object fallen 4 seconds after being dropped?

4. How far has the object fallen 6 seconds after being dropped?

5. Now pose the question in reverse by asking how many seconds the object has been falling if it travels 64 feet. You can accomplish this by "reversing direction" and replacing each operation indicated in Problem 1 with its inverse. Which operation is the inverse of squaring?

   Start with 64 ft $\rightarrow$ _____ $\rightarrow$ _____ to obtain _____ sec.

The reverse process used in Problem 5 can be completed in a more systematic way. The distance that the object has fallen is 64 feet. Replace s by 64 in the equation $s = 16t^2$ and proceed as follows:

$$64 = 16t^2$$
$$\frac{64}{16} = \frac{16t^2}{16} \qquad \text{Divide by 16.}$$
$$4 = t^2 \qquad \text{Simplify.}$$
$$\pm\sqrt{4} = t. \qquad \text{Apply square root property.}$$

Therefore, $t = 2$ or $t = -2$. Which of these values is meaningful with respect to the physical situation?

In this problem, we applied the square root property of equations.

SQUARE ROOT PROPERTY

If $x^2 = n$, where n is any *positive* number, then x has two possible values: $\sqrt{n}$ or $-\sqrt{n}$.

If $x^2 = 0$, then x has only a single value, namely $x = 0$.

If $x^2 = n$, where n is any *negative* number, then x has no real values.

---

6. Use an algebraic approach to solve each of the following (if possible).

   a. $x^2 = 9$ **b.** $8w^2 = 24$

   c. $3t^2 - 21 = 0$ **d.** $x^2 + 3 = 0$

Hang Time

The hang time t (output, in seconds) that a basketball player spends in the air during a jump depends on the height s (input, in feet) of that jump according to the algebraic rule

$$t = \tfrac{1}{2}\sqrt{s}.$$

7. To determine the hang time of a jump 1 foot above the court floor, you need to perform a sequence of two operations:

   Start with s (1 ft) $\rightarrow$ _____ $\rightarrow$ _____

   to obtain t (_____ sec in the air).

8. A good college athlete can usually jump about 2 feet above the court floor. Determine his hang time.

9. Michael Jordan is known to have jumped as high as 39 inches above the court floor. What was the hang time on that jump?

10. Now pose the question in reverse by asking how high would a jump have to be to have a hang time of 0.25 second. You can accomplish this by "reversing direction" and replacing each operation indicated in Problem 7 with its inverse. Which operation is the inverse of taking a square root?

Start with 0.25 sec $\rightarrow$ _____ $\rightarrow$ _____ to obtain _____ ft.

Once again, the reverse process can be accomplished in a more systematic way.

Substitute 0.25 second for t in the equation $t = \frac{1}{2}\sqrt{s}$ and proceed as follows:

$$0.25 = \frac{1}{2}\sqrt{s}$$
$$2(0.25) = 2 \cdot \frac{1}{2}\sqrt{s} \qquad \text{Multiply by 2.}$$
$$0.5 = \sqrt{s} \qquad \text{Simplify.}$$
$$(0.5)^2 = (\sqrt{s})^2 \qquad \text{Square both sides.}$$
$$0.25 = s \text{ ft} \qquad \text{Simplify.}$$

11. Is it reasonable for a professional basketball player to have a hang time of 3 seconds? How high would he or she have to jump?

12. Solve each of the following.

a. $\sqrt{x} = 9$ **b.** $\sqrt{t} + 3 = 10$

c. $4\sqrt{x} = 12$ **d.** $\sqrt{t+1} = 5$

e. $\sqrt{5t} = 10$ **f.** $\sqrt{2t} - 3 = 0$

13. Consider the following solution.

$$2\sqrt{x} + 10 = 0$$
$$\underline{-10 = -10}$$
$$\frac{2\sqrt{x}}{2} = \frac{-10}{2}$$
$$\sqrt{x} = -5$$
$$(\sqrt{x})^2 = (-5)^2$$
$$x = 25$$

a. Use the result, 25, to evaluate the left-hand side of the original equation, $2\sqrt{x} + 10 = 0$. Does the value of the left-hand side equal the value of the right-hand side? What does that indicate about the result $x = 25$?

b. Graph $y_1 = 2\sqrt{x} + 10$ and $y_2 = 0$. Do the graphs intersect? What does that indicate about the solution(s) to the equation $2\sqrt{x} + 10 = 0$?

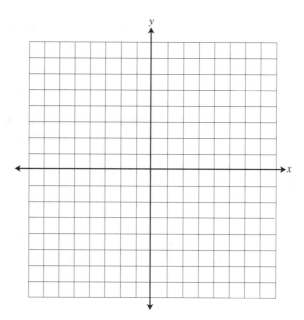

E X E R C I S E S

1. **Falling Objects** An object is dropped from the roof of a five-story building whose floors are vertically 12 feet apart. (Use $s = 16t^2$ to answer these questions.)

   a. How long will it take the object to reach the level of the fifth floor?

   b. How long will it take the object to reach the level of the third floor?

c. How long will it take the object to reach the first floor, at ground level?

2. **Moon Walk**   On the Moon, gravity is only one-sixth as strong as it is on Earth, so the distance an object will fall is one-sixth the distance it would fall on Earth in the same time. This means that the gravity (falling object) rule for the Moon is

$$s = \left(\tfrac{16}{6}\right)t^2.$$

a. How far will an object fall on the Moon in 6 seconds?

b. Determine how long it would take an object to fall 64 feet.

3. **Ball Toss**   If an object is thrown upward with an initial velocity v (measured in feet per second) from an initial height above the ground h (in feet), its height above ground s (in feet), in t seconds after it is thrown, is given by the general formula

$$s = h + vt - 16t^2.$$

A ball is tossed upward with an initial velocity of 20 feet per second from a height of 4 feet. So the general formula becomes $s = 4 + 20t - 16t^2$.

a. How high is the ball after half a second?

b. How high is it after one second?

c. How high is it after 3 seconds?

d. Estimate, using a numerical guess and check method, in how many seconds the ball reaches its maximum height of 10.25 feet.

NUMBER OF SECONDS	HEIGHT (ft)

4. **How Fast Were You Driving?**   You are probably aware that the stopping distance for a car, once the brakes are firmly applied, depends on its speed and on the tire and road conditions. The following formula expresses this relationship:

$$S = \sqrt{30fd},$$

where S is the speed of the car (in miles per hour)

f is the drag factor, which takes into account the condition of tires, pavement, and so on

d is the stopping distance (in feet), which can be determined by measuring the skid marks.

Assume that the drag factor f has the value 0.83.

a. At what speed was your car traveling if you left skid marks of 100 feet?

b. Your friend stops short at an intersection when she notices a police car nearby. She is pulled over and claims that she was doing 25 mph in a 30 mph zone. The skid marks measure 60 feet. Should she be issued a speeding ticket?

c. Determine the length of your skid if you are doing 65 miles per hour when you slam on your brakes.

STOPPING DISTANCE (ft)	SPEED (mph)

5. Solve each of the following equations (if possible) using an algebraic approach. Check your answers numerically or graphically.

   a. $3x^2 = 27$ b. $5x^2 + 1 = 16$

   c. $t^2 + 9 = 0$ d. $3r^2 + 10 = 2r^2 + 14$

   e. $\sqrt{x} - 4 = 3$ f. $5\sqrt{r} = 10$

   g. $\sqrt{x + 4} - 3 = 0$ h. $4 = \sqrt{2w}$

PROJECT ACTIVITY 2.21

Summer Job Opportunities

Topics: *Critical Thinking, Algebraic Rules*

It can be very difficult keeping up with college expenses, and so it is important for you to find a summer job that pays well. Luckily, the classified section of your newspaper lists numerous summer job opportunities in sales, road construction, and food service. The advertisements for all these positions welcome applications from college students. All positions involve the same ten-week period from early June to mid-August.

Sales

A new electronics store has opened recently. There are several sales associate positions that pay an hourly rate of $5.50 plus a 5% commission of your total weekly sales. You would be guaranteed at least 30 hours of work per week, but not more than 40 hours.

Construction

Your state's highway department hires college students every summer to help with road construction projects. The hourly rate is $11.75 with the possibility of up to 10 hours per week in overtime, for which you would be paid time and a half. Of course, the work is totally dependent on good weather, and so the number of hours that you would work per week could vary.

Restaurants

Local restaurants experience an increase in business during the summer. There are several positions for waiters and waitresses. The hourly rate is $2.90, and the weekly tip total ranges from $200 to $750. You are told that you can expect a weekly average of approximately $450 in tips. You would be scheduled to work five dinner shifts of 6.5 hours each for a total of 32.5 hours per week. Daily shift lengths will vary slightly. For example, a dinner shift may be only 5 hours on a slow night.

All of the jobs would provide an interesting summer experience. Your personal preferences might favor one position over another. Keep in mind that you have a lot of college expenses.

1. A good sales associate working 40 hours per week averages $7000 in electronics sales each week.

   a. Based on the expected weekly amount of $7000 in sales, calculate your gross weekly paycheck (before any taxes or other deductions) if you worked a full 40-hour week in sales.

   b. Use the average weekly sales figure of $7000 and write an algebraic rule for your weekly earnings s where x represents the total number of hours you would work.

   c. What would your gross paycheck for the week be if you worked 30 hours and still managed to sell $7000 in merchandise?

d. You are told that you would typically work 35 hours per week if your total electronic sales do average $7000. Calculate your typical gross paycheck for a week.

e. You calculate that to pay college expenses for the upcoming academic year, you need to gross at least $550 a week. How many hours would you have to work in sales each week? Assume that you would sell $7000 in merchandise.

2. For the construction job, you would average a 40-hour workweek.

   a. Calculate your gross paycheck for a typical 40-hour workweek.

   b. Write an algebraic rule for your weekly salary s where x represents the total number of hours worked in a week with no overtime.

   c. If the weather for a week is ideal, you can expect to work 10 hours in overtime (over and above the regular 40-hour workweek). Determine your total gross pay for a week with 10 hours of overtime.

   d. The algebraic rule in part b can be used to determine the weekly salary s when x, the total number of hours worked, is less than or equal to 40 (no overtime). Write an algebraic rule to determine your weekly salary s if x is greater than 40 hours.

   e. Suppose it turns out to be a gorgeous summer and your supervisor says that you can work as many hours as you want. If you are able to gross $800 a week, you will be able to afford to buy a computer. How many hours would you have to work each week to achieve your goal?

3. The restaurant would schedule you to work five dinner shifts of 6.5 hours each. The hourly pay rate is $2.90, and the weekly tip total for a five-day dinner shift schedule ranges from $200 to $750. You are told to expect an average of $450 in tips each week.

   a. Calculate what your gross paycheck would be for an exceptionally busy week of five 6.5-hour dinner shifts and $750 in tips.

b. Calculate what your gross paycheck would be for an exceptionally slow week of five 5-hour dinner shifts and only $200 in tips.

c. Calculate what your gross paycheck would be for a typical week of five 6.5-hour dinner shifts and $450 in tips.

d. Use $450 as your typical weekly total for tips to write an algebraic rule for your gross weekly salary s where x represents the number of hours.

e. Calculate what your gross paycheck would be for a 27-hour week and $450 in tips.

f. During the holiday week of July 4, you would be asked to work an extra dinner shift. You are told to expect $220 in tips for that night alone. Assuming a typical workweek for the rest of the week, would working that extra dinner shift enable you to gross at least $800?

4. You would like to make an informed decision in choosing one of the three positions. Based on all the information you have about the three jobs, fill in the following table.

	LOWEST WEEKLY GROSS PAYCHECK	TYPICAL WEEKLY GROSS PAYCHECK	HIGHEST WEEKLY GROSS PAYCHECK
SALES ASSOCIATE			
CONSTRUCTION WORKER			
WAITER/WAITRESS			

5. Money may be the biggest factor in making your decision. But it is summer, and it would be nice to enjoy what you are doing. Discuss the advantages and disadvantages of each position. What would your personal choice be? Why?

6. You decide that you would prefer an indoor job. Use the algebraic rules you developed for the sales job in Problem 1b and for the restaurant position in Problem 3d to calculate how many hours you would have to work on each job to receive the same weekly salary.

What Have I Learned?

1. On your last math test, the following math problem was completed incorrectly. Your instructor said that he would give you more credit if you could find your errors and correct them. He asked that you circle your errors and correct them, showing all steps next to the problem.

$$2(x - 3) = 5x + 3x - 7(x + 1)$$
$$2x - 5 = 8x - 7x + 7$$
$$2x - 5 = x + 7$$
$$3x = 12$$
$$x = 4$$

2. Does every equation have a solution? Use the following equation in your explanation.

$$3x - x + 8 = 2(x - 7)$$

3. Does every equation have only one solution? Use the following equation in your explanation.

$$6(x - 3) + 10 = 2(3x - 4)$$

4. The diameter d of a sphere having volume V is given by the cube root formula $\sqrt[3]{\frac{6V}{\pi}}$.

   Use what you have learned in this cluster to explain the steps you would use to solve this formula for V in terms of d.

How Can I Practice?

1. Determine the solution to each of the following equations.

   a. $3(2x - 4) = 12$

   b. $-25 = 5(3t + 4)$

   c. $3(x - 4) + 5 = -4$

   d. $6 - 5(2x - 3) = 1$

   e. $2(3x + 1) - 5(2x + 4) = -6$

   f. $\frac{1}{2}(3x - 2) - \frac{5}{2} = -\frac{1}{2}$

   g. $2y + 10 = 5y - 11$

   h. $2(3x + 7) - 18 = 4(5x - 1)$

   i. $3(2x + 8) = 3[2 - 4(x - 1)]$

   j. $4 - 2(p - 7) = 2p - 3(p - 4)$

   k. $8x - 3(2x - 7) - 5 = 2(x + 8)$

   l. $14x - (3x + 2) = 11x + 19$

   m. $3x^2 = 108$

   n. $7 = \frac{1}{2}x^2 - 1$

2. Solve each of the following equations for the specified variable.

   a. $4x - 2y = 5$ for y

   b. $A = \frac{1}{2}h(b + B)$ for B

   c. $\frac{y}{4} + 2x = 7$ for y

   d. $V = \pi r^2 h$ for r

   e. $V = \frac{4}{3}\pi r^3$ for r

   f. $a^2 + b^2 = c^2$ for b

3. The formula $V = \sqrt{\frac{1000p}{3}}$ shows the relationship between the velocity of the wind V (in miles per hour) to the wind pressure p (in pounds per square foot).

   a. If the pressure gauge on a bridge indicates a wind pressure p of 10 pounds per square foot, what is the velocity V of the wind?

   b. If the velocity of the wind is 50 mph, what is the wind pressure?

4. You contact the local print shop to produce a commemorative booklet for your college theater group. It is the twenty-fifth anniversary of your theater group, and you want in the booklet a short history plus a description of all the theater productions for the past 25 years. It costs $750 to prepare the type for the booklet and 25 cents for each copy produced.

a. Let C represent the total cost (output) of producing x booklets (input). Complete the following table.

NO. OF BOOKLETS, x	TOTAL COST, C
50	
100	
150	
200	

b. Write a symbolic rule that relates the total cost C in terms of the number x of booklets produced.

c. Use the rule from part c to determine the total cost of producing 500 booklets.

d. How many booklets can be produced for $1000?

e. Suppose the booklets are sold for 75 cents each. Write a symbolic rule for the total revenue R from the sale of x booklets.

f. How many booklets must be sold to break even? That is, for what value of x is the total cost of production equal to the total amount of revenue?

g. How many booklets must be sold to make a $500 profit?

5. You live 7.5 miles from work, where you have free parking. Some days, you must drive to work so that you can call on clients. On other days, you can take the bus. It costs you 20 cents per mile to drive the car and $1.50 round-trip to take the bus. Assume that there are 22 working days in a month.

a. Let x represent the number of days that you take the bus. Write an expression in terms of x that represents the number of days that you drive.

   b. Write an expression in terms of x for the cost of taking the bus.

   c. Write an expression in terms of x for the cost of driving.

   d. Write an expression in terms of x that represents the total cost of transportation.

   e. How many days can you drive if you budget $40 a month for transportation?

   f. How much should you budget for the month if you would like to take the bus only half of the time?

6. As a prospective employee in a clothing store, you are offered a choice of salary. The following table shows your options.

OPTION 1	$100 per wk	Plus 30% of all sales
OPTION 2	$150 per wk	Plus 15% of all sales

   a. Write a symbolic rule to represent the relationship between the total salary T_1 and the total amount of sales x per week for option 1.

   b. Write a symbolic rule to represent the relationship between the total salary T_2 and the total amount of sales x per week for option 2.

   c. Construct a table to investigate your total salary under each option. Use sales of $100, $300, $500, and $1000 per week.

SALES ($)	OPTION 1	OPTION 2
$100		
$300		
$500		
$1000		

d. Write an equation that you could use to determine how much you would have to sell weekly to have the same weekly salary under both plans.

e. Solve the equation in part d. Interpret your result.

f. What is the common salary for the amount of sales found in part e?

g. Check your result in part e by graphing both equations and finding the intersection. Use y_1 for T_1 and y_2 for T_2.

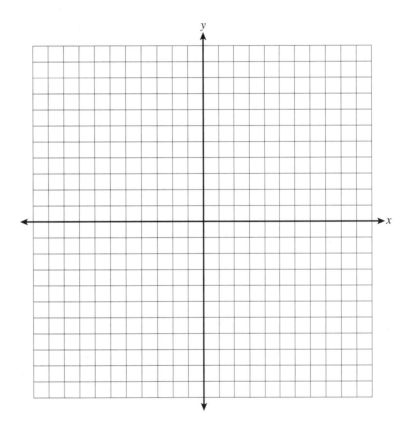

h. Decide which option you will choose and discuss why you will choose it.

Gateway Review

1. Scale each axis on the Cartesian coordinate plane below, and plot the points whose coordinates are given.

   a. $(1, 2)$ **b.** $(3, -4)$

   c. $(-2, 5)$ **d.** $(-4, -6)$

   e. $(0, 4)$ **f.** $(-5, 0)$ **g.** $(0, 0)$

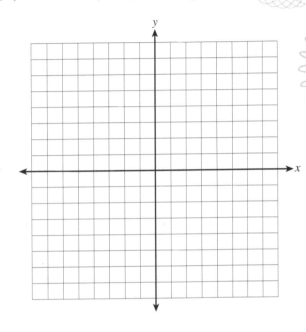

2. Let x represent the input variable. Translate each of the phrases into an algebraic expression.

   a. Add 5 to the input.

   b. Subtract the input from 18.

   c. Multiply the input by 2.

   d. Divide 4 by the input.

   e. Add 17 to the product of 3 and the input.

   f. Multiply the sum of 8 and the input by 12.

   g. Multiply the difference of 14 and the input by 11.

   h. Subtract 49 from the quotient of the input and 7.

3. Determine the following.

   a. If $y = 2x$ and $y = 18$, what is the value of x?

   b. If $y = -8x$ and $x = 3$, what is the value of y?

   c. If $y = 15$ and $y = 6x$, what is the value of x?

   d. If $y = -8$ and $y = 4 + x$, what is the value of x?

   e. If $x = 19$ and $y = x - 21$, what is the value of y?

   f. If $y = -71$ and $y = x - 87$, what is the value of x?

   g. If $y = 36$ and $y = \frac{x}{4}$, what is the value of x?

   h. If $x = 5$ and $y = \frac{x}{6}$, what is the value of y?

   i. If $y = \frac{x}{3}$ and $y = 24$, what is the value of x?

4. Tiger Woods had scores of 63, 68, and 72 for three rounds of a golf tournament.

   a. Let x represent his score on the fourth round. Write an algebraic expression to represent his average after four rounds of golf.

   b. To be competitive in the tournament, he must maintain an average of about 66. Use the expression from part a to determine what he must score on the fourth round to achieve a 66 average for the tournament.

5. Determine the following.

   a. If $y = 2x + 8$ and $y = 18$, what is the value of x?

   b. If $x = 14$ and $y = 6x - 42$, what is the value of y?

   c. If $y = -33$ and $y = -5x - 3$, what is the value of x?

   d. If $y = -38$ and $y = 24 + 8x$, what is the value of x?

e. If $x = 72$ and $y = -54 - \frac{x}{6}$, what is the value of y?

f. If $y = 66$ and $y = \frac{2}{3}x - 27$, what is the value of x?

g. If $y = 39$ and $y = -\frac{x}{4} + 15$, what is the value of x?

h. If $y = 32.56x + 27$ and $x = 0$, what is the value of y?

6. You must drive to Syracuse to take care of some legal matters. The cost of renting a car for a day is $25 plus 15 cents per mile.

   a. Identify the input variable.

   b. Identify the output variable.

   c. Using x to represent the input and y to represent the output, translate the verbal rule into a symbolic rule.

   d. Complete the following table.

INPUT, x	100	200	300	400	500
OUTPUT, y					

   e. Plot the points from your table. Then, sketch the line containing all 5 points. (Make sure you label your axes and use appropriate scaling.)

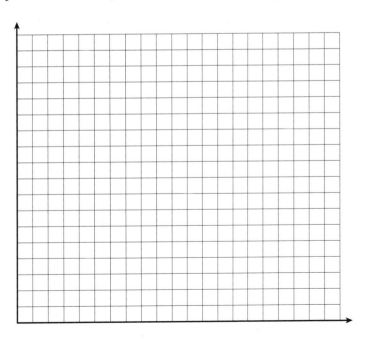

f. The distance from Buffalo to Syracuse is 153 miles. How much will it cost to travel from Buffalo to Syracuse and back? Estimate from your graph.

g. Use the symbolic rule in part c to determine the exact amount that it will cost for this trip.

h. If you have budgeted $115 for car rental for your day trip, what is the greatest number of miles you can travel in the day and not exceed your allotted budget? Estimate from your graph.

i. Use the symbolic rule from part c to determine the exact number of miles you can travel in the day and not exceed $115.

j. What are realistic replacement values for the input if you rent the car only for one day?

7. As a real estate salesperson, you earn a small salary plus a percentage of the selling price of each house you sell. If your salary is $100 a week plus 3.5% of the selling price of each house sold, what must your total annual home sales be to gross $30,000 in one year? Assume that you work 50 weeks per year.

8. Use the formula $I = Prt$ to evaluate I, given the following information.
   a. $P = \$2000, r = 5\%, t = 1$

   b. $P = \$3000, r = 6\%, t = 2$

9. Using the formula $P = 2(w + l)$, evaluate P given the following information.
   a. $w = 2.8$ and $l = 3.4$

   b. $w = 7\frac{1}{3}$ and $l = 8\frac{1}{4}$

10. You need $20,000 in your bank account to make a deposit on a home. If you make a deposit of $10,000 in a bank paying interest at the annual compounded rate of 12%, estimate how long it will take for you to acquire your home.

11. Determine numerically and algebraically which, if any, of the following expressions are equivalent.

    a. $(4x - 3)^2$ **b.** $4x^2 - 3$ **c.** $(4x)^2 - 3$

x	(4x − 3)²	(4x² − 3)	(4x)² − 3
−1			
0			
3			

12. Simplify the following using the properties of exponents.

    a. $x^3 \cdot x^5$ **b.** $(2x^2y^3)(3xy)$ **c.** $(x^4)^7$ **d.** $8(x^2)^3(y^3)^5$

    e. $(x^2y^5)(x^4)^2$ **f.** $(-s^6)^3$ **g.** $(4xyz)(x^3yz^2)(x^2y^3z^6)$

13. Simplify the following expressions.

    a. $3(x + 1)$ **b.** $-6(2x^2 - 2x + 3)$ **c.** $x(4x - 7)$

    d. $6 - (2x + 14)$ **e.** $4 + 2(6x - 5) - 19$ **f.** $5x - 4x(2x - 3)$

14. Factor completely.

    a. $4x - 12$ **b.** $18x^2 + 60x - xy$ **c.** $-12x - 20$

15. Simplify the following expressions.

    a. $4x^2 - 3x - 2 + 2x^2 - 3x + 5$

    b. $\left(3x^2 - 7x + 8\right) - \left(2x^2 - 4x + 1\right)$

16. Use the distributive property to multiply the following.

    a. $(x + 3)(x - 1)$ **b.** $(x - 2y)(2x - 3y)$ **c.** $(x + 2)\left(x^2 - 4x - 1\right)$

17. Solve the given equations and check your results.

    a. $4(x + 5) - x = 80$

    b. $-5(x - 3) + 2x = 6$

    c. $38 = 57 - (x + 32)$

    d. $-13 + 4(3x + 5) = 7$

    e. $5x + 3(2x - 8) = 2(x + 6)$

    f. $-4x - 2(5x - 7) + 2 = -3(3x + 5) - 4$

    g. $-32 + 6(3x + 4) = -(-5x + 38) + 3x$

    h. $4(3x - 5) + 7 = 2(6x + 7)$

    i. $2(9x + 8) = 4(3x + 4) + 6x$

    j. $5x^2 = 3x^2 + 8$

    k. $\sqrt{3x} + 1 = 7$

18. You have just found the perfect dress for a wedding. The price of the dress is reduced by 30%. You are told that there will be a huge sale next week, so you wait to purchase it. When you go to buy the dress, you find that it has been reduced again by 30% of the already reduced price.

    a. If the original price of the dress is $110, what is the price of the dress after the first reduction?

    b. What is the price of the dress after the second reduction?

    c. Let x represent the original price (input). Determine an expression that represents the price of the dress after the first reduction. Simplify this expression.

    d. Use the result from part c to write an expression that represents the price of the dress after the second reduction. Simplify this expression.

    e. Use the result from part d to determine the price of the dress after the two reductions if the original price is $110. How does this compare with your answer to part b?

    f. You see another dress that is marked down to $147 after the same two reductions. You want to know the original price of the dress. Write the equation and solve to determine the original price.

19. The proceeds from your college talent show to benefit a local charity totaled $950. Since the seats were all taken, you know that 500 people attended. The cost per ticket for students was $1.50, and for adults the cost was $2.50. Unfortunately, you misplaced the ticket stubs that would indicate how many students and how many adults attended. You need this information for accounting purposes and future planning.

    a. Let n represent the number of students that attended. Write an expression in terms of n to represent the number of adults who attended.

    b. Write an expression in terms of n that will represent the proceeds from the student tickets.

    c. Write an expression in terms of n that will represent the proceeds from the adult tickets.

d. Write an equation that indicates that the total proceeds from the student and adult tickets totaled $950.

e. How many student tickets and how many adult tickets were sold?

20. You have an opportunity to be the manager of a day camp for the summer. You know that your fixed costs for operating the camp are $600 per week, even if there are no campers. Each camper who attends costs the management $10 per week. The camp charges each camper $40 per week.

    Let x represent the number of campers.

    a. Write a rule in terms of x that represents the total cost of running the camp per week.

    b. Write a rule in terms of x that represents the total income (revenue) from the campers per week.

    c. Write a rule in terms of x that represents the total profit from the campers per week.

    d. How many campers must attend the camp to break even with revenue and costs?

    e. The camp would like to make a profit of $600. How many campers need to enroll to make that profit?

    f. How much money would the camp lose if only 10 campers attend?

     g. Use your grapher to graph the revenue and cost equations. Compare the break-even point from the graph (point of intersection) with your answer in part d.

21. Solve each of the following equations for the specified variable.

    a. $I = Prt$ for P **b.** $f = v + at$ for t **c.** $2x - 3y = 7$ for y

    d. $A = 2\pi r^2$ for r **e.** $V = \sqrt{\frac{110P}{3}}$ for P

22. In your science class, you are learning about pendulums. The period of a pendulum is the time it takes the pendulum to make one complete swing back and forth. The relationship between the period and the pendulum length is described by the formula $T = \frac{\pi}{4}\sqrt{\frac{l}{6}}$ where l is the length of the pendulum in inches and T is the period in seconds.

    a. The length of the pendulum inside your grandfather clock is 24 inches. What is the period?

    b. A metal sculpture hanging from the ceiling of an art museum contains a large pendulum. In science class you are asked to determine the length of the pendulum without measuring it. You use a stopwatch and determine that the period is approximately three seconds. What is the length of the pendulum in feet?

CHAPTER 3

Function Sense

Chapter 3 continues the study of relationships between input and output variables. The focus here is on functions, which are special relationships between input and output variables. You will learn how functions are represented verbally, numerically, graphically, and symbolically, and you will be introduced to some particular types of functions.

CLUSTER 1

ACTIVITY 3.1

Course Grade
Topics: *Function Representations, Function Notation*

Exploring Functions

The semester is drawing to a close, and you are concerned about your grade in your anthropology course. During the semester, you have already taken four exams and scored 82, 75, 85, and 93. Your score on exam 5 will determine your final average for the anthropology course.

1. Four possible exam 5 scores are listed in the following table. Calculate the final average corresponding to each one and record your answers.

EXAM 5 SCORE (INPUT)	FINAL AVERAGE (OUTPUT)
100	
85	
70	
60	

2. Plot the (input, output) pairs from the table in Problem 1.

Output

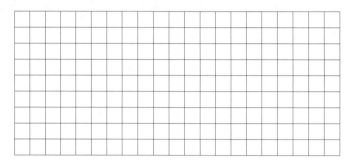

Input

An input/output situation such as the one just described is an example of a *function*.

> A **function** relates the input to the output in a particular way. It assigns a *single* output value to each input value.

3. Explain how the table of data you obtained in Problem 1 fits this definition of a function.

4. In Problem 2, you represented each input/output pair as the coordinates of a point on a graph. Draw a smooth line through the plotted points on the graph in Problem 2. Use the line to answer parts a and b.

   a. Estimate what your final average will be if you score a 95 on exam 5.

   b. Estimate what score you will need on exam 5 to earn a final average of 85.

A function describes the relationship, or correspondence, between an input variable and an output variable. It is customary to simply say that the output is a function of the input. For example, you would say that the final average is a *function* of the fifth exam score at this stage of your course.

There are several ways to represent a function. So far, you have seen that a function may be presented *numerically* by a table, as in Problem 1, and *graphically* on a grid, as in Problem 2. A function can also be given *verbally* by stating how the output value is obtained for a given input value.

5. Describe in words the final average as a function of your exam 5 score; that is, state how to obtain the final average for any given score on the fifth exam.

A fourth way to define a function is *symbolically*. The verbal rule you used to calculate your final average is given symbolically by the formula

$$A = \frac{82 + 75 + 85 + 93 + s}{5} \quad \text{or} \quad A = \frac{335 + s}{5},$$

where the variable s represents the input (fifth exam score) and the variable A represents the output (final average).

    6. a. Use the symbolic rule to determine the final average for a score of 75 on exam 5.

       b. Use the symbolic rule to determine the score you will need on exam 5 to earn a final average of 87.

Often we want to emphasize the input/output relationship of a function. We do this symbolically by writing $A(s)$ and read the symbols as "A is a function of s."

The formula is then written as

$$A(s) = \frac{335 + s}{5}.$$

If you want to determine the final average A when the fifth exam score is 100, you write $A(100)$, pronounced "A of 100." You interpret this to mean that in the formula, s is replaced with the input value 100, and the resulting expression on the right-hand side is evaluated to obtain the output value.

$$A(100) = \frac{335 + 100}{5}$$

$$A(100) = 87$$

You interpret this symbolic statement as "The final average for a fifth exam score of 100 is 87."

Notice that the symbolic way of defining a function is very useful because it clearly shows the rule for finding an output no matter what the input might be.

    7. a. In the statement $A(75) = 82$, identify the input value and output value.

       b. Interpret the practical meaning of $A(75) = 82$ in the final average situation. Write your answer as a complete sentence.

8. a. Use the symbolic formula for the final average function to determine the missing coordinate in each input/output pair of the final average function.

 i. (78, _____)   **ii.** (_____ , 67)

b. Evaluate $A(95)$ in the final average function.

c. Write a sentence that interprets the practical meaning of $A(95)$ in the final average function.

Function Notation—A Word of Caution.

In **function notation,** the parentheses *do not* indicate multiplication. The notation $f(x)$ is a way to represent the output variable when you want to call attention to the input/output relationship.

For example, $f(x) = x^2 + 3$ tells you that the function name is f, and indicates that you must square the input, then add 3 to obtain the output.

If $x = 3$, $f(3) = (3)^2 + 3 = 12$. This means that (3, 12) is one (input, output) pair associated with the function f, and (3, 12) is one point on the graph of f.

9. a. In the function defined by $H(a) = 2a + 7$, the function name is _____, and the input variable is _____. The function rule, in words, is:

b. If $a = -5$, then the corresponding output value is $H(-5) =$ _____, and thus $(-5,$ _____ $)$ is a point on the graph of H.

10. Describe the function $P(w) = 6 - 5w$ in a way similar to that outlined in Problem 9, using $w = 8$.

A function, by definition, assigns only one output value to each input value. Visually, this means that if you can find even one vertical line that intersects a graph more than once, then the graph does not represent a function. This is known as the **vertical line test.**

11. The graph of your possible final averages should pass the vertical line test. Check it out.

12. The graph pictured does not represent a function. Explain.

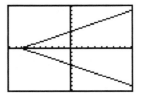

13. Which of the following graphs represent functions? Explain.

a. 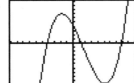 b.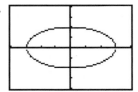

14. Give two (or more) examples of functions from your daily life. Explain how each fits the description of a function. Be sure to identify the input variable and output variable.

SUMMARY

A **function** is a rule that relates two variables. One variable is designated as the input variable and the other as the output variable. In a function, each input value is related to *one and only one* output value.

There are four ways to define the function rule: *verbally, numerically, graphically,* and *symbolically.*

Functions are often defined symbolically using the **function notation** $f(x)$, where f is the name of the function and x is the input variable. The notation $f(x)$ is often used to represent the output y for a given x. Thus, you can write $y = f(x)$ and say that the output y is a function of x.

A graph defines a function if it is not possible to draw a vertical line that intersects the graph more than once. This is known as the **vertical line test.**

E X E R C I S E S

1. In Chapter 2, you obtained the following input/output table for the history club trip, and determined the symbolic rule $y = 2x + 78$, where x represents the number of students (input) and y represents the cost of the trip (output).

NUMBER OF STUDENTS	COST OF TRIP ($)
10	98
15	108
20	118
25	128

a. Rewrite the symbolic rule in function notation that emphasizes the input/output relationship. Use n to represent the number of students and C to represent the name of the cost function for the trip.

b. Use function notation to write a symbolic statement for "The cost per student is $108 if 15 students go on the trip."

c. Choose another (input, output) pair for the cost function C. Write this pair in both ordered pair notation and in function notation.

d. Are there any restrictions on the number of students who can sign up for the trip? Explain.

e. $C(37) =$ _____

f. Describe, in words, what information about the cost of the trip is given by the statement $C(24) = 126$.

2. Which of the following relationships represent functions? Explain your answers.

a.

INPUT	−8	−3	0	6	9	15	24	38	100
OUTPUT	24	4	9	72	−14	−16	53	29	7

b.

INPUT	−8	−5	0	6	9	15	24	24	100
OUTPUT	24	4	9	72	14	−16	53	29	7

c.

INPUT	−8	−3	0	6	9	15	24	38	100
OUTPUT	24	4	9	72	4	−16	53	24	7

3. Identify the input and output variables in each of the following. Which statement is true? Give a reason for your answer.

a. Your letter grade in a course is a function of your numerical grade.

b. Your numerical grade is a function of your letter grade.

4. Let $f(x) = 2x - 1$. Evaluate $f(4)$.

5. Let $g(n) = 3n + 5$. Evaluate $g(-3)$.

6. Let $h(m) = 2m^2 + 3m - 1$. Evaluate $h(-2)$.

7. Let $p(x) = 3x^2 - 2x + 4$. Evaluate $p(5)$.

8. Let $f(x) = 2x + 3$ and $g(x) = x^2 - 1$. Evaluate $f(2) + g(2)$.

9. Let $n(x) = 5x - 2$ and $m(x) = x^2 + 5$. Evaluate $n(-1) - m(-1)$.

10. Let f be a function defined by $f(x) = x + 1$. Determine the value of x for which $f(x) = 3$.

11. Let g be a function defined by $g(x) = x - 4$. Determine the value of x for which $g(x) = 10$.

12. Let h be a function defined by $h(t) = \frac{t}{4}$. Determine the value of t for which $h(t) = 12$.

13. Let k be a function defined by $k(w) = 0.4w$. Determine the value of w for which $k(w) = 12$.

14. Which of the following graphs represents a function? Explain your answer.

a.
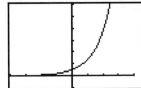

b.
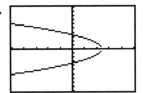

c.
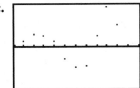

d.
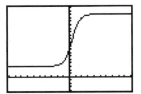

ACTIVITY 3.2

Graphs Tell Stories

Topics: *Interpretations of Graphs; Increasing, Decreasing, and Constant Functions*

"A picture is worth a thousand words" may be a cliché, but nonetheless, it is frequently true. Numerical relationships are often easier to understand when presented in visual form. Understanding graphical pictures requires practice going in both directions—from graphs to words and from words to graphs.

Graphs are always constructed so that as you read the graph from left to right, the input variable increases in value. The graph illustrates the change (increasing, decreasing, or constant) in the output values as the input values increase.

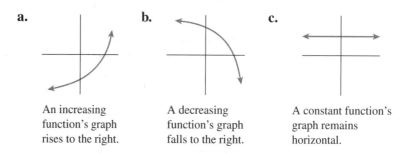

a. An increasing function's graph rises to the right.

b. A decreasing function's graph falls to the right.

c. A constant function's graph remains horizontal.

Suppose that you are a technician at the local power plant, and you have been asked to prepare a report that compares the output and efficiency of the six generators in your sector. Each generator has a graph that shows output as a function of time, over the previous week, Monday through Sunday. You take all the paperwork home for the night (your supervisor wants this report on his desk at 7:00 A.M.), and to your dismay your feisty cat scatters your pile of papers out of the neat order in which you left them. Unfortunately, the graphs for generators A through F were not labeled (you will know better next time!). You recall some information and find evidence elsewhere for the following facts.

- Generators A and D were the only ones that maintained a fairly steady output.

- Generator B was shut down for a little more than two days during midweek.

- Generator C experienced a slow decrease in output during the entire week.

- On Tuesday morning, there was a problem with generator E that was corrected in a few hours.

- Generator D was the most productive over the entire week.

   1. Identify each graph with its corresponding generator. Explain in complete sentences how you arrived at your answers.

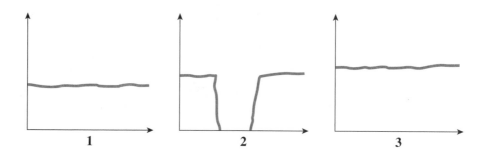

1   2   3

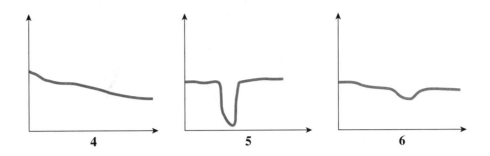

4   5   6

Graphs to Stories

The following graphs present visual images of several functions. Each graph shows
how the outputs change in relation to the input. Interpret the situation being represented. That is, describe, in words, what the graph is telling you about the situation.
Indicate what occurs when the graph reaches either a minimum (smallest) or maximum (largest) output value. In each case, provide a reasonable explanation for the
behavior you describe.

2. A person's core body temperature (°F) versus time of day

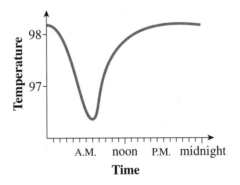

3. Performance on a simple task versus arousal level

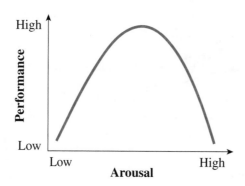

4. Net profits of a particular business versus time given quarterly

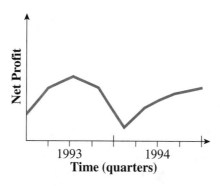

5. Annual gross income versus number of years

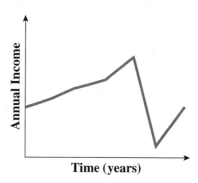

6. Time required to complete a task versus number of times task is attempted

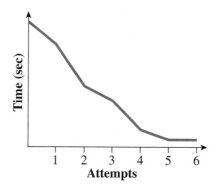

Stories to Graphs

Sketch a graph that best represents the situation described in each of the following problems. Note that in many cases, the actual values are unknown, so you will need to estimate what seems reasonable to you. Your graphs will be more qualitative than quantitative in these problems. (Remember to label your axes.)

7. You drive to visit your parents, who live 100 miles away. Your average speed is 50 mph. On arrival, you stay for 5 hours and then return home. Graph your distance from home as a function of time, from the time you leave until you return home.

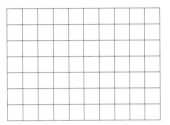

8. You just started a new job that pays minimum wage, with a small raise every 6 months. After $1\frac{1}{2}$ years, you receive a promotion that doubles your wages, but your next raise won't come for another year. Sketch a graph of your approximate wage as a function of time over your first *two and a half* years.

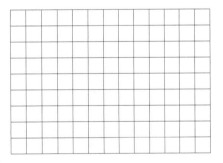

E X E R C I S E S

1. The following graph presents a visual image of a function. Interpret the situation being represented. That is, describe in words what the graph is telling you about the situation. Indicate what occurs when the function reaches either a minimum or maximum value. Provide a reasonable explanation for the behavior you describe.

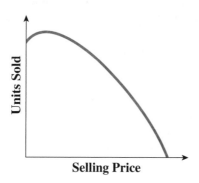

2. In baseball, a .300 batting average is considered quite good. Consider a player who struggled his first few years, but was a consistently good hitter throughout the rest of his 20 years in the majors. His lifetime average was around .300, and he retired before his skills fell off too noticeably. Sketch a graph that roughly shows his batting average as a function of time (years in the majors).

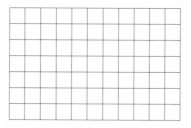

The Dangers of Asthma

Topics: *Domain, Range, Minimum and Maximum Function Values*

Asthma can be a very dangerous condition. The following graph gives the rate of asthma deaths (per 100,000) in the United States versus the year, from 1979 to 1994 (*Time*, August 7, 1995).

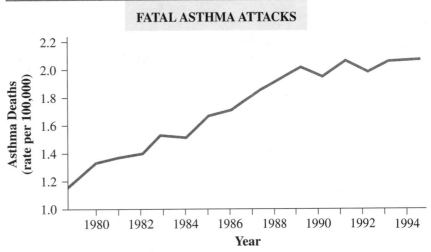

FATAL ASTHMA ATTACKS

Source: National Center for Health Statistics

1. a. Is the rate of asthma deaths a function of the year? Explain.

b. Identify the input and output variables.

2. a. What is the smallest output (minimum value) of the asthma function as indicated by the graph?

b. When did it occur?

3. a. Approximate the largest output (maximum value) of the asthma function as indicated by the graph.

b. When did it occur?

The collection of all possible input values for a function is called the **domain** of the function. The collection of all possible output values of a function is called the **range** of the function. In practice, domains and ranges are determined by the conditions imposed on the situation being studied. In that case, we speak of the practical domain and the practical range.

When the function is presented graphically, domain values are indicated on the horizontal axis and range values are represented on the vertical axis.

4. What is the domain of the asthma function as indicated by the graph?

5. What is the range of the asthma function as indicated by the graph?

6. If t represents the input in years, and $D(t)$ represents the corresponding output, approximate $D(1986)$. Write your answer in sentence form, including units of measurement.

7. a. During which year(s) were there 1.6 deaths per 100,000? Explain how you arrived at your answer.

b. If, as in Problem 6, $D(t)$ represents the output of the asthma function, write your answer to part a in function notation.

8. Determine the value(s) of t for which $D(t) = 1.8$. What does your answer represent in this situation? Is your answer an exact value or an approximation?

9. During which year(s) were there no more than 1.5 deaths per 100,000? Explain.

10. In 1994, the population of the United States was approximately 300,000,000 people. Based on the graph, estimate the number of asthma deaths in 1994.

11. Has the number of asthma deaths been steadily increasing since 1979? If not, identify the years in which the number of deaths from asthma decreased.

12. Did the number of asthma deaths increase more rapidly during 1979 or 1980? Explain.

13. Use the techniques you learned on how to read a graph and the information you found in the previous problems to write a short description of the information given in the graph of fatal asthma attacks.

14. Compare your description to those written by two other classmates. Add to your description one thing each classmate included that you did not.

The domain of a function is the set of all meaningful input values. The range of a function is the set of all meaningful output values.

A function is said to be increasing if the output increases as the input increases. The graph of an increasing function rises to the right.

A function is said to be decreasing if the output decreases as the input increases. The graph of a decreasing function falls to the right.

EXERCISES

The following graph describes the number of nuclear warhead tests that were performed between 1945 and 1993.

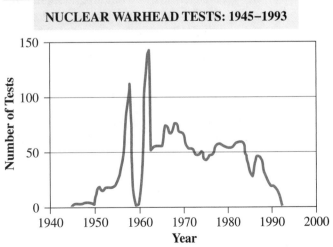

NUCLEAR WARHEAD TESTS: 1945–1993

Source: Bulletin of the Atomic Scientists

1. a. Identify the input and output variables.

b. Is the number of nuclear tests a function of the year? Explain.

c. Complete the following table with approximate values obtained from the graph.

YEAR	NUMBER OF TESTS
1950	
1960	
1970	
1980	
1990	

d. Use the graph to determine the maximum number of nuclear warhead tests and the year in which the maximum number occurred.

e. Write a summary of the information provided by the graph of nuclear warhead tests.

2. Determine the practical domain and range of each of the following functions by examining the graphs. Assume that the window encloses the practical domain and range.

a.

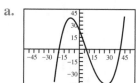

b.

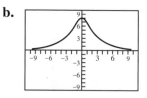

c.

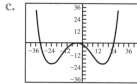

3. Consider the function g defined by the following graph.

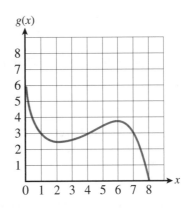

a. What is the domain of g?

b. What is the range of g?

c. For which values of x is g increasing?

d. For which values of x is g decreasing?

e. The maximum value of g is _____ and occurs when x is _____.

f. The minimum value of g is _____ and occurs when x is _____.

g. g(2) = _____

h. For what value(s) of x is g(x) = 3?

4. Sometimes a table is used to define the domain and range of a function that has a finite number of inputs and outputs. Consider the function f defined by the following table.

x	−3	−2	−1	0	1	2	3	4
f(x)	5	4	2	−1	1	3	5	6

a. What is the domain of f?

b. What is the range of f?

c. Where (over which set of consecutive x-values) is f increasing?

d. The maximum value of f is _____ and occurs when x is _____.

e. The minimum value of f is _____ and occurs when x is _____.

f. Write f as a set of eight ordered pairs.

g. $f(3) = $ _____

h. For what value of x is $f(x) = 2$?

5. Functions may be defined by an algebraic rule, sometimes with an explicit domain given. Consider the function h defined by

$$h(x) = x^2 + x + 1, \quad \text{where } -2 \le x \le 2.$$

a. Use your grapher to graph and produce a table for h.

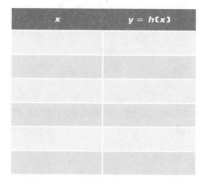

x	y = h(x)

b. What is the domain of h?

c. What is the range of h?

d. Over which values of x is h increasing?

e. Over which values of x is h decreasing?

f. The maximum value of h is _____ when x is _____.

g. The minimum value of h is _____ when x is _____.

h. $h(-1) =$ _____

i. For what value(s) of x, if any, is $h(x) = 3$?

j. For what value(s) of x, if any, is $h(x) = -2$?

6. Suppose that all you know about a function f is that $f(2) = -4$. This fact can be expressed in two different ways.

   a. If the input value for f is 2, then the corresponding output value is _____.

   b. One point on the graph of f is _____

   c. The statement above indicated that $f(x) = -4$ for $x = 2$. Is it possible for f to have an output of -4 for another value of x? Explain.

What Have I Learned?

1. Identify four different ways that a function can be defined. Give an example of each.

2. Given that $f(x) = 5x - 8$, what does $f(1)$ tell you about the graph of f?

3. If $g(t)$ is the weight (in grams) of a melting ice cube t minutes after being removed from the freezer, interpret the meaning of $g(10) = 4$.

4. Describe how you can tell from a graph when a function is increasing and when it is decreasing.

5. Use function notation to state that the point (5, 100) is on the graph of a function H.

6. If a function is defined by a graph, describe how you would determine its domain and range.

How Can I Practice?

1. Students at one community college in New York State pay $105 per credit hour when taking fewer than 12 credits, provided they are New York State residents. For 12 or more credit hours, they pay $1250 per semester.

   a. Determine the tuition cost for a student taking the given number of credit hours.

NUMBER OF CREDIT HOURS	TUITION COST ($)
3	
6	
10	
12	
16	
18	

   b. Is the tuition cost a function of the number of credit hours for the values in your completed table from part a? Explain. Be sure to identify the input and output variables in your explanation.

   c. Is the number of credit hours taken a function of tuition cost? In your explanation, be sure to identify which variables now play the role of the input and output.

   d. Use the table in part a to help graph the tuition cost at this college as a function of the number of credit hours taken.

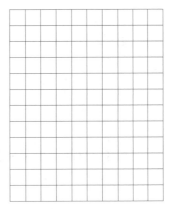

e. Suppose you have saved \$700 for tuition. Use the graph to estimate the greatest number of credit hours you can take.

f. Does the graph of your tuition cost function pass the vertical line test?

g. Let h represent the number of credit hours (input) and C represent the tuition cost (output). Write an equation for the cost of part-time tuition in terms of the number of credit hours taken.

h. Use your equation from part g to verify the tuition cost for the credit hours given in the table in part a.

2. Let $g(x) = (x + 3)(x - 2)$. Determine $g(-4)$.

3. Let $h(s) = (s - 1)^2$. Determine $h(-3)$.

4. Let $f(x) = 3x - 7$ and $g(x) = x^2 - 3$. Determine $f(-3) + g(-3)$.

5. Let $s(x) = \sqrt{x + 3}$. Determine $s(6)$.

6. Let $r(t) = t^2 - 2t + 3$ and $s(t) = 3t - 5$. Find $r(-1) \cdot s(2)$.

7. Interpret each situation being represented. That is, describe, in words, what the graph is telling you. Indicate what occurs when the function reaches either a minimum or maximum value. In each case, try to provide some reasons for the behavior you describe.

a. Hours of daylight per day versus time of year

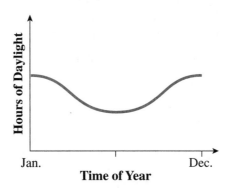

b. Population of fish in a pond versus the number of years since stocking

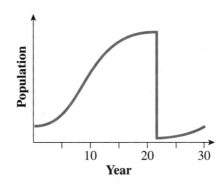

c. Distance from home (in miles) versus driving time (in hours)

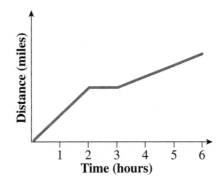

d. Amount of money saved per month versus amount of money earned

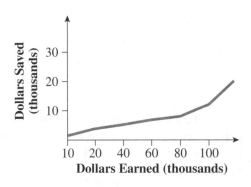

8. In the following situations, determine which variable should be the input variable and then sketch a graph that best describes the situation. Remember to label the axes with the names of the variables.

a. Sketch a graph that approximates average temperatures where you live as a function of time of year.

Input variable _____.

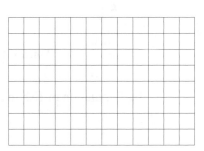

b. If you don't study, you expect to do poorly on the next test. If you study a few hours, you should do quite well, but if you study for too many more hours, your test score will probably not improve. Sketch a graph of your test score as a function of study time.

Input variable _____.

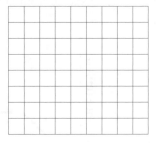

c. Two cars leave Pittsburgh traveling the same route toward Chicago. One leaves at noon and the second leaves at 2:00 P.M. The second car averages a much faster speed and catches up to the first car at 6:00 P.M. Sketch graphs for both cars on the same coordinate system, showing distance traveled as a function of time elapsed since noon.

Input variable _____.

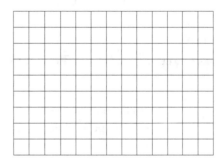

CLUSTER 2

Modeling with Functions

ACTIVITY 3.4

How Fast Did You Lose?

Topic: *Introduction to Average Rate of Change*

Suppose you are a member of a health and fitness club. The club's registered dietitian and your personal trainer have developed a special diet and exercise program for you. Data for your weight w as a function of time t over an eight-week period is given in the following table.

TIME, t (WEEKS)	0	1	2	3	4	5	6	7	8
WEIGHT, w (lb)	140	136	133	131	130	127	127	130	126

1. Plot the data points using ordered pairs of the form (t, w). For example, $(3, 131)$ is a data point that represents your weight at the end of the third week.

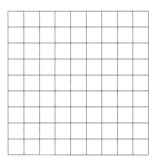

2. **a.** What was your weight at the beginning of the program?

   b. What was your weight at the end of the first week?

3. An important question in this situation is, How did your weight change from one week to the next?
   a. During which week(s) did your weight increase?

   b. During which week(s) did your weight decrease?

   c. During which week(s) did your weight remain unchanged?

4. Your weight decreased during each of the first five weeks of the program.

    a. Determine the actual change in your weight (output value) over the first five weeks of the program by subtracting your initial weight from your weight at the end of the first five weeks. (Recall the concept of change from Chapter 1.)

    b. What is the sign (positive or negative) of your answer? What is the significance of this sign?

    c. Determine the change in the input value over this time period; that is, from $t = 0$ to $t = 5$.

    d. Write the ratio of the change in weight from part a to the change in time in part c. Interpret the meaning of this ratio.

The ratio in Problem 4d above, $\frac{-13 \text{ lb}}{5 \text{ weeks}} = -2.6$ lb per week, is called the **average rate of change** of weight with respect to time over the first five weeks of the program. It can be interpreted as an average loss of 2.6 lb each week for the first five weeks of the program. Rates of change in outputs with respect to corresponding changes in inputs are so important in discussing functions that special symbolic notation has been developed to describe them.

To begin, the change in value of a variable quantity from a starting point to an ending point is denoted by Δ, the symbol for the uppercase Greek letter delta. Delta is the Greek version of d, for difference, the result of a subtraction that produces the change in value.

So, for example, the change in value of the output variable, weight (w), is written as Δw and is calculated by subtracting the initial value of w, often denoted by w_1, from the final value of w, denoted by w_2. Symbolically, this change is represented by $\Delta w = w_2 - w_1$.

    5. Write your answer to Problem 4a using the Δ notation.

Similarly, the change in value of the input variable, time (t), is written as Δt and is calculated by subtracting the initial value of t, denoted by t_1, from the final value of t, denoted by t_2. Symbolically, this change is represented by $\Delta t = t_2 - t_1$.

6. Write your answer to Problem 4c using the Δ notation.

The value of Δw describes how much weight you have gained or lost, but it does not tell how quickly you shed those pounds. That is, a loss of 3 pounds in one week is more impressive than a loss of 3 pounds over a month's time.

The ratio $\frac{\Delta w}{\Delta t}$, called the **average rate of change** of w over the given time interval, describes the **rate** at which you have lost weight over this period of time. Its units are output units *per* input unit—in this case, pounds per week.

7. a. Write your answer to Problem 4(d) using the Δ notation.

    b. On the graph in Problem 1, connect the points (0, 140) and (5, 127) with a line segment. Does the line segment rise, fall, or remain horizontal as you follow it from left to right?

    c. Write a statement relating the sign of your answer in part a to your answer in part b.

8. a. Determine the average rate of change of your weight over the time period from $t = 5$ to $t = 7$ weeks. Write your answer using the Δ notation and include the appropriate sign and units.

    b. Interpret the rate in part a with respect to your diet.

    c. On the graph in Problem 1, connect the points (5, 127) and (7, 130) with a line segment. Does the line segment rise, fall, or remain horizontal as you follow it from left to right?

    d. Write a statement relating the sign of your answer in part a to your answer in part c.

9. a. At what rate is your weight changing during the sixth week of your diet; that is, from $t = 5$ to $t = 6$?

b. Interpret the rate in part a with respect to your diet.

c. Connect the points (5, 127) and (6, 127) on the graph with a line segment. Does the line segment rise, fall, or remain horizontal as you follow it from left to right?

d. Write a statement relating your answers to parts a and c.

10. a. What is the average rate of change of your weight over the period from $t = 4$ to $t = 7$ weeks?

b. Explain how the rate in part a reflects the progress of your diet over those three weeks.

SUMMARY

The ratio $\frac{\Delta y}{\Delta x}$ is the average rate of change of output y with respect to input x, where

$$\Delta y = y_2 - y_1 \quad \text{and} \quad \Delta x = x_2 - x_1.$$

The units of the rate are output units per input unit.

E X E R C I S E S

1. The following table presents the median age (output) of a U.S. male at the time of his first marriage as a function of the year (input).

INPUT, (YEAR)	1900	1910	1920	1930	1940	1950	1960	1970	1980	1990
OUTPUT, (AGE)	25.9	25.1	24.6	24.3	24.3	22.8	22.8	23.2	24.7	26.1

Exercise numbers appearing in color are answered in the Selected Answers section of this book.

    a. During which decade(s) did the age at first marriage increase for males?

    b. During which decade(s) did the age at first marriage decrease?

    c. During which decade(s) did the age at first marriage remain unchanged?

    d. During which decade(s) was the change in first-marriage age the greatest?

2. Graph the data from Exercise 1 and connect consecutive points with straight line segments. Then answer the following questions using the graph.

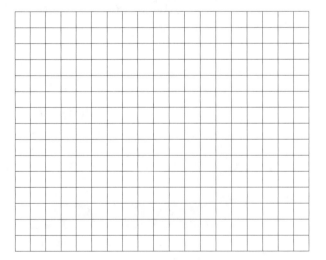

    a. During which decade(s) is your graph rising?

    b. During which decade(s) is your graph falling?

    c. During which decade(s) is your graph horizontal?

d. During which decade(s) is your graph steepest?

e. Compare the answers of the corresponding parts (a–d) of Exercises 1 and 2. What is the relationship between the sign of the output change for a given input interval and the direction (rising, falling, horizontal) of the graph in that interval?

3. Refer to the data in Exercise 1 to determine the average rate of change per decade of first-marriage age for males from 1900 through 1950. Compare this average rate of change to the average rate of change prior to World War II (1940) and with that from 1950 to 1970.

4. The National Weather Service recorded the following temperatures one February day in Chicago.

TIME OF DAY	10 A.M.	12 noon	2 P.M.	4 P.M.	6 P.M.	8 P.M.	10 P.M.
TEMPERATURE (°F)	30	35	36	36	34	30	28

a. Determine the average rate of change (including units and sign) of temperature with respect to time over the entire 12-hour period given in the table.

b. Over which period(s) of time is the average rate of change zero? What, if anything, can you conclude about the actual temperature fluctuation within this period?

c. What is the average rate of change of temperature with respect to time over the evening hours from 6 P.M. to 10 P.M.? Interpret this value (including units and sign) in a complete sentence.

d. Write a brief paragraph describing the temperature and its fluctuations during the 12-hour period in the table.

ACTIVITY 3.5

**Nuclear Arsenals
Shrinking**
Topic: *Rate of Change
Application*

Since 1945, the world has lived with the threat of nuclear war. The accompanying table and graph describe the buildup and subsequent decline of nuclear warheads around the world.

The table in the margin indicates that the number of nuclear warheads in 1960 was 20,430, and the number of warheads in 1965 was 39,050. This can be expressed symbolically as $N(1960) = 20,430$ and $N(1965) = 39,050$. The change in the number of nuclear warheads during the five-year period from 1960 to 1965 is therefore

$$N(1965) - N(1960) = 39,050 - 20,430$$
$$= 18,620 \text{ nuclear warheads.}$$

This indicates an increase of 18,620 warheads over this five-year period. Therefore, the number of warheads increased at an average rate of $18,620 \div 5 = 3724$ warheads per year.

**WORLD NUCLEAR
ARSENALS: 1945–1993**

Year, Y	Nuclear Warheads, N
1945	2
1950	303
1955	2,490
1960	20,430
1961	25,700
1962	30,405
1963	34,080
1964	37,015
1965	39,050
1966	40,330
1967	41,685
1968	41,055
1969	39,600
1970	39,695
1971	41,365
1972	44,020
1973	47,745
1974	50,840
1975	52,325
1976	53,255
1977	54,980
1978	56,805
1979	59,120
1980	61,480
1981	63,055
1982	64,770
1983	66,980
1984	67,865
1985	68,590
1986	69,480
1987	68,835
1988	67,040
1989	63,650
1990	60,240
1991	55,775
1992	52,875
1993	49,910

Source: Bulletin of the Atomic Scientists

1. Use the table to determine the change in the number of nuclear warheads from 1965 to 1970 and calculate the average rate of change of nuclear warheads during that period. (Remember that your answer must include sign and units and be stated in a complete sentence.)

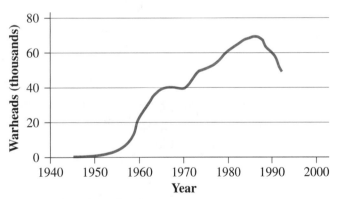

Source: Bulletin of the Atomic Scientists

The preceding graph indicates that the change in the number of nuclear warheads from 1985 to 1990 is approximately

$$N(1990) - N(1985) = 60,000 - 70,000$$
$$= -10,000 \text{ nuclear warheads.}$$

This is a decrease of 10,000 warheads over this five-year period. Therefore, the number of warheads decreased at an average rate of $10,000 \div 5 = 2000$ warheads per year.

2. Use the graph to determine the change in the number of nuclear warheads from 1975 to 1980 and the average rate of change during this period.

3. Use the table or graph to determine the period of years in which the number of nuclear warheads increased. Over which period of years did the number of nuclear warheads decrease?

4. The following graph gives information concerning the number of U.S. and Soviet nuclear warheads as a function of the year. Translate the graph into tabular form on the accompanying table.

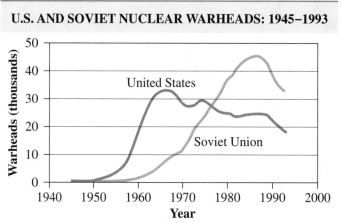

U.S. AND SOVIET NUCLEAR WARHEADS: 1945–1993

Source: Bulletin of the Atomic Scientists

YEAR	U.S. WARHEADS (IN THOUSANDS)	SOVIET WARHEADS (IN THOUSANDS)
1950		
1960		
1970		
1980		
1990		

5. In what year did the Soviet Union catch up to the United States in the number of warheads? How many nuclear warheads did both countries have at that time? Explain the basis for your answer.

6. Over which period of years did the U.S. nuclear warhead arsenal decrease? What feature of the graph or table demonstrates this decrease?

7. Use the graph to estimate during which five-year period the United States had the most rapid buildup of nuclear warheads. What feature of the graph illustrates this rapid buildup?

8. What was the greatest number of nuclear warheads possessed by the Soviet Union? In what year did this maximum occur? Explain how you can read this from the graph.

9. In 1990, what percentage of nuclear warheads were from countries other than the United States or Soviet Union? Explain the basis for your answer.

EXERCISES

1. The following table presents the Dow Jones industrial average (DJIA) at the start of each year from 1980 to 1990. The DJIA is a measure of the vigor of the stocks traded on the New York Stock Exchange, and its output values are measured in points.

YEAR	1980	1981	1982	1983	1984	1985	1986	1987	1988	1989	1990
DJIA	839	964	875	1047	1259	1212	1547	1896	1939	2169	2753

a. What was the average rate of change (include units) over the ten-year period described in the table?

b. During which year did the DJIA increase the most?

c. What would a negative average rate of change represent in this situation?

Does this occur for the data in this table? When?

d. What would a zero average rate of change represent in this situation?

Does this occur for the data in this table? When?

e. Over which year will the graph of this function rise most steeply?

f. Over which year will the graph of this function rise least steeply?

g. Over which year will the graph of this function fall most steeply?

2. If a variable increases from a value of $10 to a value of $50 over a ten-day period, what is the average rate of increase?

3. If $f(x) = 5x - 3$, evaluate the following.

a. $\dfrac{f(6) - f(0)}{6 - 0}$

b. $\dfrac{f(0) - f(6)}{0 - 6}$

4. Medicare is a government program that affects us all in some way. As U.S. demographics change and greater numbers of senior citizens join the Medicare rolls each year, the expense and quality of service have come under increased scrutiny. In Chapter 2, you examined the graph resulting from a study of Medicare expenditures (costs) over the past 30 years. That graph is reproduced here, along with another graph from a study of enrollment numbers in Medicare.

MEDICARE BY THE NUMBERS

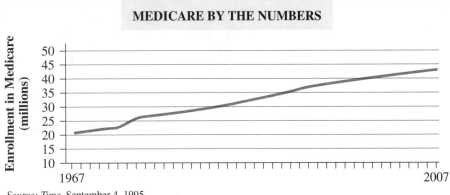

Source: *Time*, September 4, 1995

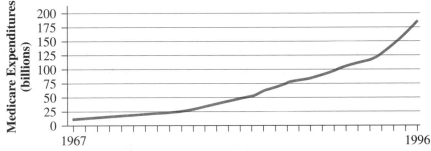

Source: *Time*, September 4, 1995

a. Identify the input and output variables in each study.

b. Which function contains the ordered pair with input 1980 and output 28,000,000? Explain.

c. Let *t* represent the year (input) and $S(t)$ represent the enrollment (output) in the Medicare enrollment function.

   i. Determine the value of $S(1967)$.

      ii. What is the domain of S, as pictured?

      iii. What is the range of S, as pictured?

    d. Let t represent the year (input) and $P(t)$ represent the expenditures (output) in the Medicare expenditures function.

      i. Explain in words the meaning of $P(1984) = 65$.

      ii. What is the domain of P, as pictured?

      iii. What is the range of P, as pictured?

    e. Write a mathematical statement, using function notation, that says, "Medicare expenditures for 1977 amounted to $25,000,000,000."

    f. What is the average rate of increase in Medicare expenditures from 1967 to 1996? (Remember to include units in your answer.)

    g. What is the average rate of increase in Medicare enrollment over the same time period? (Remember to include units in your answer.)

Inflation

Topics: *Rate of Change,*
Exponential Function

Inflation causes your dollar to buy less in the future. In other words, the loaf of bread you buy today for $1 will cost more next year because of inflation. The inflation rate is usually expressed as an annual percentage rate. If the current inflation rate is 10%, then you can expect your $1 loaf of bread to cost 10 cents more next year.

1. At the current inflation rate of 10%, how much will a $20 pair of shoes cost next year?

2. If you assume that inflation stays at 10% per year for the next decade, you can calculate the cost of a currently priced $5 pizza for each of the next ten years. Complete this table. Round to the nearest cent.

YEARS FROM NOW, t	COST OF PIZZA, $c(t)$
0	$5.00
1	$5.50
2	
3	
4	
5	
6	
7	
8	
9	
10	

3. Plot the data you found in Problem 2. Use years from now as the input and cost of pizza as the output.

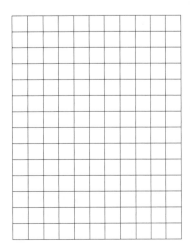

4. Determine the rate of change in the first year (from $t = 0$ to $t = 1$), the fifth year (from $t = 4$ to $t = 5$), and the tenth year (from $t = 9$ to $t = 10$). Describe what these results mean for the cost of your pizza over the next ten years.

The cost of a pizza is a function of the number of years from now. In Problem 2, you could have calculated each output value by multiplying the previous output by 1.10, the growth factor (see Activity 1.15). Thus, after ten years, you would have multiplied the original cost, \$5, the output, by 1.10 *ten times*. Symbolically, this is written $5(1.10)^{10}$. Hence, the cost of pizza can be modeled algebraically as $c(t) = 5(1.10)^t$, where c represents the cost and t represents the number of years from now.

> A function, in this case $c(t) = 5(1.10)^t$, where the variable appears as an exponent is called an **exponential function**.

5. a. What will be the cost for a pizza after 20 years?

b. Use the cost function to calculate the cost of a pizza after $5\frac{1}{2}$ years.

6. The cost function $C(t) = 5(1.10)^t$ is graphed below for t between -10 and 10. Graph this function on your grapher. Use the trace feature of your grapher to examine the coordinates of some points on your graph. Does your original plot agree with the graph shown here?

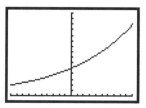

7. Is the entire graph that you see really relevant to the original problem? Explain. Resize your window to include only the first quadrant.

8. a. How many years will it take for the price of a pizza to double? Explain how you found your answer.

b. How many years will it take for the price of a pizza to triple? Explain how you found your answer.

EXERCISES

1. Use your grapher to graph the functions $f(x) = \left(\frac{1}{2}\right)^x$ and $g(x) = 2^x$ in the window Xmin $= -2$, Xmax $= 2$, Ymin $= -2$, Ymax $= 5$. Compare the graphs. List similarities and differences.

2. Suppose the inflation rate is 7% per year.

a. What exponential function could we use to model the cost of $45 sneakers after t years of 7% inflation?

b. What would the cost of the sneakers be in ten years if the inflation rate stayed at 7%?

3. An exponential function may be increasing or decreasing. Determine which is the case for each of the following functions. Explain how you determined your answers.

   a. $f(x) = 5^x$　　　　　　　　　**b.** $g(x) = \left(\frac{1}{2}\right)^x$

   c. $h(t) = 1.5^t$　　　　　　　　**d.** $k(p) = 0.2^p$

4. Compare the rate of increase of the functions $f(x) = 3^x$ and $g(x) = 3x$ from $x = 0$ to $x = 5$ by the following methods.

   a. Compare the output values in the following table.

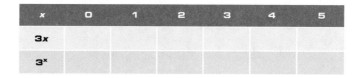

x	0	1	2	3	4	5
3x						
3ˣ						

   b. Examine the lines shown in the graph.

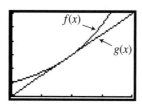

   c. Calculate the average rate of change for each function from $x = 0$ to $x = 5$.

ACTIVITY 3.7

Apple Orchard

Topics: *Interpolation and Extrapolation, Functions of the Form* $y = \frac{k}{x}$

Your apple orchard has done well this year, and it's time to harvest. In previous years, you hired workers to pick the fruit, and you plan to do so again this year. To better plan for shipping to the market, you would like to estimate how long it will take to pick all your apples. Last year you hired 12 workers and it took them 30 hours to harvest all the apples. Two years ago, 20 workers needed 18 hours. And three years ago, 8 workers needed 45 hours to accomplish the same job.

1. Make a table for the given data. Then, plot the ordered pairs (number of workers, number of hours) on a coordinate system, labeling your axes with an appropriate scale. Use the three plotted points to sketch a graph that you think best describes the function.

NUMBER OF WORKERS	NUMBER OF HOURS

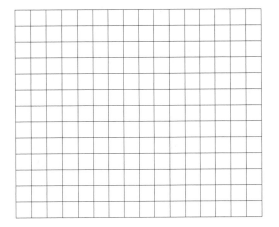

2. In the context of this activity, which quantity is your input variable and which is the output variable?

3. What do you think is a reasonable domain and range for this function?

4. Use your graph to predict how long it would take 15 workers to pick all your apples. Your estimate is an example of **interpolation** (estimating within the bounds of the original set of data pairs).

5. Use your graph again to predict how long it would take 4 workers to do the job. Your estimate is an example of **extrapolation** (estimating beyond the bounds of the original set of data pairs).

6. From the graph, you can also predict the number of workers needed to complete the job in a certain number of hours.

   a. How many workers would you need to complete the job in 20 hours?

   b. Is this an example of interpolation or extrapolation?

7. How many workers would you need to complete the job in 5 hours?

8. Is it possible to use the graph to predict how long it will take 50 workers to do the job? Explain.

9. If w represents the number of workers, then $h(w)$ is the number of hours it will take w workers to pick all the apples. Determine each of the following.

   a. $h(12)$ b. $h(20)$ c. $h(8)$

10. You can use unit analysis (also known as dimensional analysis) to determine the formula for the function $h(w)$. The amount of work required to do a job is often measured in worker-hours (also called man-hours). Worker-hours are calculated by multiplying the number of workers by the total time to do the job. This assumes that each worker works for the entire time required for his or her group to get the job done. Are you making the same assumption in this problem? For each year, how many worker-hours are required to harvest all your apples?

11. Use your results in Problem 10 to write a formula for h as a function of w.

12. Use your function formula from Problem 11 to determine $h(4)$, $h(15)$, and $h(50)$. How do these predictions compare with your previous estimates?

13. Use your function formula to solve $h(w) = 5$ and $h(w) = 20$ for the number of workers w. How do these predictions compare with your estimates in Problems 6 and 7?

14. Use your grapher to graph this function. What window best shows the part of the graph relevant to this problem?

15. Is zero in the domain of this function? Explain.

SUMMARY

For a function defined by $y = \frac{k}{x}$, where $k > 0$, as the input x increases, the output y decreases.

The input value $x = 0$ is not in the domain of this function.

Estimating an output value for an input value *within* the bounds of the given data is called **interpolation.**

Estimating an output value for an input value *outside* the bounds of the given data is called **extrapolation.**

EXERCISES

1. You are looking for a new car, and you think you have found a great deal. The Sell-a-Lot dealership will sell you the car you want in exchange for your old car plus $6000 financed at 2% interest. The deal sounds good to you, but you aren't really sure what the 2% interest means. Your major concern is your monthly payment, so you ask your salesperson to explain. She says that roughly speaking, if you finance for 24 months, your payments will be about $320; if you finance for 30 months, your payments will be roughly $260; and if you finance for 36 months, your payments will be roughly $230. However, she says that you can finance your loan for any monthly time period between 12 and 60 months.

    a. Make a table of the data. Then plot the ordered pairs on a coordinate system, clearly labeling your axes with an appropriate scale.

MONTHS, m	MONTHLY PAYMENT, p

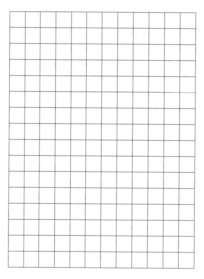

    b. Use your graph to estimate your monthly payment if you finance your car for 27 months.

    c. Use your graph to estimate your monthly payment if you finance your car for 40 months.

d. About which of these estimates do you feel more confident? Explain.

e. Use unit analysis to determine a formula that approximates the relationship between m and p found in your table. Add the graph of this equation to the grid in part a and label it clearly.

f. You would like to keep your monthly payments to about $200 per month. Use your formula to determine for how many months you would finance your car.

g. When you finally meet with your salesperson to work out the details, she tells you that if you finance your car for 27 months, your payments will be $289.67. Are you surprised? Explain, referring to the concepts of interpolation and extrapolation.

h. When you ask the salesperson about monthly payments of $200, she informs you that you would need to finance your car for 46 months to reach a monthly payment of $200.72. Are you surprised? Explain, referring to the concepts of interpolation and extrapolation.

2. a. Solve $xy = 120$ for y. Is y a function of x? What is its domain?

b. Solve $xy = 120$ for x. Is x a function of y? What is its domain?

3. Solve the following equations. Check your solutions either numerically or graphically.

   a. $30 = \frac{120}{x}$

   b. $2 = \frac{250}{x}$

   c. $-8 = \frac{44}{x}$

   d. $25 = \frac{750}{x + 2}$

ACTIVITY 3.8

Electricity Charges

Topics: *Linear Function, Vertical Intercept, Rate of Change*

The bimonthly cost of electricity is a function of the amount of electricity used. Your power company currently charges $20.60 as a fixed bimonthly basic service charge, not including usage. The rate charged for usage is 10.6245 cents per kilowatt-hour (kWh).

1. Complete the following table of electricity costs. Round to the nearest cent.

KILOWATT HOURS, k	COST, $C(k)$
600	$84.35
650	$89.66
700	
750	$100.28
800	$105.60
850	
900	
950	$121.53
1000	

2. If C represents the total bimonthly cost of electricity and k the amount of electricity used, write a symbolic rule for C as a function of k.

3. Plot the points from the table in Problem 1 and connect them with a smooth line. Remember to label and scale the axes.

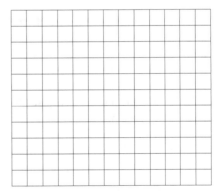

4. Determine the value of $C(875)$ and interpret its meaning in the context of this activity.

5. How much electricity could you use if you could afford to pay $150 for electricity? (Determine your result to two decimal places.)

The **vertical intercept** $(0, b)$ is the point where the graph of a function meets the vertical axis.

---

6. **a.** Estimate the vertical intercept from your graph of the cost function if possible. What is the practical meaning of the vertical intercept in this situation?

   b. Determine the vertical intercept using the equation for the cost function (see Problem 2).

7. **a.** What is the rate of change of cost C as k increases from $k = 600$ to $k = 700$?

   b. What is the rate of change of cost C as k increases from $k = 750$ to $k = 950$?

   c. Does the rate of change have any practical meaning in this situation? Explain.

8. The electricity cost function is an example of a **linear function.** Can you justify the name?

The linear function will be studied in depth in Chapter 4.

What Have I Learned?

1. What must be true about the average rate of change between any two points on the graph of an increasing function?

2. Suppose you are told that the average rate of change of a function is always negative. What can you conclude about the graph of the function and why?

3. Describe a step-by-step procedure for finding the average rate of change between any two points on the graph of a function. Use (85, 350) and (89, 400) in your explanation.

4. Can the graph of a function have more than one vertical intercept? Explain.

5. Give an example of a function from your major field of study or area of interest.

6. What is the mathematical definition of a function?

7. In what ways are interpolation and extrapolation the same?

8. In what ways are interpolation and extrapolation different?

9. Which is more reliable, interpolation or extrapolation?

How Can I Practice?

1. Let $f(x) = -3x + 4$. Find $f(-5)$.

2. Let $m(x) = 2x^2 + 6x - 7$. Find $m(3)$.

3. Let $f(x) = 2x - 9$ and $p(x) = x^2 - x - 1$. Evaluate: $\frac{f(-1)}{p(-1)}$.

4. Let f be defined by the ordered pairs $\{(2, 3), (0, -5)\}$.

   a. List the set of numbers that constitute the domain of f.

   b. List the set of numbers that constitute the range of f.

5. If $h(x) = 3x - 1$, determine the range of the function when the domain is

$$D = \{-3, -2, -1, 0, 1, 2\}.$$

6. You decide to lose weight and will cut down on your calories to lose 2 pounds a week. Suppose that your present weight is 180 pounds. Sketch a graph covering 12 weeks showing your projected weight loss. Describe your graph. If you stick to your plan, how much will you weigh in six months? Is this realistic for you?

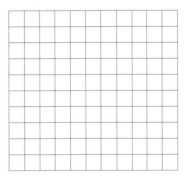

7. A taxicab driver charges a flat rate of $2.50 plus $1.50 per mile. The fare F (in dollars) is a function of the distance driven, x (in miles). The driver wants to display a table for her customers to show approximate fares for different locations within the city.

   a. Write a symbolic rule for f in terms of x.

   b. Create a table using the domain (measured in miles): $\{0.25, 0.5, 0.75, 1, 1.5, 2, 3, 5, 10\}$.

x									
$f(x)$									

c. Plot the data points or graph this function on your grapher.

d. Determine from the graph what happens to the fare as the number of miles increases.

e. What is the average rate of change of the fare as the distance traveled increases from 1 to 10 miles?

f. What is the average rate of change of the fare as the distance traveled increases from 0.5 to 3 miles?

g. What do you observe about the rates in parts e and f?

8. Sketch a graph that represents the following situation: The stock market closed on Monday at 7300. The market gained 200 points on Tuesday, then lost 100 points on Wednesday and then remained unchanged on Thursday. Finally, on Friday, the market was up by 50 points.

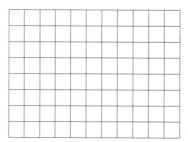

a. What is the average rate of change of the market from Monday's to Friday's closing?

b. What do you observe about the graph when the average rate of change is negative?

c. What do you observe about the graph when the average rate of change is zero?

9. Your union contract provides that your salary will increase by 3% in each of the next five years.

   a. If you make $20,000 to start, complete the following table to determine what your salary will be in five years.

YEARS, y	SALARY, $s(y)$
0	$20,000
1	
2	
3	
4	
5	

   b. Represent your salary symbolically as a function of years.

   c. Plot the data from your table, or graph this function on your grapher. Identify the type of function you have graphed.

   d. Refer to Activity 3.6, Inflation. Write in your own words what is happening to your salary over the next five years.

   e. If you continue to receive a 3% raise each year, how much will you be earning in ten years?

10. The function $N(p) = \frac{1200}{p}$ is used to model the relationship between the price p (in dollars) of a candy bar and the number N of bars that will be sold.

   a. How many bars will be sold if the price is 50 cents?

   b. What price should be charged if you wish to sell 4000 bars?

   c. Use your grapher to verify your answers to parts a and b graphically. Use the window: Xmin = 0, Xmax = 2, Xscl = .25, Ymin = 0, Ymax = 5000, Yscl = 200.

   d. What is the domain for this function, based on the algebraic definition?

   e. What would you estimate the practical domain to be for this function?

11. a. Solve $xy = 25$ for y. Is y a function of x? What is its domain?

   b. Solve $xy = 25$ for x. Is x a function of y? What is its domain?

Solve the following equations. Check your solutions either numerically or graphically.

12. $24 = \frac{36}{x}$ 13. $2500 = \frac{50}{x}$

14. $18 = \dfrac{1530}{x + 2}$

15. $-5 = \dfrac{120}{x - 5}$

16. Consider the function defined by the following table.

x	5	8	12	15	20	30
f(x)	288	180	120	96	72	48

a. Plot the ordered pairs on the grid, clearly labeling your axes with an appropriate scale.

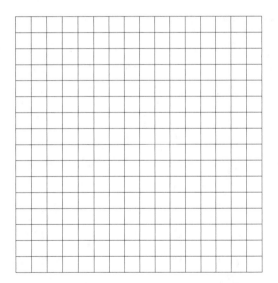

b. Check the product of each input and its corresponding output to find a formula that describes the relationship between x and $f(x)$ found in your table. Add the graph of this equation to the grid in part a and label it clearly.

c. Use your graph to estimate that value of x for which $f(x) = 200$.

d. Use the formula found in part b to calculate the value of x for which $f(x) = 200$. How does your result compare to your result in part c?

17. You wake up on Monday morning and listen to the news and weather, noting that the temperature is 25°F, a balmy February day in Oswego, New York. However, the announcer adds that the wind speed makes it feel like −4°F. Curious about the underlying mathematical model, you research windchill and discover the following chart.

Wind Speed (mph)

Air Temperature (°F)	4	5	10	15	20	25	30	35	40	45	50
35	35	32	22	15	11	8	5	3	2	1	1
30	30	27	16	9	4		−3	−5	−6	−7	−7
25	25	21	9	2	−4	−7	−10	−12	−14	−15	−15
20	20	16	3	−5	−11		−18	−20	−22	−23	−23
15	15	11	−3	−12	−18	−22	−26	−28	−30	−31	−31
10			−9	−18	−25		−33	−36	−38		
5	5	0	−15	−25	−32	−37	−41	−44	−46	−47	−47
0	0	−5	−21	−32	−39	−45	−49	−51	−54	−55	−55
−5	−5	−10	−28	−39	−46	−52	−56	−60	−61	−63	−63
−10	−10	−15	−34	−45	−53	−59	−64	−67	−69	−71	−71
−15	−15	−21	−40	−52	−61		−72	−75	−77	−79	−79
−20	−20	−26	−46	−58	−67	−74	−79	−83	−85	−87	−87
−25	−25	−31	−52	−66	−75	−82	−87	−91	−93	−95	−95
−30	−30	−36	−58	−72	−82	−89	−94	−98	−101	−103	−103
−35	−35	−42	−65	−79	−89	−97	−102	−106	−109	−111	−111
−40	−40	−47	−71	−86	−96		−110	−114	−117	−119	−119
−45	−45	−52	−77	−92	−103	−111	−117	−122	−125	−127	−127
−50	−50	−58	−83	−99	−111	−119	−125	−130	−133	−135	−135

a. You will notice that the chart has several values missing. Fill them in with reasonable estimates. Explain how you obtained these estimates.

b. What wind speed accounts for 25°F feeling like −4°F? Explain how you obtained your answer. Remember to label your results with appropriate units.

c. What windchill temperature is produced by an air temperature of 0°F with a wind speed of 25 mph? Explain how you obtained your answer.

d. If the wind speed is 15 mph, what temperature produces a windchill of −45°F? How did you arrive at your answer?

e. If the wind speed is 10 mph, graph the input/output ordered pairs (air temperature, windchill) that define windchill as a function of air temperature. Label the axes.

According to the graph, as air temperature increases, what happens to wind chill temperature? Is this reasonable in the context of this problem?

f. On the same set of axes, repeat part e with a wind speed of 30 mph. Compare the graphs and describe what you observe.

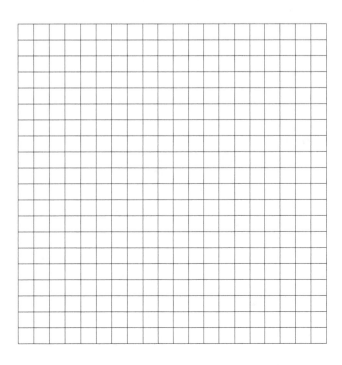

g. If the air temperature is 20°F, graph the input/output ordered pairs (wind speed, windchill) that define windchill as a function of wind speed. Label the axes and then describe your graph in words.

h. On the same set of axes, repeat part g with an air temperature of 30°F. Compare the graphs and describe what you observe.

Gateway Review

1. Let $f(x) = (-x)^2$. Find $f(2)$.

2. Let $f(x) = -x^2$. Find $f(2)$.

3. Let $f(x) = (-x)^2$. Find $f(-2)$.

4. Let $f(x) = -x^2$. Find $f(-2)$.

5. Let $h(x) = 5x - 3$. Find $h(-10)$.

6. Let $g(n) = 4n^2 - 3n + 8$. Find $g(-7)$.

7. Let $p(m) = -6m^2 + 2m - 12$. Find $p(2)$.

8. Let $m(k) = k^2 - 3k + 5$ and $n(j) = 3j^2 + 4j + 6$. Find $m(4) + n(4)$.

9. Let $f(x) = 2x^2 - 5$ and $g(y) = 3y + 8$. Find $f(-3) - g(-3)$.

10. Let $f(x) = -x^2 + 3x + 1$ and $g(k) = -8k + 4$. Find $\frac{f(3)}{g(3)}$.

In Exercises 11 and 12, interpret the situation being represented. That is, describe in words what the graph is telling you. Indicate what occurs when the graph reaches either a minimum or maximum value. In each case, try to provide reasons for the behavior you describe.

11. Cost of mailing a package versus the weight of the package

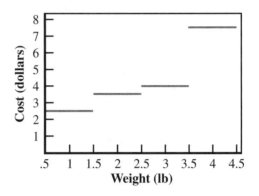

12. Distance from your home versus time traveled

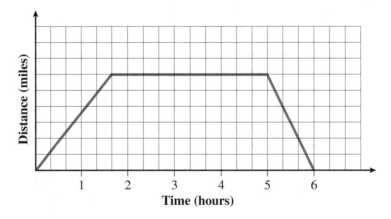

Sketch a graph that best describes the situation in Exercises 13 and 14.

13. You climb a hill situated a little above sea level and climb to the top of a cliff, taking a short rest midway. Then you dive into the sea. Describe your height above sea level as a function of time.

14. You toss an apple upward toward your friend who is across the room. Describe the height of the apple as a function of the time from when it leaves your hand.

The first 3 minutes of an international long-distance phone call costs $3.32, whether the call lasts 3 minutes or not. After the first 3 minutes, the charges are $0.89 per each additional minute or part of a minute. Use this information to answer Exercises 15–28.

15. Complete the following table, showing your phone bill as a function of the minutes you spend on the phone.

NUMBER OF MINUTES, n	COST OF CALL, $C(n)$ ($)
0	
1	
2	
3	
4	
5	
10	
12	
15	
20	

16. Explain how you calculated the cost of talking long distance for 10 minutes.

17. Use n for the input variable (duration of call in minutes) and $C(n)$ for the output variable (cost of phone call) to write symbolic statements for when:

    a. You do not talk at all, since no one answered the phone.

    b. You talk for only the first 3 minutes.

    c. You talk for more than 3 minutes.

18. What is the domain of the phone call function?

19. What is the minimum cost of a phone call (if there is a minimum)?

20. What is the maximum cost of a phone call (if there is a maximum)?

21. For what values of n is the phone call function increasing?

22. For what values of n is the phone call function decreasing?

23. For what values of n is the phone call function constant?

24. Determine the rate of change of the phone call function of C for $n = 1$ to $n = 3$.

25. Determine the rate of change of the phone call function any time after the first 3 minutes, using any two appropriate points from the table.

26. What does the sign of the rate of change in Exercise 25 imply?

27. Repeat Exercise 25 using two different points. How does your answer compare with that of Exercise 25?

28. From your solution in Exercise 27, what can you say about the rate of change of C for $n > 3$?

29. Use the following graph to answer parts a and b. State the months in which the maximum and minimum values occur for each of the given periods.

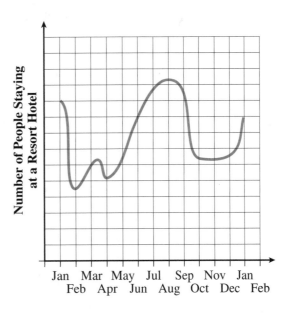

a. January to May

b. May to September

30. The following graph shows electricity use in kilowatt-hours as a function of the time of year.

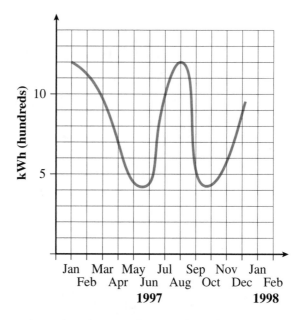

a. Extrapolate what the use of electricity might be (in kilowatt-hours) for January 1, 1998.

b. Estimate the use of electricity for March 1, 1997.

Linear Functions

Introduction to Linear Functions

The Snowy Tree Cricket

Topics: *Linear Functions, Constant Rate of Change, Slope*

One of the more familiar late-evening sounds during the summer is the rhythmic chirping of a male cricket. Of particular interest is the snowy tree cricket, sometimes called the temperature cricket. It is very sensitive to temperature, speeding up or slowing down its chirping as the temperature rises or falls. The following data show how the number of chirps per minute of the snowy tree cricket is related to temperature.

t, TEMPERATURE (°F)	$N(t)$, NUMBER OF CHIRPS PER MINUTE
55	60
60	80
65	100
70	120
75	140
80	160

1. **a.** Determine the average rate of change of the number of chirps per minute with respect to temperature as the temperature increases from 55°F to 60°F.

   b. What are the units of measure of this rate of change?

2. **a.** How does the average rate of change found in Problem 1 compare with the average rate of change as the temperature increases from 65°F to 80°F?

b. Check the average rate of change of number of chirps per minute with respect to temperature for the given temperature intervals. Add several more of your own choice. List all your results in the table.

TEMPERATURE INCREASES	AVERAGE RATE OF CHANGE (NUMBER OF CHIRPS PER MINUTE WITH RESPECT TO TEMPERATURE)
From 55° to 60°F	
From 65° to 80°F	
From 55° to 75°F	

c. What can you conclude about the average rate of increase in the number of chirps per minute for any particular increase in temperature?

3. For any 7° increase in temperature, what is the expected increase in chirps per minute?

If the rate of change in output with respect to input remains constant (stays the same) for *any* two points in a data set, the points will lie on a straight line. That is, the output is a **linear** function of the input. Conversely, if the points of a data set lie on a straight line when graphed, the rate of change of output with respect to input is constant for any two data points.

4. Plot the data points (temperature, chirps per minute) from the table on page 404. What type of graph is suggested by the pattern of points?

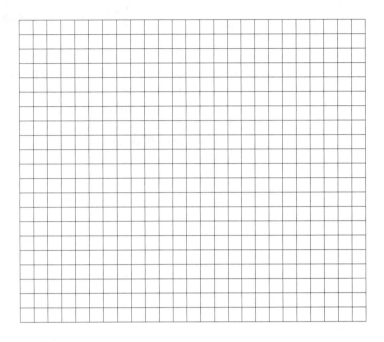

5. From the graph in Problem 4, would you conclude that the number of chirps per minute is a linear function of the temperature? Explain.

In the language of graphing, the **average rate of change** of output with respect to input is called the **slope** of the line segment connecting the two points.

6. What is the slope of the line in the snowy tree cricket situation?

7. a. Because the slope of this line is positive, what can you conclude about the direction of the line as the input variable (temperature) increases in value?

   b. What is the practical meaning of slope in this situation?

The slope of a line indicates its steepness; the larger the absolute value of its slope, the steeper the line. Since slope is really an average rate of change, it can be calculated in exactly the same way that an average rate of change is determined.

Slope is denoted by the letter m and may be briefly described here as,

$$m = \text{slope} = \frac{\text{change in output}}{\text{change in input}} = \frac{\text{vertical change}}{\text{horizontal change}}.$$

Geometrically, slope is the ratio of two distances. In going *from* one point *to* another point on the same line, the vertical distance between the points (the rise) is divided by the horizontal distance between the points (the run). The direction of the movement, whether in a positive or negative direction along each coordinate axis, determines the *signs* of the vertical and horizontal distances.

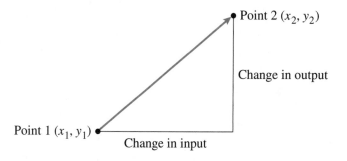

Point 2 (x_2, y_2)

Change in output

Point 1 (x_1, y_1)

Change in input

Here the vertical change from point 1 to point 2 is positive since the vertical movement is upward. Similarly, the horizontal movement here is to the right, so the horizontal distance is also positive.

From its geometric meaning, slope may be described as

$$m = \text{slope} = \frac{\text{rise}}{\text{run}} = \frac{\text{distance up } (+) \text{ or down } (-)}{\text{distance right } (+) \text{ or left } (-)}.$$

8. Use any two points plotted in Problem 4 to geometrically determine the slope of the line.

9. Crickets are usually silent when the temperature falls below 55°F. What is a possible practical domain for the snowy tree cricket function?

A linear function is one whose average rate of change of output with respect to input from any one data point to any other data point is always the same (constant) value.

The slope of a line segment joining two points (x_1, y_1) and (x_2, y_2) is denoted by m and can be calculated using the formula $m = \dfrac{y_2 - y_1}{x_2 - x_1}$.

The graph of every linear function with positive slope is a line rising to the right. The graph of every linear function with negative slope is a line falling to the right.

E X E R C I S E S

1. Consider the following data regarding the growth of the U.S. national debt from 1940 to 1990.

NUMBER OF YEARS SINCE 1940	0	10	20	30	40	50
NATIONAL DEBT (IN BILLIONS OF DOLLARS)	51	257	291	381	909	3113

   a. Compare the average rate of increase from 1940 to 1950 with that from 1980 to 1990. Is the rate of change constant? Explain, using the data.

   b. Plot the data points. If the points are connected to form a smooth curve, is the graph a straight line? Are the input and output variables in this problem related linearly?

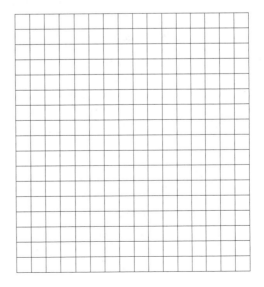

c. What must be true about the rate of change between any two data points in order for all the data points to lie on a straight line? Compare the graph in part b with the graph of the snowy tree cricket data in Problem 4.

2. Your friend's parents began to give her a weekly allowance on her fourth birthday. On each birthday her allowance increased, so that her weekly allowance in dollars was equal to her age in years.

   a. List your friend's allowance for the ages in the table.

AGE (years)	ALLOWANCE ($)
4	
5	
6	
7	

   b. Is her allowance a linear function of her age? Explain.

3. Calculate the average rate of change between consecutive data points to determine whether the output in each table is a linear function of the input.

   a.
INPUT	−4	0	3	5
OUTPUT	−23.8	1	19.6	32

   b.
INPUT	2	7	10	12
OUTPUT	0	−10	−16	−18

   c.
INPUT	−5	0	5	8
OUTPUT	−45	−5	35	59

While on a trip, you notice that the video screen on the airplane, in addition to showing movies and news, records your altitude (in kilometers) above the ground. As you start your descent (at time $t = 0$), you record the following data.

TIME, t (min)	0	2	4	6	8	10
ALTITUDE, $A(t)$ (km)	12	10	8	6	4	2

1. **a.** What is the rate of change in the altitude of the plane from 2 to 6 minutes into the descent? Pay careful attention to the sign of this rate of change.

   b. What are the units of measurement of this rate of change?

2. **a.** Determine the rate of change over several other input intervals.

   b. What is the significance of the signs of these rates of change?

   c. Based on your calculation in Problems 1a and 2a, do you think that the data lie on a single straight line? Explain.

   d. What is the practical meaning of slope in this situation?

3. By how much does the altitude of the plane change for each 3-minute change in time during the descent?

4. a. Plot the data points and verify that the points lie on a line. What is the slope of the line?

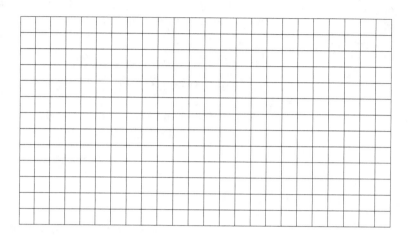

b. Explain to a classmate the method you used to determine the slope in part a.

c. Describe another method to determine the slope (other than the one you used in part a).

5. The slope of the line for the descent function in this activity is negative. What does this tell you about how the outputs change as the input variable (time) increases in value?

6. a. How many minutes since the time your descent began will it take the plane to reach the ground?

b. Use a straightedge to connect the data points on your graph in Problem 4. Extend the line so that it crosses both axes.

A **horizontal intercept** of a graph is a point where the graph crosses the horizontal axis. It is clearly identified by its second (vertical) coordinate, which is always equal to 0. The ordered pair notation for a horizontal intercept has the form $(a, 0)$.

c. Estimate the horizontal intercept of the line.

d. How is the horizontal intercept related to the answer you obtained in part a? That is, what is the practical meaning of the horizontal intercept?

A **vertical intercept** of a graph is a point where the graph crosses the vertical axis. It is clearly identified by its first (horizontal) coordinate, which is always equal to 0. The ordered pair notation for a vertical intercept has the form $(0, b)$.

e. Determine the vertical intercept of the descent function. What is the practical meaning of this intercept?

7. a. You found a promotion for unlimited access to the Internet for $20 per month. Complete the table of values, where t is the number of hours a subscriber spends on-line during the month and c is the monthly access cost for that subscriber.

t, TIME (hours)	1	2	3	4	5
c, COST ($)					

b. Sketch a graph of the data points. What is the slope of the line drawn through the points?

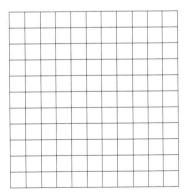

c. What single word best describes a line with zero slope?

d. Describe another situation (or give a data collection) in which the average rate of change is zero.

8. You earn $100 each week as a part-time aide in your college's health center to cover your weekly expenses while going to school. Complete the following table, where x represents your weekly salary and y represents your weekly expenses, for a typical month.

x, WEEKLY SALARY ($)	100	100	100	100
y, WEEKLY EXPENSES ($)	50	70	90	60

a. Sketch a graph of the data points. Do the points lie on a line? Explain.

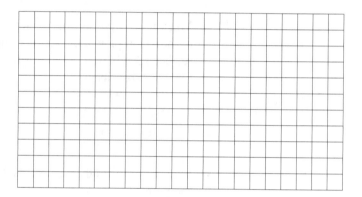

b. Can you determine a numerical value for the slope of the line in part a?

c. Write down your slope formula. What must be true about the change in the input variable for the slope formula to result in a numerical value?

d. What type of line results whenever the slope is undefined, as in this case?

e. Is y a function of x? Explain.

SUMMARY

The horizontal intercept and the vertical intercept are two points of special importance on the graph of a function.

- The horizontal intercept is the point at which the graph crosses the horizontal axis. Its ordered pair notation is $(a, 0)$; that is, the second coordinate is equal to zero.
- The vertical intercept is the point at which the graph crosses the vertical axis. Its ordered pair notation is $(0, b)$; that is, the first coordinate is equal to zero.

The slope of a horizontal line is zero.

The slope of a vertical line is undefined; it has no numerical value.

EXERCISES

1. In a science lab, you collect the following sets of data. Which of the four data sets are linear functions? If linear, determine the slope.

TIME (sec)	0	10	20	30	40
TEMPERATURE (°C)	12	17	22	27	32

TIME (sec)	0	10	20	30	40
TEMPERATURE (°C)	41	23	5	−10	−20

TIME (sec)	3	5	8	10	15
TEMPERATURE (°C)	12	16	24	28	36

TIME (sec)	3	9	12	18	21
TEMPERATURE (°C)	25	23	22	20	19

2. **a.** Suppose you are a member of a health and fitness club. A special diet and exercise program has been developed for you by the club's registered dietitian and your personal trainer. You weigh 181 pounds and would like to lose 2 pounds every week. Complete the following table of values for your hoped-for weight each week.

N, NUMBER OF WEEKS	0	1	2	3	4
W(N), DESIRED WEIGHT (lb)					

b. Plot the data points.

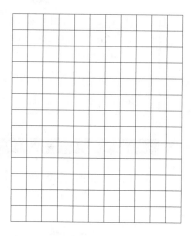

c. Explain why your desired weight is a linear function of time. What is the slope of the line containing the five data points?

d. What is the practical meaning of slope in this situation?

e. How long will it take to reach your ideal weight of 168 pounds?

3. Your aerobics instructor informs you that to receive full physical benefit from exercising, your heart rate must be maintained at a certain level for at least 12 minutes. The proper exercise heart rate for a healthy person, called the target heart rate, is determined by the person's age. The relationship between these two quantities is illustrated by the data in the following table.

A, AGE (years)	20	30	40	50	60
R(A), TARGET HEART RATE (beats per minute)	140	133	126	119	112

a. Do the data in the table indicate that the target heart rate is a linear function of age? Explain.

b. What is the slope of the line for these data? What are the units?

c. What are suitable replacement values (domain) for age, A?

d. Plot the data points and sketch the line through them. Label both axes starting with zero.

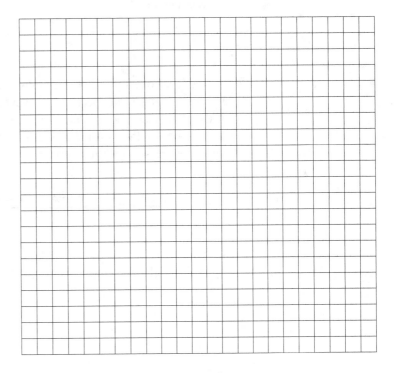

e. Extend the line to locate the horizontal and vertical intercepts. Do these intercepts have a practical meaning in the problem? Explain.

4. a. Determine the slope of each of the following lines.

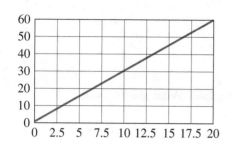
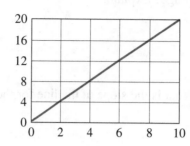
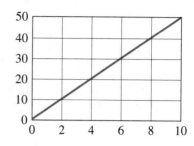

b. At first glance, the three graphs in part a may appear to represent the same line. Do they?

5. The concept of slope arises in many practical applications. When designing and building roads, engineers and surveyors need to be concerned about the *grade* of the road. The grade, usually expressed as a percent, is one way to describe the steepness of the finished surface of the road. A 5% grade means that the road has a slope (rise over run) of $0.05 = \frac{5}{100}$.

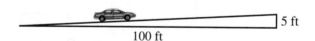

100 ft

5 ft

a. If a road has a 5% upward grade over a 1000-foot run, how much higher will you be at the end of that run than at the beginning?

b. What is the grade of a road that rises 26 feet over a distance of 500 feet?

c. Suppose you are traveling on a road with a 2.5% downward grade. What horizontal distance do you travel if you have dropped 5 feet in elevation?

6. Each question refers to the accompanying graph. The graphed line in each grid represents the distance a car travels as a function of time (in hours).

a. How fast is the car traveling?

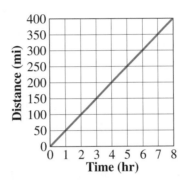

b. How can you determine visually which car is going faster? Calculate the speed of each car.

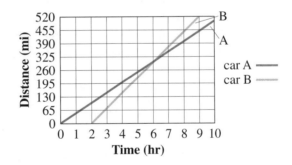

c. Describe the car's movement.

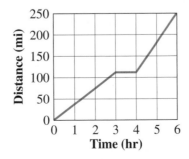

Charity Event
Topics: *Deriving a Linear Function Rule from Contextual Information, Slope-Intercept Form of a Linear Function, Characteristics of Linear Functions*

Professor Abrahamsen's social psychology class is organizing a campus entertainment night to benefit charities in the community. You are a member of the budget committee for this class project. The committee suggests an admission donation of $10 per person for food, nonalcoholic beverages, and entertainment. The committee members expect that each student in attendance will purchase a raffle ticket for $1. Faculty members volunteer to emcee the event and perform comedy sketches. Two student bands are hired at a cost of $200 each. Additional expenses include $1000 for food and drinks, $200 for paper products, $100 for posters and tickets, and $500 for raffle prizes. The college is donating the use of the gymnasium for the evening.

1. Determine the total fixed costs for the entertainment night.

2. a. Determine the **total revenue** (gross income *before* expenses are deducted) if 400 students attend and each buys a raffle ticket.

   b. Determine the **profit** (net income *after* expenses are deducted) if 400 students attend and each buys a raffle ticket.

3. The total revenue (gross income) for the event depends on the number n of students who attend. Write an expression in terms of n that represents the total revenue if n students attend and each buys a raffle ticket.

4. Write a symbolic function rule defining the profit p in terms of the number n of students in attendance.

5. List some suitable replacement values (domain) for the input variable n. Is it meaningful for n to have a value of $\frac{1}{2}$ or -3?

6. a. Use the symbolic rule in Problem 4 to determine the profit if 100 students attend.

b. What is the practical meaning of the negative value for profit in part a?

7. If the gymnasium holds a maximum of 650 people, what is the maximum amount of money that can be donated to charity?

8. a. Suppose that the members of the class want to be able to donate $1000 to community charities. How many students must attend the entertainment night in order to have a profit of $1000?

b. How many students must attend in order to be able to donate at least $2000 to the local charities?

9. a. Complete the following table of values.

n, NUMBER OF STUDENTS	0	50	100	200	300	400
p(n), PROFIT ($)						

b. How much does the profit change as the number of students in attendance increases from 300 students to 400 students?

c. Is the change in profit constant for any increase of 100 students?

d. Determine the average rate of change between consecutive data pairs in the table.

e. Is profit a linear function of the number of students attending? Explain.

 10. a. Sketch a graph of the profit function. Use a similar window to sketch the profit function on your grapher.

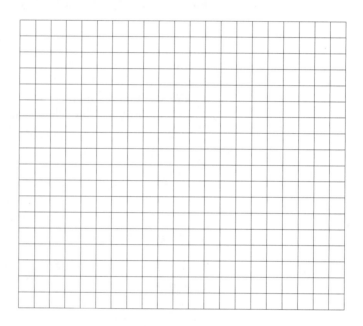

b. Verify your answers to Problems 6 and 8 using the graph.

11. a. Is your graph a straight line? Why should you have expected that the plotted points would all lie on the same line?

b. Determine the slope of the line.

c. What are the units of measurement of the slope? What is the practical meaning of the slope in this situation?

d. What is the vertical intercept of the line?

e. What is the practical meaning of the vertical intercept in this situation?

The profit function has a symbolic form that is representative of *all* linear functions. That is,

> The symbolic form of a linear function consists of the sum, or difference, of two terms:
>
> a *variable term* (the input variable together with its coefficient)
>
> and a *constant term* (a fixed number).

The symbolic form or rule in this situation is $p = 11n - 2200$.

12. a. Identify the variable term in this symbolic rule. What is its coefficient?

b. Identify the constant term in this symbolic rule.

c. What characteristic of the linear function graph does the coefficient represent?

d. What characteristic of the linear function graph does the constant term represent?

The answers to Problem 12c and d generalize to *all* linear functions.

> The coefficient of the variable term in a linear function is the *slope* of the line; the constant term (including its sign) gives the location of the *vertical intercept* of the line.

Recall that the letter m denotes the slope of a line and $(0, b)$ is the ordered pair that represents its vertical intercept. Therefore, the symbolic rule for y, a linear function of x, is given by

$$y = mx + b$$

and is called the **slope-intercept** form of the equation of a line.

> **NUMERICAL, GRAPHICAL, AND ALGEBRAIC CHARACTERISTICS OF LINEAR FUNCTIONS**
>
> - The average rate of change between any two input/output pairs is always the same constant value.
> - Equally spaced input values produce equally spaced output values.
> - The graph of every linear function is a line whose slope, m, is precisely the constant rate of change of the function.
> - The symbolic rule for a linear function, also called the slope-intercept form of the line, is given by $y = mx + b$, where m is the slope and $(0, b)$ is the y- (vertical) intercept of the line.

EXERCISES

Suppose the budget committee decides to increase the admission fee to $12 per person. It is still expected that the students will each purchase a raffle ticket for $1.

1. Write a new symbolic rule for profit in terms of the number of tickets sold.

2. Sketch a graph of the function.

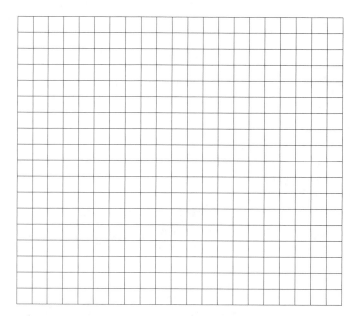

3. What is the slope of the line? What is the practical meaning of the slope?

4. What is the vertical intercept of the line? What is the practical meaning of the intercept in this situation?

5. If the gymnasium holds a maximum of 900 people, what is the maximum amount of money that can be given to charity?

6. What attendance is needed to make a profit of $1000?

7. How many students must attend so that the class project will not incur a loss? This number is called the **break-even number.**

ACTIVITY 4.4

Super Bowl Commercials

Topics: *Recognizing Linear Functions, Determining Horizontal Intercepts, Graphing Linear Equations Using Slope and Intercepts, Interpreting Slopes and Points Contextually*

The cost of a 30-second commercial during the Super Bowl has steadily increased since 1977. The following symbolic rule models the functional relationship between the cost of a 30-second commercial and the number of years since 1977:

$$C(t) = 48t + 104,$$

where $C(t)$ is the cost in thousands of dollars and t is the number of years since 1977.

1. Do you recognize the form of the symbolic rule for this function: the sum of a variable term and a constant term? What type of function does this model describe?

2. **a.** Determine the slope of this linear function.

   b. What are the units of measurement of the slope? What is the practical meaning of the slope in this situation?

3. **a.** Determine the vertical (y) intercept of this linear function.

   b. What is the practical meaning of the vertical intercept in this situation?

4. What is a reasonable domain for the function?

One way to graph a linear function by hand is to first plot the vertical intercept and then to make use of the slope.

5. Follow the steps given here for the slope-intercept method to graph the Super Bowl cost function on the grid on the following page.
   - Plot (0, 104) on the vertical axis.
   - Write the slope, 48, in fraction form as $\frac{48}{1} = \frac{\text{change in cost}}{\text{change in year}}$.
   - Starting at (0, 104), move 48 numerator units up and then move 1 denominator unit to the right. Mark the point you have reached.
   - The coordinates of the point you have reached are (1, 152).
   - Use a straightedge to draw the line through (0, 104) and (1, 152).

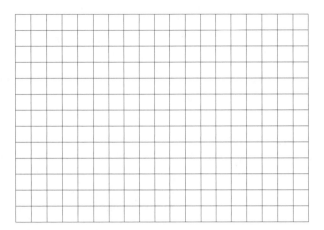

6. Starting at (0, 104), explain how to use the slope to calculate the coordinates of the point (1,152) without actually moving on the graph.

7. Interpret the practical meaning of the ordered pair (1, 152) in terms of the Super Bowl commercial situation.

8. Use the slope once again to reach a third point on the line. Interpret the practical meaning of this new ordered pair in terms of the Super Bowl situation.

9. Use the given symbolic rule to determine the cost of a 30-second commercial in the year 2002. What ordered pair best conveys the same information?

10. Use the symbolic rule to approximate in what year the cost of a 30-second commercial will reach $2 million. What ordered pair on the graph best conveys the same information?

A second way to graph a linear function is to plot its vertical and horizontal intercepts, and then use a straight edge to sketch the line containing these two points.

11. Suppose you need to graph the linear function $y = 2x + 6$. Follow the steps given here to use the intercepts to graph this function.

    a. Write the ordered pair that represents the vertical intercept.

    b. Recall that the horizontal intercept is a point at which the graph crosses the horizontal axis. Its vertical coordinate is equal to 0. Determine the horizontal intercept by setting y equal to 0 in the symbolic rule, and then solving the equation for x. Write the ordered pair that represents the horizontal intercept.

    c. Plot the vertical and horizontal intercepts on the grid below. Then use a straightedge to draw the line containing these points.

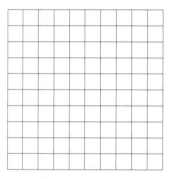

12. Identify the slope, vertical intercept, and horizontal intercept of each linear function in the following table.

LINEAR FUNCTION RULE	SLOPE	VERTICAL INTERCEPT	HORIZONTAL INTERCEPT
$y = 3x - 7$			
$y = -2x + 3$			
$y = 15 + 2x$			
$y = 10x$			
$y = 5$			
$3x + 4y = 12$			

EXERCISES

In Exercises 1–6, determine the slope, y-intercept, and x-intercept of each line. Then sketch each graph on graph paper, labeling and verifying the coordinates of each intercept. Use your grapher to check your results.

1. $y = 3x - 4$

2. $y = -5x + 2$

3. $y = 8$

4. $y = \dfrac{x}{2} + 5$

Hint: Solve for y first in Exercises 5 and 6.

5. $2x - y = 3$

6. $3x + 2y = 1$

7. Graph the following functions in the order given. Use your grapher to verify your answer. In what ways are the graphs similar? In what ways are they different?

a. $y = x - 4$ **b.** $y = x - 2$

c. $y = x$ **d.** $y = x + 2$

e. $y = x + 4$

8. Graph the following functions in the order given. Use your grapher to verify your answer. In what ways are the graphs similar? In what ways are they different?

a. $y = -4x + 2$ **b.** $y = -2x + 2$

c. $y = 2$ **d.** $y = 2x + 2$

e. $y = 4x + 2$

9. What is the equation of the linear function with slope 12 and y-intercept $(0, 3)$?

10. **a.** Suppose you start with $20 in your savings account and add $10 every week. What is the rate at which the amount in your account is changing from week to week?

   b. Write an equation that models your savings as a function of time (in weeks).

11. **a.** What is the slope of the line that goes through the points $(0, 5)$ and $(2, 11)$?

   b. What is the equation (symbolic rule) of the line through these two points?

12. **a.** What is the slope of the line that goes through the points $(0, -4)$ and $(2, 8)$?

   b. What is the equation of the line through these two points?

13. **a.** What is the slope of the line that goes through the points $(0, 120)$ and $(3, 165)$?

   b. What is the equation (symbolic rule) of the line containing these two points?

14. a. What is the slope of the line that goes through the points $(0, -43.5)$ and $(-1, 13.5)$?

b. What is the equation of the line through these two points?

15. Determine the horizontal and vertical intercepts of each of the following. Use the intercepts to sketch a graph of the function.

a. $y = -3x + 12$

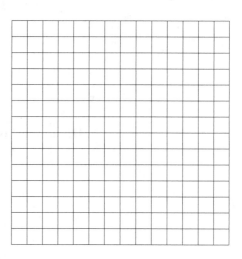

b. $y = \dfrac{1}{2}x + 6$

What Have I Learned?

1. a. Does the slope of the line having the equation $4x + 2y = 3$ have a value of 4? Why or why not?

    b. Solve the equation in part a for y so that it is in the form $y = mx + b$.

    c. What is its slope?

2. A line is given by the equation $y = -4x + 10$.

    a. Determine its horizontal and vertical intercepts algebraically from the equation.

    b. Use your grapher to confirm these intercepts.

3. Explain the difference between a line with zero slope and a line with an undefined slope.

4. Describe how you recognize that a function is linear when it is given

    a. graphically.

b. symbolically.

c. numerically in a table.

5. Do vertical lines represent functions? Explain.

How Can I Practice?

1. A function is linear because the rate of change of the output with respect to the input from point to point is constant. Use this idea to determine the missing input (x) and output (y) values in each table, assuming that each table represents a linear function.

a.

x	y
1	4
2	5
3	

b.

x	y
1	4
2	6
3	

c.

x	y
1	4
2	9
3	

d.

x	y
−1	3
0	5
	7
4	

e.

x	y
2	11
3	8
4	
	−1

f.

x	y
−2	−5
0	−8
	−11
4	

g. Explain how you used the idea of constant rate of change to determine the values in the tables.

2. Determine a symbolic rule of the form $y = mx + b$ for each function in Problem 1.

3. The pitch of a roof is an example of slope in a practical setting. The roof slope is usually expressed as a ratio of rise over run. For example, in the building shown, the pitch is 6 to 24, usually stated as "3 in 12." Also, a pitch of 6 to 24 is $\frac{1}{4}$ when expressed as a fraction.

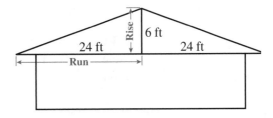

   a. If a roof has a pitch of 5 to 16, how high will the roof rise over a 24-foot run?

   b. If a roof's slope is 0.25, how high will the roof rise over a 16-foot run?

   c. What is the slope of a roof that rises 12 feet over a run of 30 feet?

4. Determine whether any of the following tables contain input and output data that represent a linear function. In each case, give a reason for your answer.

   a. You make an investment of $100 at 5% interest compounded semiannually. The following table represents the amount of money you will have at the end of each year.

TIME (years)	AMOUNT ($)
1	105.06
2	110.38
3	115.97
4	121.84

   b. A cable-TV company charges a $45 installation fee and $28 per month for basic cable service. The table values represent the total usage cost since installation.

NUMBER OF MONTHS	6	12	18	24	36
TOTAL COST ($)	213	381	549	717	1053

   c. For a fee of $20 a month, you have unlimited video rental. Values in the table represent the relationship between the number of videos you rented each month and the monthly fee.

NUMBER OF RENTALS	10	15	12	9	2
COST ($)	20	20	20	20	20

5. After stopping your car at a stop sign, you accelerate at a constant rate for a period of time. The speed of your car is a function of the time since you left the stop sign. The following table shows your speed in each of the first 7 seconds.

t TIME (sec)	s SPEED (mph)
0	0
1	11
2	22
3	33
4	44
5	55
6	55
7	55

a. Graph the data using ordered pairs of the form (t, s).

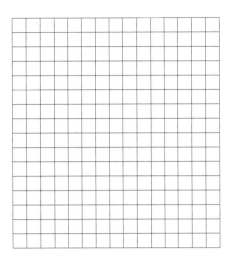

b. For what values of t is the graph increasing?

c. What is the slope of the line segment during the period of acceleration?

d. What is the practical meaning of the slope in this situation?

e. For what values of t is the speed a constant? What is the slope of the line connecting the points of constant speed?

6. a. The three lines shown in the following graphs appear to be different. Calculate the slope of each line.

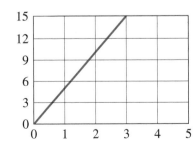 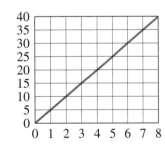 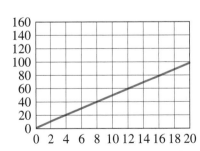

b. Do the three graphs represent the same linear function? Explain.

7. **a.** Determine the slope of the line through the points $(2, -5)$ and $(2, 4)$.

b. Determine the slope of the line $y = -3x - 2$.

c. Determine the slope of the line $2x - 4y = 10$.

d. Determine the slope of the line from the graph.

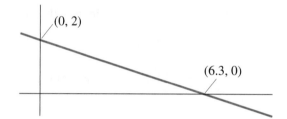

(0, 2)

(6.3, 0)

8. Determine the vertical and horizontal intercepts for the graph of each of the following.

    a. $y = 2x - 6$ **b.** $y = -\frac{3}{2}x + 10$

    c. $y = 10$

9. Determine the equation of each line.

   a. The line passes through the points $(2, 0)$ and $(0, -5)$.

   b. The slope is 7 and the line passes through the point $\left(0, \frac{1}{2}\right)$.

   c. The slope is 0 and the line passes through the point $(2, -4)$.

10. Sketch a graph of each of the following. Use your grapher to verify.

   a. $y = 3x - 6$

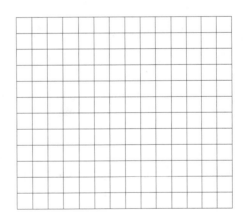

   b. $2x + y = 10$

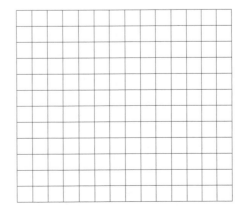

11. Write each equation in slope-intercept form to discover what the graphs have in common. Use your grapher to verify.

   a. $y = 3x - 4$ **b.** $y - 3x = 6$ **c.** $3x - y = 0$

12. Write each equation in slope-intercept form to discover what the graphs have in common. Use your grapher to verify.

   a. $y = -2$ **b.** $y - 3x = -2$ **c.** $x = y + 2$

13. a. Determine an equation of the line containing the points $(3, 4)$ and $(3, -1)$. What is the slope?

   b. Determine the vertical and horizontal intercepts, if any, of the graph of the line in part a.

   c. Does the graph of the line in part a represent a function? Explain.

Problem Solving with Linear Functions

ACTIVITY 4.5

Predicting Population

Topics: *Deriving Linear Equations from Two Points, Functions in Slope-Intercept Form*

According to the U.S. Census Bureau, the population of the United States was approximately 132 million in 1940 and 151 million in 1950.

For this activity, assume that the rate of change of population with time is a constant value over the decade from 1940 to 1950; that is, a linear relationship exists between the input variable t and the output variable P.

1. Write the data as ordered pairs of the form (t, P), where t is the number of years since 1940 and P is the corresponding population, in millions. Plot the data points and draw a straight line through them. Label your horizontal axis from 0 to 25 and your vertical axis from 130 to 180.

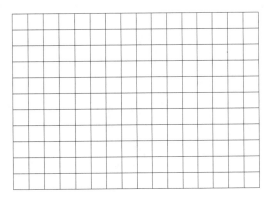

2. **a.** What is the average rate of change of population from $t = 0$ (1940) to $t = 10$ (1950)?

   b. What is the vertical intercept of this line? What is the practical meaning of the vertical intercept in this situation?

3. What is the slope of the line connecting the two points in Problem 1? What is the practical meaning of the slope in this situation?

4. Use the slope and vertical intercept from Problems 3 and 2 to write an equation for the line.

5. Assume that the rate of change you found in Problem 2 stays the same through 1960. Use the equation you found in Problem 4 to predict the population of the United States in the year 1960. Also estimate the population in 1960 from the graph.

6. The actual population of the United States in 1960 was approximately 179 million. What is the *relative error* (error ÷ actual population) in your prediction?

7. What do you think was the cause of your prediction error?

8. Suppose you want to develop a population model based on more recent data. The population of the United States was approximately 226 million in 1980 and 249 million in 1990. Plot these data points using ordered pairs of the form (t, P), where t is the number of years since 1980 (now, $t = 0$ corresponds to 1980). Draw a line through the points.

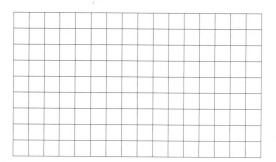

9. a. Determine the slope of the line in Problem 8. What is the practical meaning of the slope in this situation? How does this slope compare with the slope in Problem 3?

   b. In which decade, 1940–1950 or 1980–1990, did the U.S. population increase more rapidly? Explain your answer in terms of slope.

c. Determine the vertical intercept of the line in Problem 8.

d. Write the equation of the line in Problem 8.

10. Use the equation, also called a linear model, developed in Problem 9 to predict the population in the year 2010. What assumptions are you making about the rate of change of the population in this prediction?

11. According to the linear model in Problem 9, in what year will the population be 350 million?

EXERCISES

1. **a.** According to the U.S. Census Bureau, the population of California in 1980 was approximately 23.7 million and was increasing at a rate of approximately 617,000 people per year. Let P represent the California population (in millions) and t represent the number of years since 1980. Complete the following table.

t	P (IN MILLIONS)
0	
1	
2	

b. Why can you assume that the California population growth is linear with respect to time? What is the slope and vertical intercept of the graph of the population data?

c. Write a linear function rule for P in terms of t.

d. Use the linear model to estimate what the population of California was in 1997.

e. Population data from the State of California Department of Finance is used by California state agencies in developing their programs and policies. The Department of Finance estimated that the population of California in 1997 was approximately 33 million. Determine the relative error (as a percent) between your prediction in part d and the estimate actually used by the state of California.

f. Use the linear model from part c to predict the population of California in 2008.

g. Do you think your prediction in part f will be too high, close, or too low? Explain.

2. a. In 1990, the rate of change of the world population was approximately 1 million people every four days. If the population of the world in 1990 was estimated to be 5.3 billion, write a symbolic rule to model the population P (in billions) in terms of t, where t is the number of years since 1990 ($t = 0$ corresponds to 1990).

b. Use the linear model to predict the world population in the year 2020.

c. According to the model, when will the population of the world be double the 1990 population?

ACTIVITY 4.6

Housing Prices
Topic: *Deriving Linear
Functions When Given Two
Points*

You have been aware of a steady increase in housing prices in your neighborhood since 1990. The house across the street sold for $125,000 in 1993, and then sold again in 1997 for $150,000. This data can be written in table format, where the input represents the number of years since 1990 and the output represents the sale price of a typical house in your neighborhood.

NUMBER OF YEARS SINCE 1990, x	HOUSING PRICE (THOUSANDS OF $), y
3	125
7	150

1. Plot the two points on the grid below and sketch the line containing them. Extend the line so that it intersects the vertical axis. Scale your input axis to include the period of years from 1990 to 2000. Scale your output axis by increments of 25, starting from 0 and continuing through 250.

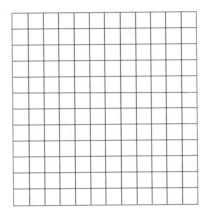

2. Determine the slope of the line. What are its units of measurement? What is the practical meaning of the slope in this situation?

3. Estimate the vertical intercept from the graph. What is the practical meaning of the vertical intercept in this situation?

4. Use the slope and your estimate of the vertical intercept to write a linear function rule for housing price, y, in terms of x, number of years since 1990.

5. Test the accuracy of your function rule by checking whether the coordinates of each plotted point satisfy the equation.

6. Starting at the point (3, 125) use the slope $\frac{6.25}{1}$ to determine the coordinates of the vertical intercept.

You can use a straightforward algebraic method to find the exact vertical intercept when two other points on the line are known. The first step is to calculate the slope, which you did in Problem 2. Therefore, the equation of the line will be of the form

$$y = 6.25x + b$$

where b is still unknown. Since the coordinates of every point on a line must satisfy the equation of the line, the ordered pair (3, 125) must satisfy this equation. So, you can substitute 3 for x and 125 for y and solve the resulting equation for b.

$$125 = 6.25(3) + b$$
$$b = 106.25$$

Therefore, the equation of the line is $y = 6.25x + 106.25$.

7. Use the ordered pair (7, 150) in the method described above to determine an equation of the line. Compare your equation with the one obtained above.

8. Summarize the algebraic procedure for determining an equation of a line given two points, neither of which is the vertical intercept. Illustrate your step-by-step procedure using the points (15, 62) and (21, 80).

1. The amount of federal income tax paid by an individual single taxpayer is a function of adjusted gross income. The following table represents the amount of 1997 federal tax for various adjusted gross incomes.

i, GROSS INCOME ($)	15,000	15,500	16,000	16,500	17,000	17,500	18,000
t, 1997 FEDERAL TAX ($)	2254	2329	2404	2479	2554	2629	2704

a. Plot the data points, with adjusted gross income *i* as input and 1997 tax *t* as output. Explain why the relationship is linear. Scale your input axis from $0 to $24,000 and your output axis from $0 to $3000.

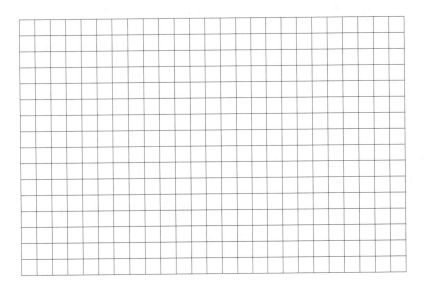

b. Determine the slope of the line. What is the practical meaning of the slope?

c. What is the vertical intercept? Does it make sense?

   d. Write a symbolic rule to model this situation. Use the variable i to represent the adjusted gross income and the variable t to represent the federal tax owed.

   e. What was the federal tax owed by a college student having an adjusted gross income of $4200?

   f. What was the adjusted gross income of a single person who paid $1680 in federal taxes for 1997?

In Exercises 2–9, determine the equation of the line with the given slope passing through the given point. Use your grapher to confirm your answer.

2. $m = 3$, through the point $(2, 6)$ 3. $m = -1$, through the point $(5, 0)$

4. $m = 7$, through the point $(-3, -5)$ 5. $m = 0.5$, through the point $(8, 0.5)$

6. $m = 0$, through the point $(5, 2)$ 7. $m = -4.2$, through the point $(-4, 6.8)$

8. $m = 25$, through the point $(12, 455)$

9. $m = -\frac{2}{7}$, through the point $\left(5, \frac{4}{7}\right)$

In Exercises 10–15 determine the equation of the line through the given points. Check your equation numerically or graphically on a grapher.

10. $(2, 6)$ and $(4, 16)$ 11. $(-5, 10)$ and $(5, -10)$

12. (3, 18) and (8, 33)

13. (0, 6) and (−10, 0)

14. (10, 2) and (19, 5)

15. (3.5, 8.2) and (2, 7.3)

ACTIVITY 4.7

Oxygen for Fish

Topics: *Scatterplots, Linear Regression Models*

Fish need oxygen to live, just as you do. The amount of dissolved oxygen in water is measured in parts per million (ppm). Trout need a minimum of 6 ppm to live. There are many variables that determine the amount of dissolved oxygen in a stream. One very important variable is the water temperature. To investigate the effect of temperature on dissolved oxygen, you take a water sample from a stream and measure the dissolved oxygen as you heat the water. Your results are as follows.

t, TEMPERATURE (°C)	11	16	21	26	31
d, DISSOLVED OXYGEN (ppm)	10.2	8.6	7.7	7.0	6.4

1. Plot the data points as ordered pairs of the form (t, d). Scale your input axis from 0°C to 40°C and your output axis from 0 ppm to 15 ppm. The resulting graph of points is called a **scatterplot.**

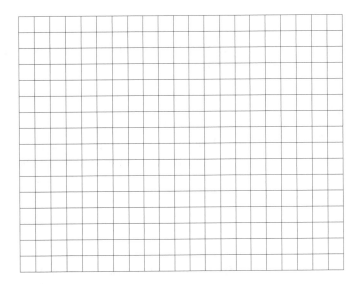

2. Does there appear to be a linear relationship between the temperature t and the amount of dissolved oxygen d? Is there an exact linear fit? That is, do all the points lie on the same line?

3. Use a straightedge to draw a single line that you believe best represents the linear trend in the data. The resulting line is commonly called a **line of best fit.** The method you are using here to draw the line, informally called the "eyeball" method, is one way to estimate a line of best fit. The line you drew and its symbolic rule is a linear model for the given set of data. In Problem 9 you will use the more objective **least squares method** to determine a line of best fit.

4. Use two points on your line (preferably, as far apart as possible) to estimate the slope of the line you drew in Problem 3. What is the practical meaning of the slope in this situation?

5. What is the vertical intercept of your line? Does this point have any practical meaning in this situation?

6. What is the equation of your linear model?

7. Use the information in the opening paragraph and your linear model to approximate the maximum temperature at which trout can survive.

8. An estimate of how well a linear model represents a given set of data is called a **goodness-of-fit measure.** To measure the goodness-of-fit of your linear model, use the linear rule you derived in Problem 6 to complete the following table.

t, INPUT	ACTUAL OUTPUT	$d(t)$ MODEL'S OUTPUT	ACTUAL VALUE − MODEL VALUE	|ACTUAL VALUE − MODEL VALUE|
11	10.2			
16				
21				
26				
31				

Find the sum of the absolute values of the differences found in the last column. This sum is called the **error.** The smaller the error, the better the fit. Compare the size of your error with those of your classmates.

Appendix

9. The *method of least squares* is an objective procedure for determining a line of best fit. This method results in the equation of a line called a *regression line*. Your graphing calculator uses this procedure to produce the equation of a regression line. Use your grapher's statistics menu to determine the equation for the regression line in this situation.

10. Determine the goodness-of-fit measure for the least squares regression line in Problem 9.

t, INPUT	ACTUAL OUTPUT	d(t) MODEL'S OUTPUT	ACTUAL VALUE − MODEL VALUE	|ACTUAL VALUE − MODEL VALUE|
11	10.2			
16				
21				
26				
31				

11. Compare the error of your line of best fit with the error of the least squares regression line.

E X E R C I S E S

During the spring and summer, a concession stand at a community Little League base-ball field sells soft drinks and other refreshments. To prepare for the season, the concession owner refers to the previous year's files, in which he had recorded the daily soft drink sales (in gallons) and the average daily temperature (in degrees Fahrenheit). These data are shown in the table.

t, TEMPERATURE (°F)	g, SOFT DRINK SALES (gal)
52	35
55	42
61	50
66	53
72	66
75	68
77	72
84	80
90	84
94	91
97	95

1. Plot the data points as ordered pairs of the form (t, g).

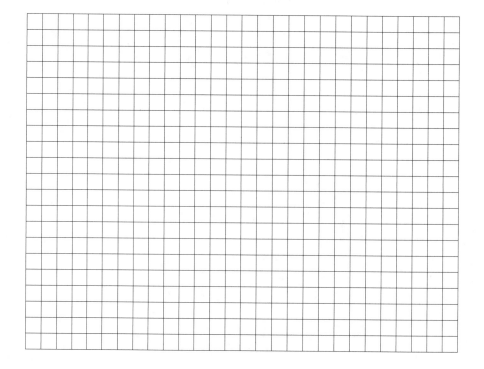

2. Does there appear to be a linear relationship between the temperature t and the soft drink sales g? Is there an exact linear fit?

3. a. Use a straightedge to draw a line of best fit that you believe best represents the linear trend in the data.

   b. Use the coordinates of two points on your line to determine the slope of your line of best fit.

   c. What is the practical meaning of the slope in this situation?

4. What is the vertical intercept of your line of best fit? Does this number have any practical meaning in this situation?

5. What is the equation of your line of best fit?

6. To measure the goodness-of-fit of the line found in Exercise 5, complete the following table and compute the error.

t, INPUT	ACTUAL OUTPUT	$g(t)$ MODEL'S OUTPUT	|ACTUAL VALUE − MODEL VALUE|
52			
55			
61			
66			
72			
75			
77			
84			
90			
94			
97			

7. Use your grapher's statistics menu to determine the equation for the regression line in this situation.

8. What is the goodness-of-fit measure (the error) for the regression line in Exercise 7? (Proceed as you did in Exercise 6.)

t, INPUT	ACTUAL OUTPUT	$g(t)$ MODEL'S OUTPUT	ACTUAL VALUE − MODEL VALUE

9. Compare the error of your line of best fit with the error of the least squares regression line.

What Have I Learned?

1. How would you write the equation of a line if you knew only the slope and the vertical intercept? Use an example to demonstrate.

2. Demonstrate how you would change the equation of a linear function such as $5y - 6x = 3$ into slope-intercept form.

3. What assumption are you making when you say that the cost c of a rental car (in dollars) is a linear function of the number n of miles driven?

4. When a scatterplot of input/output values from a data set suggests a linear relationship, you can determine a line of best fit. Why might this line be useful in your analysis of the data?

5. Explain how you would use the "eyeball" method to determine a line of best fit for a set of data. How would you estimate the slope and y-intercept?

6. Suppose a set of data pairs suggests a linear trend. The input values range from a low of 10 to a high of 40. You use your grapher to calculate the regression equation in the form $y = ax + b$.

    a. Will that equation provide a good prediction of the output value for an input value of $x = 20$? Explain.

    b. Will that equation provide a good prediction of the output value for an input value of $x = 60$? Explain.

How Can I Practice?

In Problems 1–9, determine the slope and the vertical and horizontal intercepts of each line.

    1. $y = 2x + 1$ **2.** $y = 4 - x$

    3. $y = -2$ **4.** $-\frac{3}{2}x - 5 = y$

    5. $y = \frac{x}{5}$ **6.** $y = 4x + \frac{1}{2}$

7. $2x + y = 2$

8. $-3x + 4y = 12$

9. $x - 6y = -2$

Determine an equation of each line in Problems 10–13.

10. The slope is 9 and the vertical intercept is $(0, -4)$.

11. The line passes through the points $(0, 4)$ and $(-5, 0)$.

12. The slope is $\frac{5}{3}$ and the line passes through the point $(0, -2)$.

13. The slope is zero and the line passes through the origin.

14. Identify the input and output variables and write a linear function in symbolic form for each of the following situations. Then give the practical meaning of the slope and vertical intercept in each.

    a. You must make a down payment of $50 and pay $10 per month for your new computer.

    b. You pay $16,000 for a new car and its book value decreases by $1500 each year.

Graph the equations in Problems 15 and 16, and determine the slope and vertical intercept of each graph. Describe the similarities and differences in the graphs.

15. a. $y = -x + 2$ **b.** $y = -4x + 2$

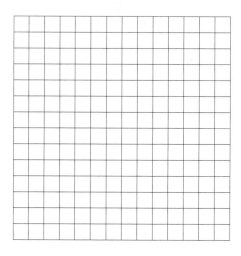

16. a. $y = 3x - 4$ **b.** $y = 3x + 5$

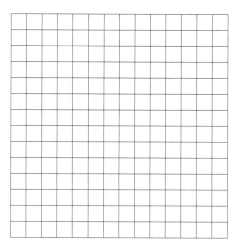

17. a. Refer back to Activity 4.1, The Snowy Tree Cricket, on page 403, to write a linear function rule that gives the number of chirps per minute in terms of temperature.

b. Use your linear model to determine the number of chirps per minute if the temperature is 62°F.

c. Use your linear model to determine the temperature if the crickets chirp 190 times per minute.

18. a. Refer back to Activity 4.2, Descending in an Airplane, on page 409. Write a linear function rule that gives the altitude A in terms of time t, where t is measured from the moment the plane begins its descent.

b. Use your linear model to determine how far the plane has descended after 5 minutes.

c. Use your linear model to determine how long it will take the plane to reach the ground.

19. Suppose you are driving on Interstate 90 in Montana at a constant speed of 75 mph.

a. Write a function rule that represents the total distance traveled in terms of the number of hours driven at 75 mph.

b. Sketch a graph of the function. What is the slope and vertical intercept of the line? What is the practical meaning of the slope?

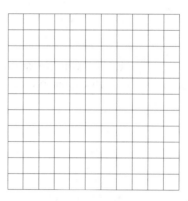

c. How long would you need to drive at 75 mph to travel a total of 400 miles?

d. Suppose you start out at 10:00 A.M. and drive for 3 hours at a constant speed of 75 mph. You are hungry and stop for lunch. One hour later you resume your travel, driving steadily at 60 mph until 6 P.M. when you reach your destination. How far have you traveled today? Sketch a graph that shows the distance traveled as a function of time.

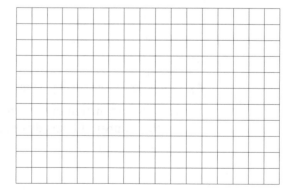

20. Determine the equation of each line in parts a through f.

a. The slope is 3, and the *y*-intercept is 6.

b. The slope is -4, and the y-intercept is -5.

c. The slope is 2, and it passes through $(0, 4)$.

d. The slope is 4, and it passes through $(6, -3)$.

e. The slope is -5, and it passes through $(4, -7)$.

f. The slope is 2, and it passes through $(5, -3)$.

21. Determine the equation of the line passing through each pair of points.

a. $(0, 6)$ and $(4, 14)$ **b.** $(-2, -13)$ and $(0, -5)$

c. $(5, 3)$ and $(-1, 3)$ **d.** $(-9, -7)$ and $(-7, -3)$

e. (6, 1) and (6, 7)   f. (2, 3) and (2, 7)

22. Determine the equation of the line shown on the following graph.

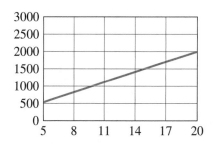

23. Determine the x-intercept of each line given in Problem 20c–f.

24. On an average winter day, the Auto Club receives 125 calls from persons who need help starting their cars. The number varies, however, depending on the temperature. Here are some data giving the number of calls as a function of the temperature (in degrees Celsius).

TEMPERATURE (°C)	NUMBER OF AUTO CLUB SERVICE CALLS
−12	250
−6	190
0	140
4	125
9	100

a. Use your grapher to sketch a scatterplot of the data from the table.

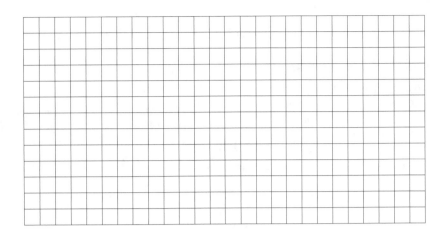

b. Use your grapher to determine the equation of the regression line for the data in the preceding table.

c. Use your regression equation from part b, $y = -7.11x + 153.9$, to determine how many service calls the Auto Club can expect if the temperature drops to $-20°C$.

d. For what temperature would the regression equation predict that the Auto Club should expect no calls?

Systems of Linear Equations

ACTIVITY 4.8

Business Checking Account

Topic: *Solving 2 × 2 Linear Systems Numerically, Graphically, and by Substitution*

In setting up your part-time business, you have two choices for a checking account at your local bank.

	MONTHLY FEE	TRANSACTION FEE
REGULAR	$11.00	$0.17 for each transaction
BASIC	$8.50	$0.22 for each transaction in excess of 20

1. If you anticipate making about 50 transactions each month, which checking account will be the most economical?

2. Complete the following table for each account, showing the monthly cost for 50, 100, 150, 200, 250, and 300 transactions. Estimate the number of transactions for which the cost of the two accounts comes the closest.

NUMBER OF TRANSACTIONS	50	100	150	200	250	300
COST OF REGULAR						
COST OF BASIC						

3. For each account, plot the ordered pairs (x, y) from the table in Problem 2, where x represents the number of transactions and y the cost. Sketch a graph through the points for each account, and estimate the coordinates of the point where the lines intersect. What is the significance of this point?

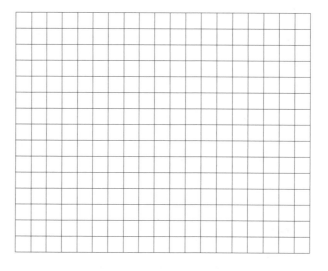

Two equations that relate the same variables are called a **system of equations.** The solution of a system of equations is the set of all ordered pairs that satisfy both equations. One method for solving a system of two linear equations is to graph both equations on the same grid (same coordinate axes). If the two lines intersect, the solution to the system is the coordinates of the point of intersection. This method is referred to as solving graphically.

4. Express the total monthly cost of the regular account as a function of the number of transactions. That is, find the equation where x is the number of transactions and y is the cost. Use your grapher to graph the function and verify that it is the same as the line you graphed in Problem 3.

5. The function for the basic account is not expressed as easily symbolically, because the transaction fee does not apply to the first 20 transactions. Assuming that you will write at least 20 checks per month, determine the equation for this function when $x > 20$. As in Problem 4, graph this function and verify that it agrees with your original plot.

Appendix

The two function rules you determined in Problems 4 and 5 can be considered together as a system of equations. The solution of this system would represent the specific number of transactions that produce identical costs in both accounts. You can always determine an exact solution by solving the system of equations algebraically. In Problem 6, you will explore one method for solving systems of equations algebraically—the **substitution method.** Another algebraic method for solving systems of linear equations, the **addition method,** can be found in Appendix C.

Because you want to determine the number of transactions that result in the same monthly cost for each account, the value of the output variable y will be the same in both accounts. Replace (or *substitute*) the y variable in one equation with its algebraic expression in x from the other equation. When the equation of each line is given in $y = mx + b$ form, this is equivalent to setting the two function rules equal to one another. You should obtain a single equation with the input variable x appearing on both sides.

6. a. Write the equations from Problem 4 and Problem 5. Impose the requirement that both costs (y-values) should be the same by setting the algebraic expressions in x on the right side of each function rule equal to one another.

   b. Solve the resulting equation for x. What does this value of x represent?

   c. Substitute this x-value into either function rule to determine the corresponding cost.

7. Summarize your results by describing under what circumstances the basic account is preferable to the regular account.

 8. If your grapher has a table feature, enter the two cost functions into the "$y = $" menu and search the table for the input value that produces two identical outputs. What is that value?

E X E R C I S E S

1. Suppose two industrious friends are selling lemonade in their neighborhood on hot summer days. They model their costs (in cents) as $C = 2x + 100$ and the revenue (in cents) from the sales as $R = 12x$, where x represents the number of glasses of lemonade sold.

   a. Interpret the practical meaning of the slope in each equation.

   b. What is the practical meaning of the vertical intercept in the cost equation?

   c. How many glasses of lemonade must they sell to break even?

Determine the exact solution to each system of equations in Exercises 2–5 algebraically using substitution. Use the table feature or graphing capability of your calculator to verify these solutions.

2. $p = q - 2$

   $p = -1.5q + 3$

3. $n = -2m + 9$

   $n = 3m - 11$

4. $y = 1.5x - 8$

   $y = -0.25x + 2.5$

5. $z = 3w - 1$

   $z = -3w - 1$

Does every system of equations have exactly one solution? Attempt to solve the systems in Exercises 6 and 7 algebraically. Explain your results using a graphical interpretation.

6. $y = -3x + 2$

   $y = -3x + 3$

7. $3x - y = 1$

   $-6x + 2y = -2$

You are employed by a company that manufactures solar collector panels. To remain competitive, the company must consider many variables and make many decisions. Two major concerns are those variables and decisions that affect operating expenses (or costs) of making the product and those that affect the gross income (or revenue) from selling the product.

Costs such as rent, insurance, and utilities for the operation of the company are called *fixed costs*. These costs generally remain constant over a short period of time and must be paid whether or not any items are manufactured. Other costs, such as materials and labor, are called *variable costs*. These expenses depend directly on the number of items produced.

1. Suppose the records of the company show that fixed costs over the past year have averaged $8000 per month. In addition, each panel manufactured costs the company $95 in materials and $55 in labor. Write a symbolic rule in function notation for the total cost C of producing n solar collector panels in one month.

2. A marketing survey indicates that the company can sell all the panels it produces if the panels are priced at $350 each. The revenue (gross income) is the amount of money collected from the sale of the product. Write a symbolic rule in function notation for the revenue R from selling n solar collector panels in one month.

3. Sketch a graph of the cost and revenue functions using the same set of coordinate axes. Use your grapher and then copy the graphs onto the following grid.

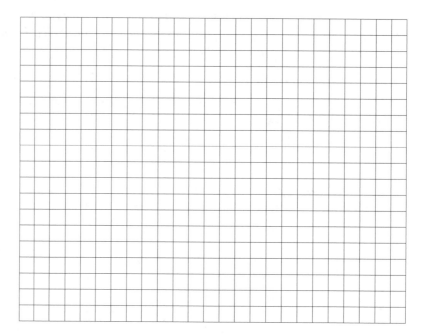

4. The point at which the cost and revenue functions are equal is called the *break-even point*. The exact coordinates of the break-even point can be found algebraically.

    a. What system of equations must be solved to determine the break-even point for your company?

    b. Solve the system to determine the break-even point algebraically.

    c. Does your graph confirm the algebraic solution in part b?

5. For what values of n is $R(n) > C(n)$? What do these values represent in this situation?

6. *Profit* is defined as revenue minus cost. Write a symbolic rule for the profit P made by selling n solar panels in one month.

$+$

7. a. Sketch a graph of the profit function.

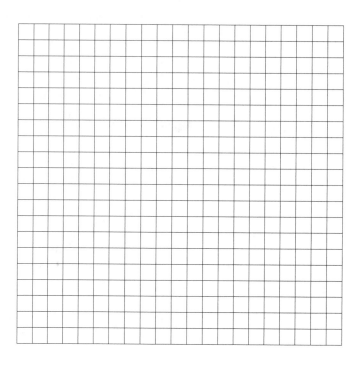

b. Determine the vertical and horizontal intercepts of the graph. What is the meaning of each intercept in this situation?

c. What is the slope of the line? What is the practical meaning of slope in this situation?

8. The profit function for a competitor is $P(n) = 175n - 7500$, where n is the number of panels produced and sold per month. Sketch a graph of the competitor's profit function on the same coordinate axes as your company's profit function. Solve the resulting system. What does the point of intersection represent in this situation?

E X E R C I S E S

1. A company will break even when its revenue R exactly equals its cost C. C and R depend on x, the number of items sold.

   Suppose you are the manager of a small company producing designer-cut interlocking pavers for driveways. You sell the pavers in bundles of 1 gross (144) each at $200 for each bundle. The total cost in dollars, $C(x)$, of producing x bundles of pavers is modeled by $C(x) = 160x + 1000$.

   a. Write a symbolic rule for the revenue function in dollars, $R(x)$, from the sale of the pavers.

   b. Determine the slope and the y-intercept for the cost function. Explain the practical meaning of each in this situation.

   c. Determine the slope and the y-intercept for the revenue function. Explain the practical meaning of each in this situation.

   d. Graph the two functions from parts b and c on the same set of axes. Estimate the break-even point from the graph. Express your answer as an ordered pair, giving units. Check your estimate of the break-even point by graphing the two functions on your grapher.

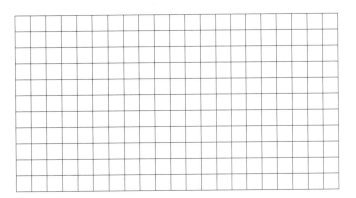

e. Determine the exact break-even point algebraically. If your algebraic solution does not agree (approximately) with your answer from the graph in part d, explain why.

f. How many bundles of pavers would you have to sell for your company to break even?

g. What is the total cost to the company when you break even? Verify that the cost and revenue values are equal at the break-even point.

h. For what values of x will your revenue exceed your cost?

i. As manager, what factors do you have to consider when deciding how many pavers to make?

j. If you knew you could sell only 30 bundles of pavers, would you make them? Consider how much it would cost you and how much you would make. What if you thought you could sell only 20?

ACTIVITY 4.10

Supply and Demand

Topics: *Solving* 2 × 2
Linear Systems, Inequalities,
Equilibrium Point

In business, the price p of an item and the number n of items produced and sold are closely related. From the manufacturer's point of view, the higher the price, the more the manufacturer is willing to supply (produce). Suppose that in a particular situation, the relationship between n and p is linear. Using n as the input and p as the output, the function that represents this relationship is called a *supply curve*.

1. Your company can afford only enough material to supply 30 solar panels per month if they sell for $225 each. However, if the panels sell for $450 each, a total of 100 panels can be supplied. Write a symbolic linear rule for the supply function.

From the consumer's point of view, the higher the price p, the fewer items consumers will purchase. Assume, as in the case of the supply curve, that the relationship between the number of items n purchased by the consumer and the price p is linear. The function that represents this relationship is called a *demand curve*.

2. Suppose your company is deciding on a price p for each panel. The results from market research indicate that 200 panels per month will sell if the price is $80 per panel. If the price per panel is $450, only 20 panels will sell. Write a linear function rule for the demand function, using n as the input variable. What is the significance of the negative slope in this situation?

If the supply and demand graphs are drawn on the same coordinate axes, the intersection is called the *equilibrium point*. This point represents the values of n and p for which supply equals demand.

3. Use the demand and supply functions developed in Problems 1 and 2 to write a system of equations that can be used to determine the equilibrium point. Represent the system and point of intersection graphically using your graphing calculator and copy the graph onto the following grid. What is the equilibrium point?

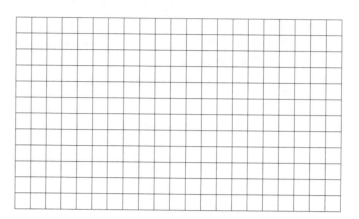

EXERCISES

1. You like to drive to a special store to buy your favorite gourmet jelly beans for a low price of $1.30 a pound. However, your best friend points out that you are spending about $3 on gas every time you drive to that store and that you might as well buy your jelly beans at your local supermarket for $2.10 a pound. You decide to show him that you are getting enough of those jelly beans to make it worth the trip. How many pounds do you need to buy? Use a system of equations to decide. Verify your answer using your grapher.

2. You want to hire someone to prune your trees and shrubs. One service you call charges a $15 consultation fee plus $8 an hour for the actual work. A neighborhood gardener says she does not include a consulting fee, but she charges $10 an hour for her work.

   a. Whom would you hire for a small (2 to 3 hours) job?

   b. When, if at all, would it be more economical to hire the other service? Set up and solve a system of equations to answer this question.

3. You and your friend are going rollerblading at a local park. There is a 5-mile path along the lake that begins at the concession stand. You rollerblade at a rate of 10 mph and your friend rollerblades at 8 mph. You start rollerblading at the concession stand. Your friend starts farther down the path, 0.5 mile from the concession stand. How long will it take you to catch up to your friend? How far will you have rollerbladed in that amount of time? Use your grapher verify your results. (Recall that Distance = Rate · Time.)

4. When warehouse workers use hand trucks to load a boxcar, it costs the management $40 for labor for each boxcar. After management purchases a forklift for $2000, it only costs $15 for labor to load each boxcar. How many boxcars must be loaded before a profit is realized on the purchase of the forklift?

What Have I Learned?

1. What is meant by *a solution to a system of linear equations*? How is a solution represented algebraically? Graphically?

2. Briefly describe three different methods for solving a system of linear equations.

3. Describe a situation in which you would be interested in determining a break-even point. Explain how you would determine the break-even point mathematically. Interpret the break-even point with respect to the situation you describe.

4. Typically, a linear system of equations has one unique solution. Under what conditions is this not the case?

How Can I Practice?

1. Which input value results in the same output value for $y_1 = 2x - 3$ and $y_2 = 5x + 3$?

2. Which point on the line given by $4x - 5y = 20$ is also on the line $y = x + 5$?

3. Solve: $y = 2x + 8$
   $y = -3x + 3$

4. You sell centerpieces for $19.50 each. Your fixed costs are $500 per month, and each centerpiece costs $8 to produce. What is your break-even point?

5. Suppose you wish to break even (in Problem 4) by selling only 25 centerpieces. At what price would you need to sell each centerpiece?

6. Solve: $y = -25x + 250$
   $y = 25x + 300$

7. Use your grapher to solve this system: $4m - 3n = -7$
$$2m + 3n = 37$$

8. Solve: $y = 6x - 7$
$$y = 6x + 4$$

9. Solve: $u = 3v - 17$
$$u = -4v + 11$$

10. Use your grapher to estimate the solution to the following system:

$$342x - 167y = 418$$
$$-162x + 103y = -575$$

11. The formula $A = P + Prt$ represents the value A of an investment of P dollars at a yearly simple interest rate r for t years.

Use a system of equations to determine the time it takes an investment of $100 at 8% to reach the same value as an investment of $120 at 5%.

Gateway Review

1. Determine the slope and vertical intercept of the line whose equation is $4y + 10x - 16 = 0$.

2. Determine the equation of the line through the points $(1, 0)$ and $(-2, 6)$.

3. Estimate the slope of the following line, and use the slope to determine an equation for the line.

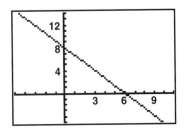

4. Match the graphs with the given equations. Assume that x is the input variable and y is the output variable. Each tick mark on the axes represents one unit.

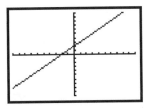 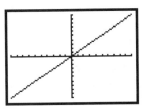 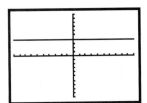

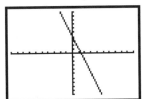 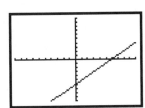 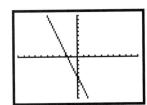

    a. $y = x - 6$ **b.** $y = 4$ **c.** $y = x + 2$

    d. $y = -3x - 5$ **e.** $y = x$ **f.** $y = 4 - 3x$

5. What is the equation of the line passing through the points $(0, 5)$ and $(2, 11)$? Write the final result in slope-intercept form, $y = mx + b$.

6. The equation of a line is $5x - 10y = 20$. Determine the horizontal and vertical intercepts.

7. Determine the horizontal and vertical intercepts of the line whose equation is $-3x + 4y = 12$. Use the intercepts to graph the line.

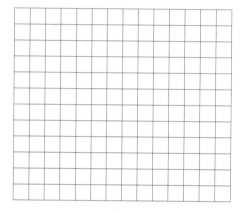

8. Determine the equation of the line that passes through the point $(0, -4)$ and has slope $\frac{5}{3}$.

9. Consider the lines represented by the following pair of equations: $3x + 2y = 10$ and $x - 2y = -2$.

   a. Solve the system algebraically.

   b. Determine the point of intersection of the lines by solving the system graphically.

c. Determine the point of intersection of the lines by solving the system using the tables feature of your calculator, if available.

10. Solve the system of equations:

$$b = 2a + 8$$
$$b = -3a + 3$$

11. Solve the system of equations:

$$x = 10 - 4y$$
$$5x - 2y = 6$$

12. Corresponding values for p and q are given in the following table.

p	4	8	12	16
q	95	90	85	80

a. Write a symbolic rule for q as a linear function of p.

b. Write a symbolic rule for p as a linear function of q.

13. The following table was generated by a linear function. Determine the symbolic rule that defines this function.

x	95	90	85	80	75
y	55.6	58.4	61.2	64.0	66.8

14. ABC Rent-a-Car offers cars at $50 a day and 20 cents a mile. Its competition rents cars for $60 a day and 15 cents a mile.

   a. For each company, write a formula giving the cost $C(x)$ of renting a car for a day as a function of the distance x traveled.

   b. Sketch the graphs of both functions on the same axes.

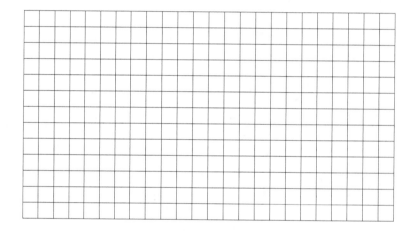

   c. Use the graphs to determine when the two daily costs would be the same.

   d. You intend to make a return trip to Washington, D.C., from New York City. Which company would be cheaper to rent from?

15. Consider a graph of Fahrenheit temperature, °F, as a linear function of Celsius temperature, °C. You know that 212°F and 100°C represent the temperature at which water boils. You also know that 32°F and 0°C each represent the freezing point of water.

   a. What is the slope of the line through the points representing the boiling and freezing points of water?

   b. What is the equation of the line?

c. Use the equation to determine the Fahrenheit temperature corresponding to 40°C.

d. Use the equation to determine the number of degrees Celsius corresponding to 170°F.

16. Your power company currently charges $20.63 per month as a fixed basic service charge plus a variable amount based on the number of kilowatt-hours used. The following table displays several sample monthly usage amounts and the corresponding bill for each.

a. Complete the table of electricity cost, rounding to the nearest cent.

KILOWATT HOURS, k	COST, $C(k)$ ($)
600	84.35
650	89.66
700	
750	100.28
800	105.59
850	
900	
950	121.52
1000	

b. Sketch a graph of the electricity cost function, labeling the axes.

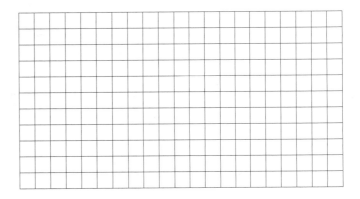

c. Write an equation for your total electric bill $C(k)$ as a function of the number of kilowatt-hours used, k.

d. Determine the total cost if 875 kilowatt-hours of electricity are used.

e. Approximately how much electricity did you use if your total monthly bill is $150?

17. Suppose that the median price P of a home rose from $50,000 in 1970 to $125,000 in 2000. Let t be the number of years since 1970.

   a. Assume that the increase in housing prices has been linear. Determine an equation for the line representing price P in terms of t.

   b. Use the equation from part a to complete the following table, where P represents the price in thousands of dollars.

t	P
0	50
10	
20	
30	125
40	

   c. Interpret the practical meaning of the values in the last line of the table.

18. For tax purposes, you may have to report the value of assets, such as a car. The value of some assets depreciates, or drops, over time. For example, a car originally costing $10,000 may be worth only $5000 a few years later. The simplest way to calculate the value of an asset is by using *straight-line depreciation,* which assumes that the value is a linear function of time. If a $950 refrigerator depreciates completely (to zero value) in 10 years, determine a formula for its value as a function of time.

19. A coach of a local basketball team decides to analyze the relationship between his players' heights and weights.

PLAYER	HEIGHT (in.)	WEIGHT (lb)
1	68	170
2	72	210
3	78	235
4	75	220
5	71	175

a. Use your grapher to determine the equation of the regression line for these data.

b. Player 6 is 80 inches tall. Predict his weight.

c. If a player weighs 190 pounds, how tall would you expect him to be?

CHAPTER 5

Quadratic Functions

Introduction to Quadratic Functions

ACTIVITY 5.1

Newton's Apple

Topic: *Introducing Quadratic Functions*

Your friend takes you to the top of a building to demonstrate a physics property discovered by Sir Isaac Newton. He tosses an apple into the air. The apple's distance from the ground as a function of time is given by

$$s(t) = -16t^2 + 20t + 176,$$

where t is the time in seconds since the apple is released and $s(t)$ is the height (in feet) of the apple above the ground.

1. Complete the following table.

TIME, t (sec)	HEIGHT $s(t)$ (ft)
0	
1	
2	
3	
4	
5	

2. a. Plot the points from the table and draw a smooth curve through the points.

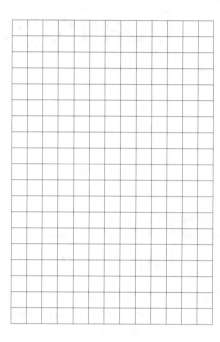

b. Does this graph represent the path of the apple after it is tossed? If not, what does the graph represent?

c. Is the apple falling at a constant rate? Explain.

 3. a. Use your grapher to produce a graph of this function. Use the table of values from Problem 1 to help you choose a window. Compare the graph that you see on your screen with the graph you drew in Problem 2.

b. Is the entire graph that you see on your grapher or the grid in Problem 2 relevant to the situation of tossing an apple off the roof of a building? Explain.

c. Use the trace feature to examine some points on the graph.

4. What is the practical domain for this function?

5. Estimate a practical range for this function. Explain your choice.

6. a. Does Newton's distance function, $s(t) = -16t^2 + 20t + 176$, have a minimum output value in the practical domain?

b. Does $s(t) = -16t^2 + 20t + 176$ have a maximum output value in the practical domain?

 c. Use your grapher to estimate the maximum value in the practical domain.

d. What is the practical meaning of the maximum point in this situation?

7. What is the horizontal intercept of the graph? What does it represent in this situation?

8. What is the vertical intercept of the graph? What does it represent in this situation?

> Any function defined by $f(x) = ax^2 + bx + c$, where a, b, and c are numbers and $a \neq 0$, is a **quadratic function.** The graph of a quadratic function is a U-shaped curve called a **parabola.**

9. The function you have been studying in this activity, $s(t) = -16t^2 + 20t + 176$, is a quadratic function of its input variable t. Identify the values of constants a, b, and c for this function.

E X E R C I S E S

1. Refer to the accompanying graph to answer the following questions.

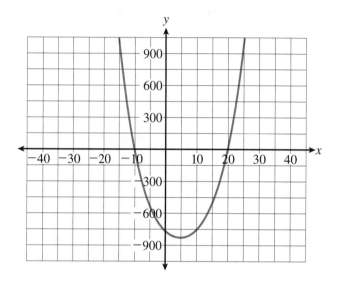

a. Determine the horizontal intercept(s).

b. Determine the vertical intercept.

c. Estimate the maximum and minimum output values on the interval $-10 \leq x \leq 25$.

2. Consider the function defined by $f(x) = x^2 - 2x - 3$. For this exercise, the domain of the function is the set of real numbers from -2 to 4.

a. Complete the table.

x	f(x)
−2	
−1	
0	
1	
2	
3	
4	

b. Plot the data from the table on the following grid and draw a smooth curve through the points. Use your grapher to verify your sketch.

c. What is the range of this function for the given domain?

d. What is the vertical intercept of this function?

e. What is (are) the horizontal intercept(s)?

f. Where (that is, for which set of input values) is the function increasing? Decreasing?

g. What is the significance of the point $(1, -4)$?

3. From their algebraic definitions, determine which of the following functions are quadratic functions. Verify your results by graphing each of the six functions in the standard window of your calculator. What distinctive shape characterizes the quadratic function graphs?

a. $y = x^2 - 3$ **b.** $y = 2x - 3$

c. $y = 4 - 3x$

d. $y = 4x^2 - 3x + 2$

e. $y = 10 - x^2$

f. $y = x^3 - x^2 - 5$

4. The general form of a quadratic function is $y = ax^2 + bx + c$. For each function that you identified as quadratic in Exercise 3, identify the values of the coefficients a, b, and c.

Rockets in Flight

Topics: *Solving Quadratic Equations by Factoring, Zero Product Rule*

Your model rockets are very popular with the children in the neighborhood. To better understand the flight of the rockets, you have found a function that describes their height as a function of time:

$$H(t) = -16t^2 + 400t,$$

where $H(t)$ is the height in feet and t is the number of seconds since launch. Note that the rockets are shot straight up into the air.

1. Complete the following table.

t (sec)	0	5	10	15	20	25
$H(t)$ (ft)						

2. What is the practical domain for this function?

3. Plot the data from the table on the following grid and sketch a smooth curve through the points. Use your grapher to verify your sketch.

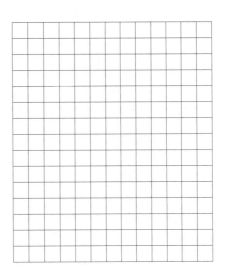

4. **a.** Use the graph from Problem 3 to determine the vertical intercept.

   b. Use the graph from Problem 3 to determine the horizontal intercepts.

   c. What do each of the intercepts of the graph represent in this situation?

d. Estimate the coordinates of the maximum point in the practical domain. What does the maximum point represent in this situation?

5. Complete the following table. What does the table indicate is the greatest (maximum) height the rockets reach? Does the table support your result in Problem 4d?

t	11.0	11.5	12.0	12.5	13.0	13.5	14.0
$H(t)$							

For this particular quadratic function, the horizontal intercepts (the t-intercepts) were easy to determine because they happen to occur at whole-number values. Likewise, the maximum height occurs at a point that is relatively easy to find. In most realistic applications of quadratic functions, however, these important points do not have such simple coordinates. In the rest of this chapter, you will explore several algebraic methods that will enable you to precisely determine any point of interest for any quadratic function. In the remainder of this activity, you will develop an algebraic procedure to determine the horizontal intercepts of a parabola.

6. What is the output value for any function at a horizontal intercept (t-intercept)?

7. Write an equation that can be used to find the t-intercepts for the function $H(t) = -16t^2 + 400t$.

8. There is a very useful principle of algebra that will allow you to solve the equation you wrote in Problem 7. It is known as the **Zero Product Rule.** If you know that the product of two factors is zero, then what must be true about at least one of the factors?

9. To apply the Zero Product Rule here, you must first rewrite the equation $-16t^2 + 400t = 0$ with the non-zero side in *factored form.* To factor the expression $-16t^2 + 400t$, factor out the greatest common factor.

10. Now use the Zero Product Rule to rewrite the factored form of the equation as two distinct equations. That is, set each factor equal to zero. Solve each one separately to obtain the solutions to the original equation $-16t^2 + 400t = 0$.

11. a. What do the solutions in Problem 10 represent graphically?

b. What do the solutions represent in the context of this situation?

THE ZERO PRODUCT RULE

Whenever the product of two numbers is equal to zero, at least one of the two factors must be equal to zero. This is expressed symbolically by:

$$\text{If } ab = 0, \text{ then either } a = 0 \text{ or } b = 0.$$

To use the Zero Product Rule in solving an equation, the equation must be written in a special form:

- One side must be written as a product of two or more factors.

- The other side must be zero.

---

12. To solve the equation $2x^2 + 5x = 13x$ using the Zero Product Rule, proceed as follows.

a. Rewrite the equation as an equivalent equation in which the right side is zero by subtracting $13x$ from both sides. This new form of the equation, with one side equaling zero, is called the **standard form** of the equation.

b. Factor the left side of the standard form equation.

c. Apply the Zero Product Rule to obtain two distinct linear equations, and then solve these two equations.

d. Check your solutions by substituting each one into the *original* equation. Also use your grapher to confirm these solutions.

The procedure outlined in Problem 12 is an algebraic approach that uses factoring to solve an equation of the form $ax^2 + bx + c = 0$. An equation of this form is called a **quadratic equation.**

PROCEDURE TO SOLVE QUADRATIC EQUATIONS BY FACTORING

1. Write the quadratic equation $ax^2 + bx + c = 0$ in standard form, (i.e., with one side zero).

2. Divide each term in the equation by the greatest common numerical factor.

3. Factor the nonzero side.

4. Apply the Zero Product Rule to obtain two linear equations.

5. Solve each linear equation.

6. Check your solutions in the original equation.

EXERCISES

1. Suppose your friend's rocket has a more powerful booster, so it travels faster and higher. The rocket is shot straight upward. Its height is described by the function $H(t) = -16t^2 + 480t$. Recall that t is given in seconds and $H(t)$ in feet.

a. Write an equation that can be used to find the horizontal intercepts of the graph of $H(t)$.

b. Solve the equation obtained in part a by factoring.

c. What do the solutions signify in terms of the rocket's flight?

d. What is the practical domain for this function?

 e. Use the practical domain to help set a window for your grapher. Then graph the function in this window.

f. Use the graph to determine how high this rocket will go, and how long after launch it will reach its maximum height.

2. Use the Zero Product Rule to solve each of the following quadratic equations by factoring. Remember to check your solutions numerically and graphically.

a. $x^2 + 10x = 0$ 　　　　　　　　　**b.** $2x^2 = 25x$

c. $15y^2 + 75y = 0$ 　　　　　　　　**d.** $2(x - 5)(x + 1) = 0$

e. $24t^2 - 36t = 0$ 　　　　　　　　　**f.** $12x = x^2$

g. $2w^2 - 3w = 5w$ 　　　　　　　　**h.** $14p = 3p - 5p^2$

3. Given two solutions, write a possible quadratic equation with those solutions. Represent your equation first in factored form, and then expand into standard form.

a. $x = 1, \ x = 3$ 　　　　　　　　　　**b.** $x = 0, \ x = -4$

c. $x = 5, \ x = -5$ 　　　　　　　　　**d.** $x = 2, \ x = 2$

ACTIVITY 5.3

More Rockets in Flight

Topics: *Solving Quadratic Equations, Factoring Trinomials of the Form* $x^2 + bx + c$

In the preceding activity, you were able to answer several questions about your model rocket's flight by analyzing the function $H(t) = -16t^2 + 400t$. There are additional questions of interest to ask.

1. Suppose you want to know exactly how long it takes for your rocket to reach a height of 1600 feet. Use a guess, check, and repeat approach (or the table feature of a graphing calculator) to solve this problem.

2. There are two instants when the rocket is exactly 1600 feet above the ground. What is the practical meaning of each?

3. You can also estimate the time your rocket is 1600 feet high from the parabola you graphed in Activity 5.2, Problem 3. Draw a horizontal line at $H = 1600$ and estimate the time from the points of intersection. Use your grapher to verify the times.

You were able to estimate the times when the rocket is 1600 feet high by using two different methods: a numerical guess and check and a graphical approach. In this situation, at least one of these methods led to an exact solution. A third approach, based on the Zero Product Rule, will always determine the exact solution.

4. What equation do you need to solve to determine the times when the rocket is 1600 feet high?

5. Recall the procedure you used in the previous activity to solve a quadratic equation by factoring. Here you will follow those same steps. First, rewrite the equation so that it is in standard form (right side zero). Then divide every term in the equation by the greatest common numerical factor so that the coefficient of the t^2 term is $+1$.

The resulting expression on the left does not have a common variable factor. This expression, called a **trinomial** because of its three terms, is of the general form $1x^2 + bx + c$. Here we have replaced the t with the more common variable x, and the b and c represent specific constant values. You will now explore how to factor an expression of the form $x^2 + bx + c$.

6. You already know the solutions to the equation in Problem 5 from your numerical and graphical methods. Use these two solutions to write the factored form of a quadratic equation with these solutions.

7. Multiply the two binomial factors you found in Problem 6 and compare your product with the trinomial expression in Problem 5.

The heart of the solution strategy is the ability to factor a trinomial, $x^2 + bx + c$. You can then apply the Zero Product Rule to complete the solution process. Factoring a quadratic trinomial is the reverse process of multiplying two binomial expressions. (Recall multiplying binomials in Activity 2.17.) For example, examine the multiplication of $x + 4$ by $x + 2$. (You may recall the FOIL method from Chapter 2.)

$$(x + 4)(x + 2) = x^2 + 2x + 4x + 8$$

$$ F \quad O \quad I \quad L$$

$$= x^2 + 6x + 8$$

The product is the trinomial $x^2 + 6x + 8$. Note that the constant term, 8, is the *product* of 4 and 2. The coefficient of the x-term, 6, is the *sum* of 4 and 2. Therefore, the only possible factors of $x^2 + 6x + 8$ are $(x + 4)(x + 2)$, because 4 and 2 is the only integer pair whose product is 8 and whose sum is 6.

> To factor a trinomial of the form $x^2 + bx + c$, determine the integer pair whose product is c (the constant term) and whose sum is b (the coefficient of the x-term). Note that not every trinomial of this form can be factored into binomials with integer constants.

8. To factor the trinomial $x^2 + 3x + 2$, you need to find the integer pair whose product is 2 and whose sum is 3.

a. Determine the factors of 2 whose sum is 3.

b. Use your result from part a to factor $x^2 + 3x + 2$. Check your answer by multiplying the factors.

9. a. In Problem 8, there were not many choices for the factors of 2. Larger numbers generally create more possibilities. For the trinomial $x^2 + 7x + 12$, list all the possible ways to factor 12 into the product of two whole numbers.

b. Which factor pair also adds up to 7? Use those two factors to rewrite the trinomial as the product of two binomials, and check your answer by multiplying.

c. Use the result obtained in part b and the Zero Product Rule to solve $x^2 + 7x + 12 = 0$.

10. If necessary, rewrite the quadratic equation in standard form and then solve by factoring. Remember to check your solutions by substituting back into the original equation.

a. $x^2 + 7x + 10 = 0$ **b.** $x^2 + 13x + 12 = 0$

c. $x^2 + 14x + 24 = 0$ **d.** $x^2 + 25x + 24 = 0$

e. $x^2 + 9x = -20$ **f.** $x^2 + 8x = -7$

You may have noticed that all the coefficients in the standard form quadratic equations of Problem 10 are positive. The rocket height equation in Problem 5 $\left(t^2 - 25t + 100 = 0\right)$ involves a negative coefficient. Such equations can be factored in the same way.

For example, to factor $x^2 - x - 12$, you need to find factors of -12 that sum to -1.

11. a. List all the possible ways to factor -12 into the product of two integers. What must be true of the signs of the two factors?

b. Which factor pair also sums to -1? Use that factor pair to rewrite $x^2 - x - 12$ as the product of two binomials, and check your answer by multiplying.

c. Solve the equation $x^2 - x - 12 = 0$ and check your solution.

12. Solve the following quadratic equations by factoring. Remember to check your solutions by substituting back into the original equation.

a. $x^2 - 2x - 15 = 0$ **b.** $x^2 + 3x - 18 = 0$

c. $2x^2 - 8x - 42 = 0$ **d.** $t^2 - 5t = 24$

e. $x^2 - 7x + 10 = 0$ **f.** $-3t^2 + 36t = 81$

13. Try to solve the quadratic equation $x^2 + 3x - 8 = 0$ by factoring. What do you notice? Solve the equation using a numerical or graphical approach.

E X E R C I S E S

1. Solve the following quadratic equations by factoring. Check your solutions either numerically or graphically.

    a. $x^2 + 7x + 6 = 0$ **b.** $x^2 - 10x - 24 = 0$

    c. $y^2 + 11y = -28$ **d.** $5x^2 - 75x + 180 = 0$

2. Your friend's rocket has a more powerful engine. Its height (in feet) is described by the function $H(t) = -16t^2 + 432t$. Determine the times (in seconds) at which the rocket achieves the following heights. Solve the corresponding equations algebraically. Then check your answer numerically (and graphically if you have a graphing calculator).

    a. When is the rocket 800 feet high?

b. When is the rocket 1760 feet high?

c. When is the rocket 2240 feet high?

d. When is the rocket 2720 feet high?

e. At what times does your friend's rocket touch the ground?

f. Use the symmetric pattern of parts a-e to help you determine the maximum height that the rocket reaches, and when it occurs. Use your grapher to verify your answer.

 3. Solve the following quadratic equations by factoring if possible. Otherwise, find an approximate solution either numerically or graphically.

a. $x^2 + 2x - 5 = 0$ **b.** $x^2 - 6x + 7 = 0$

c. $x^2 - 2x = 15$ **d.** $2x^2 - 6x + 10 = 0$

ACTIVITY 5.4

Maximum Enclosure

Topics: *Quadratic Formula, Axis of Symmetry, Vertex*

Your pet ducks need a pen to protect them from the dogs that run loose in your neighborhood. You find 100 feet of fencing that will be perfect for the job. In planning the construction of the pen, you decide to make it rectangular, and you want the dimensions to give the ducks the largest possible grazing area.

1. Assuming that you will use all 100 feet of fencing, what arithmetic relationship must the length and width satisfy? Make a table with some possible dimensions and calculate the area of the resulting enclosure.

LENGTH (ft)	WIDTH (ft)	AREA (sq ft)
35	15	525
	5	

2. Did you expect that different dimensions would produce different areas? Do you think that the largest area you found in Problem 1 is the best you can do? How can you know for certain?

By describing the situation algebraically, you can determine what the largest possible area is precisely.

3. Let *x* represent the length of the pen. Write an expression in terms of *x* that will describe its width.

4. Use the fact that the area of a rectangle is given by its length times its width to express the area, $A(x)$, as a function of *x*, the length.

5. a. Use your function rule from Problem 4 to complete the following table. What is the practical domain and range?

x, LENGTH (ft)	A(x), AREA OF RECTANGLE (sq ft)
5	
15	
20	
25	
30	
35	
45	

b. According to the data in the table, what dimensions result in the maximum area? Does this agree with your answer in Problem 2?

6. Plot the points or use your grapher to produce the graph of your function. Use the graph to determine the maximum area and the *x*-value that produces it.

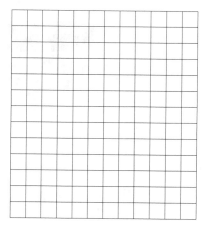

7. You should recognize the shape of the graph of this function. What is the name of this curve?

VERTEX AND AXIS OF SYMMETRY

The turning point on a parabola is called its **vertex.** A vertical line drawn through the vertex cuts the parabola into two halves that are mirror images of each other. This line, which cuts the parabola at its vertex but is not part of the parabola itself, is called the **axis of symmetry** of the parabola.

8. What is an equation of the axis of symmetry for the parabola you graphed in Problem 6?

9. From your table in Problem 5, write down each pair of points that yields the same area. Verify that the axis of symmetry lies exactly halfway between the *x*-values in every pair.

10. Now suppose you need an area of exactly 400 square feet. You want to determine the dimensions of such a rectangle, still using your 100 feet of fencing. From your graph (or table), estimate what you think the dimensions should be.

11. a. What equation must you solve to determine the dimensions of a pen whose area is exactly 400 square feet?

b. Solve the equation by factoring.

12. a. Write an equation to determine the dimensions of the pen whose area is exactly 504 square feet.

b. Solve this equation by factoring.

c. Explain some difficulties that you can encounter in solving quadratic equations by factoring.

13. a. Write an equation to determine the dimensions of a pen whose area is exactly 300 square feet.

b. In attempting to solve this equation by factoring, you must factor 300 into a product of two negative integers whose sum is -50. List all such integer pairs of factors and verify that none add up to -50.

c. Explain some additional difficulties that you can encounter in attempting to solve quadratic equations by factoring.

d. Use your graph to estimate the dimensions that will result in an area of 300 square feet.

The trinomial in Problem 13 was not factorable. Does this mean that you cannot solve the quadratic equation algebraically? No. There is another algebraic method that can be used to solve *any* quadratic equation of the form $ax^2 + bx + c = 0$.

SOLVING QUADRATIC EQUATIONS USING THE QUADRATIC FORMULA

The **quadratic formula** gives the solutions to a quadratic equation that has been written in standard form, $ax^2 + bx + c = 0$. The solutions are

$$x = \frac{-b \pm \sqrt{b^2 - 4ac}}{2a}$$

where a, b, and c are the numerical values of the coefficients in the equation. The $\pm$ symbol is a shorthand notation for two distinct solutions: one obtained by adding the two terms in the numerator, the other by subtracting.

---

14. Verify your solution to the quadratic equation in Problems 11 and 12 using the quadratic formula. Remember to write each equation in standard form, $ax^2 + bx + c = 0$. Identify and record the values of the coefficients a, b, and c for use in the formula.

$$ax^2 + bx + c = 0: \quad a = \underline{\hspace{1cm}} \quad b = \underline{\hspace{1cm}} \quad c = \underline{\hspace{1cm}}$$

$$x = \frac{-b + \sqrt{b^2 - 4ac}}{2a} \qquad x = \frac{-b - \sqrt{b^2 - 4ac}}{2a}$$

15. a. Now return to your equation in Problem 13a. Solve by using the quadratic formula. Use your calculator to determine the decimal approximation of your solutions. Verify the reasonableness of your answers by comparing with your estimate in Problem 13d.

b. Notice that there are two different solutions to the equation. How do these solutions relate to the axis of symmetry of the parabola?

If you separate the quadratic formula into two terms,

$$x = \frac{-b}{2a} \pm \frac{\sqrt{b^2 - 4ac}}{2a},$$

the first term will always be a number that is midway between the two solutions. Hence, it gives you the location of the axis of symmetry.

The axis of symmetry of the parabola whose equation is $y = f(x) = ax^2 + bx + c$ can always be found in the first term of the quadratic formula: $x = \frac{-b}{2a}$. Because the vertex of the parabola lies on its axis of symmetry, the x-coordinate of the vertex is also $x = \frac{-b}{2a}$. The y-coordinate of the vertex can be calculated by evaluating the function at this value. That is, the y-coordinate of the vertex is given by $f\left(\frac{-b}{2a}\right)$.

16. a. Determine the vertex of the parabola defined by the area function $A(x)$ in Problem 4.

b. Compare the axis of symmetry defined by the x-coordinate of the vertex with your result in Problem 8.

c. Compare the y-coordinate of the vertex with the maximum area you calculated in Problem 5b.

E X E R C I S E S

Algebraically determine the axis of symmetry and the coordinates of the vertex for each parabola in Exercises 1–8. Calculate the x-intercepts of each graph by setting $y = 0$ and solving the resulting equation using the quadratic formula. Confirm that the axis of symmetry lies midway between the x-intercepts. (Use your grapher to check your results graphically.)

1. $y = x^2 + 2x - 15$ **2.** $y = 4x^2 + 32x + 15$

3. $y = -2x^2 + x + 1$ **4.** $y = x^2 + x - 3$

5. $y = -x^2 + 10x + 9$

6. $y = 3x^2 - 2x - 1$

7. $y = 5x^2 + 10x + 5$

8. $y = x^2 + 2x + 2$

9. You may have noticed that some parabolas open upward (the vertex is at the bottom of the curve) and some parabolas open downward (the vertex is at the top of the curve). Examine the parabolas in Exercises 1–8 and their respective equations, specifically the coefficients. What characteristic of the equation determines whether a parabola opens upward or downward? Test your conjecture with a few quadratic functions of your own choosing.

10. In Exercises 1–8, you saw that a quadratic equation can have two solutions, only one solution, or possibly no solutions. What is the connection between the number of solutions of a quadratic equation, $ax^2 + bx + c = 0$, and the graph of the corresponding parabola, $y = ax^2 + bx + c$?

11. A ball is thrown up in the air. Its height above the ground t seconds later is described by the quadratic function $H(t) = -16t^2 + 48t + 6$.

   a. How high does the ball go, and exactly when does it reach this maximum height?

   b. When does the ball hit the ground?

12. Create a quadratic equation that has two solutions, but that cannot be solved by factoring.

ACTIVITY 5.5

Roots in the Garden

Topics: *Pythagorean Theorem, Solving Quadratic Equations by Using Square Roots*

Recall, from Chapter 1, the Pythagorean theorem: In any right triangle, where a and b are the lengths of the legs and c is the length of the hypotenuse, $c^2 = a^2 + b^2$.

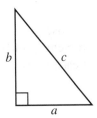

1. Given a right triangle with legs 5 inches and 12 inches, determine the hypotenuse.

To determine the hypotenuse c of the given triangle, you had to take the square root of both sides of the equation. Taking the square root is one way to solve some quadratic equations.

2. **a.** Is the solution you found in Problem 1 the only solution to the quadratic equation, taken out of context? (*Hint:* How many solutions can a quadratic equation have?)

   b. What must you do when taking a square root to be sure that you obtain all the solutions to the quadratic equation?

   c. Are both solutions to the quadratic equation also solutions to Problem 1? Explain.

Suppose you have a 10-by-10-foot square garden plot. This year you want to increase the area of your plot and still keep its square shape.

3. Use the inner square of the sketch to represent your existing plot and the known dimensions. Use the variable x to represent the increase in length to each side. Write an expression for the length of a side of the larger square.

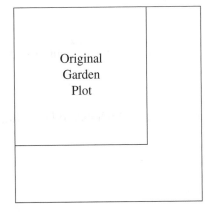

Original
Garden
Plot

4. Express the area A of your new larger plot as the square of a binomial in function notation.

5. Suppose you decide to double the original area. Use the area expression in Problem 4 to write an equation to determine the required increase in the length of each side.

6. Solve for x by first taking the square root of both sides of your equation.

7. What is the length of each side of your new garden, to the nearest tenth of a foot? Check your answer by calculating the area of this new plot.

8. a. Use the quadratic formula to solve the quadratic equation in Problem 5. Remember to write the equation in the standard form $ax^2 + bx + c = 0$ by expanding the squared term.

b. How does your answer compare to the answer in Problem 6?

c. Which method of solution is more efficient for this situation?

9. a. Use your grapher to graph the area function of Problem 4 along with the constant function for the required area, $y_2 = 200$. Determine the intersection of the parabola and the horizontal line.

b. Which point of intersection of the parabola and the line represents the relevant solution to the garden problem? Explain.

c. How does the result from part b compare to the result from Problem 7?

10. Write the quadratic equation from Problem 5 in standard form and graph. Where on the graph do you find the solution to your equation? Verify that your graphical solution is the same as the solution obtained by the quadratic formula in Problem 8.

EXERCISES

In Exercises 1–8, solve the quadratic equations by the square root method and at least one other way. Check your solutions numerically.

1. $x^2 = 4$

2. $x^2 = 36$

3. $x^2 = -25$

4. $(x - 4)^2 = 49$

5. $(x + 5)^2 = 64$

6. $(x - 7)^2 = 37$

7. $(2x - 3)^2 = 81$

8. $(x - 3)^2 = -5$

9. A 16-foot-long ladder is leaning against a wall, so that the bottom of the ladder is 5 feet from the wall, measuring along the floor (see diagram). You want to determine how much farther the bottom of the ladder must be moved away from the wall so that the top of the ladder is exactly 10 feet above the floor.

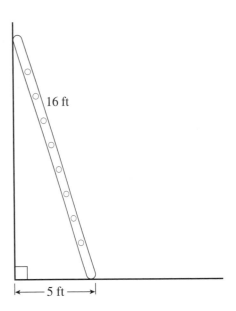

   a. Use the Pythagorean theorem to write a quadratic equation in terms of x for this situation.

   b. Do not multiply out the squared binomial expression in x but keep it on the left-hand side. Combine the constant numbers on the right side of the equation.

c. Solve for x using the square root method.

10. You change your mind about increasing the size of your garden and wish to decrease it to only 50 square feet (from the original 10-by-10-foot plot in the activity). To keep your plot in a square shape, by how much must you decrease the length of the original side? (*Hint:* Use the area function from Problem 4.)

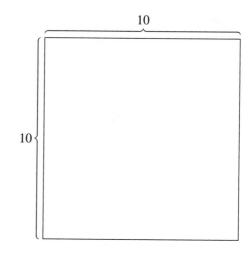

What Have I Learned?

1. List the methods commonly used to solve quadratic equations. (You have used five such methods in this chapter.) Briefly describe each method, highlighting the things you want to be sure to remember.

2. Which algebraic method of solving quadratic equations will *always* work? Explain.

3. What is the graph of a quadratic function called?

4. Draw a parabola in a Cartesian coordinate system, and label the following features (if they apply).

   a. Minimum point

   b. Maximum point

   c. Vertical intercept

   d. Horizontal intercept(s)

   e. Axis of symmetry

   f. Vertex

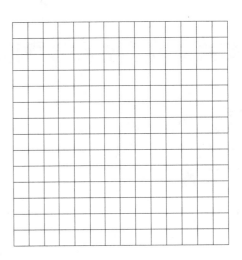

5. Does the graph of every parabola have *x*-intercepts? Explain.

6. How many vertical intercepts are possible for a quadratic function?

7. a. What does the formula $x = -\frac{b}{2a}$ represent with respect to a quadratic function and its associated parabola?

   b. Where do you obtain the values of a and b?

8. The vertex of a parabola is the point $(3, 1)$. Use the symmetry of a parabola to complete the following table.

x	1	2	3	4	5
y	5	2	1		

How Can I Practice?

1. Complete the following table by determining if the parabola opens upward or downward, identifying the values of a, b, c, and the y-intercept, and writing the equation of the axis of symmetry.

QUADRATIC FUNCTION	a	b	c	UP OR DOWN	y-INTERCEPT	AXIS OF SYMMETRY
$y = x^2$						
$f(x) = -x^2 - 2x$						
$H(t) = -16t^2 + 400t$						
$s(t) = -16t^2 + 20t + 176$						
$y = 2x^2 + 5$						

2. For each of the quadratic functions, $f(x)$, $g(x)$, $h(x)$, $p(x)$, $q(x)$, and $r(x)$, determine the direction in which the parabola opens, the axis of symmetry, and the vertical and horizontal intercepts. Then use this information to match each function with its corresponding graph (a–f). Indicate your answer by writing the function name on the line provided. Verify, using your grapher if you have access to one.

$$f(x) = x^2 \qquad\qquad h(x) = -x^2 \qquad\qquad r(x) = x^2 + 4x + 4$$

$$g(x) = x^2 - 8x \qquad\qquad p(x) = -x^2 + 3x \qquad\qquad q(x) = x^2 + 3$$

a.   b. 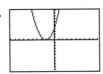   c.

d. 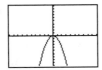   e.   f.

a. _____   b. _____   c. _____

d. _____   e. _____   f. _____

3. **a.** Use at least two algebraic methods to solve $(x - 3)^2 = 4$.

   b. Which method do you prefer? Explain.

4. **a.** How do you determine the axis of symmetry of the parabola whose equation is $f(x) = ax^2 + bx + c$?

   b. How do you use the knowledge of the axis of symmetry to calculate the y-coordinate of the vertex?

   c. What is the axis of symmetry of the graph of the function defined by $f(x) = -x^2 - 6x - 4$?

d. Determine the coordinates of the vertex of the function in part c.

5. For each of the following quadratic functions, determine the direction in which the parabola opens, the axis of symmetry, and the vertical and horizontal intercepts. Then use this information to sketch the functions by hand. Use your calculator to verify your graph.

   a. $y = x^2 + 3$ **b.** $y = -x^2 - 4$

   c. $y = 2x^2 + 6x$ **d.** $y = x^2 - 6x + 9$

6. Solve the following equations by factoring, if possible. Use your grapher to verify your solutions.

   a. $x^2 - 9x + 20 = 0$ **b.** $x^2 - 9x + 14 = 0$

   c. $x^2 - 11x = -24$ **d.** $m^2 + m = 6$

   e. $2x^2 - 4x - 6 = 0$ **f.** $3t^2 + 21t - 18 = 0$

g. $-7x + 10 = -x^2$

h. $-9a - 12 = -3a^2$

i. $2x^2 + 3x = 0$

j. $2x^2 = 12x$

k. $v^2 + 8v = 0$

7. Use the quadratic formula to solve the following equations. Use your grapher to check your solutions.

a. $x^2 = -5x + 6$

b. $6x^2 + x = 15$

c. $x^2 - 7x - 18 = 0$

d. $3x^2 - 8x + 4 = 0$

e. $4x^2 + 28x - 32 = 0$ f. $2x^2 + 4x - 3 = 0$

g. $x^2 + 7 = 6x$ h. $x^2 + 6x = -6$

i. $x^2 + 2 = 4x$ j. $2x - 6 = -x^2$

Fractions

Proper and Improper Fractions

A fraction in the form $\frac{a}{b}$ is called **proper** if a and b are counting numbers and a is less than b $(a < b)$. If a is greater than or equal to b $(a \geq b)$ then $\frac{a}{b}$ is called **improper**.

NOTE: a can be zero but b cannot, since you may not divide *by* zero.

Example 1: $\frac{2}{3}$ is proper since $2 < 3$.

Example 2: $\frac{7}{5}$ is improper since $7 > 5$.

Example 3: $\frac{0}{4} = 0$.

Example 4: $\frac{8}{0}$ has no numerical value and, therefore, is said to be **undefined.**

Reducing a Fraction

A fraction, $\frac{a}{b}$, is in **lowest terms** if a and b have no common divisor (other than 1).

Example 1: $\frac{3}{4}$ is in lowest terms. The only common divisor is 1.

Example 2: $\frac{9}{15}$ is not in lowest terms. 3 is a common divisor of 9 and 15.

To **reduce a fraction $\frac{a}{b}$ to lowest terms,** divide both a and b by a common divisor until the numerator and denominator no longer have any common divisors.

Example 1: Reduce $\frac{4}{8}$ to lowest terms.

Solution: $\dfrac{4}{8} = \dfrac{4 \div 4}{8 \div 4} = \dfrac{1}{2}$

Example 2: Reduce $\frac{18}{14}$ to lowest terms.

Solution: $\dfrac{18}{14} = \dfrac{18 \div 2}{14 \div 2} = \dfrac{9}{7}$

Mixed Numbers

A **mixed number** is the sum of a whole number plus a proper fraction.

Example: $3\frac{2}{5} = 3 + \frac{2}{5}$

Changing an Improper Fraction, $\frac{a}{b}$, to a Mixed Number

1. Divide the numerator, a, by the denominator, b.
2. The quotient becomes the whole number part of the mixed number.
3. The remainder becomes the numerator, and b remains the denominator, of the fractional part of the mixed number.

Example 1: Change $\frac{7}{5}$ to a mixed number.

Solution: $7 \div 5 = 1$ with a remainder of 2.

So $\frac{7}{5} = 1 + \frac{2}{5} = 1\frac{2}{5}$

Example 2: Change $\frac{22}{6}$ to a mixed number.

Solution: First reduce $\frac{22}{6}$ to lowest terms.

$$\frac{22}{6} = \frac{22 \div 2}{6 \div 2} = \frac{11}{3}.$$

$11 \div 3 = 3$ with a remainder of 2.

So $\frac{11}{3} = 3 + \frac{2}{3} = 3\frac{2}{3}$

Changing a Mixed Number to an Improper Fraction

1. To obtain the numerator of the improper fraction, multiply the denominator by the whole number and add the original numerator to this product.
2. Place this sum over the original denominator.

Example: Change $4\frac{5}{6}$ to an improper fraction.

Solution: $6 \cdot 4 + 5 = 29$

So $4\frac{5}{6} = \frac{29}{6}$

E X E R C I S E S

1. Reduce these fractions, if possible.

   a. $\frac{12}{16}$ **b.** $\frac{33}{11}$ **c.** $\frac{21}{49}$ **d.** $\frac{13}{31}$

2. Change each improper fraction to a mixed number. Reduce if possible.

   a. $\frac{20}{16}$ **b.** $\frac{34}{8}$ **c.** $\frac{48}{12}$ **d.** $\frac{33}{5}$

3. Change each mixed number to an improper fraction.

   a. $7\frac{2}{5}$ **b.** $8\frac{6}{10}$ **c.** $9\frac{6}{7}$ **d.** $11\frac{3}{4}$

Finding a Common Denominator of Two Fractions

A **common denominator** is a number that is divisible by each of the original denominators.

The **least common denominator** is the smallest possible common denominator.

A common denominator can be obtained by multiplying the original denominators. Note that this product may not necessarily be the least common denominator.

Example: Find a common denominator of $\frac{5}{6}$ and $\frac{7}{9}$.

Solution: $6 \cdot 9 = 54$, so 54 is a common denominator.
However, 18 is the least common denominator.

Equivalent Fractions

Equivalent fractions are fractions with the same numerical value.

To obtain an equivalent fraction, multiply or divide both numerator and denominator by the same nonzero number.

NOTE: Adding or subtracting the same number to the original numerator and denominator does *not* yield an equivalent fraction.

Example: Find a fraction that is equivalent to $\frac{3}{5}$ whose denominator is 20.

Solution: $\dfrac{3}{5} = \dfrac{3 \cdot 4}{5 \cdot 4} = \dfrac{12}{20}$

NOTE: $\dfrac{3 + 15}{5 + 15} = \dfrac{18}{20} = \dfrac{9}{10} \neq \dfrac{3}{5}$

Comparing Fractions: Determining Which is Larger. Is $\frac{a}{b} < \frac{c}{d}$ or is $\frac{a}{b} > \frac{c}{d}$?

1. Obtain a common denominator by multiplying b and d.
2. Write $\frac{a}{b}$ and $\frac{c}{d}$ as equivalent fractions, each with denominator $b \cdot d$.
3. Compare numerators. The fraction with the larger (smaller) numerator is the larger (smaller) fraction.

Example: Determine whether $\frac{3}{5}$ is less than or greater than $\frac{7}{12}$.

Solution: A common denominator is $5 \cdot 12 = 60$.

$$\frac{3}{5} = \frac{3 \cdot 12}{5 \cdot 12} = \frac{36}{60}, \qquad \frac{7}{12} = \frac{7 \cdot 5}{12 \cdot 5} = \frac{35}{60}$$

Since $36 > 35$, $\qquad \frac{3}{5} > \frac{7}{12}$

EXERCISES

Compare the two fractions and indicate which one is larger.

1. $\frac{4}{7}$ and $\frac{5}{8}$ **2.** $\frac{11}{13}$ and $\frac{22}{39}$ **3.** $\frac{5}{12}$ and $\frac{7}{16}$ **4.** $\frac{3}{5}$ and $\frac{14}{20}$

Addition and Subtraction of Fractions with the Same Denominators

To add (or subtract) $\frac{a}{c}$ and $\frac{b}{c}$:

1. Add (or subtract) the numerators, a and b.

2. Place the sum (or difference) over the common denominator, c.

3. Reduce to lowest terms.

Example 1: Add $\frac{5}{16} + \frac{7}{16}$

Solution: $\dfrac{5}{16} + \dfrac{7}{16} = \dfrac{5+7}{16} = \dfrac{12}{16} \qquad \dfrac{12}{16} = \dfrac{12 \div 4}{16 \div 4} = \dfrac{3}{4}$

Example 2: Subtract $\frac{19}{24} - \frac{7}{24}$

Solution: $\dfrac{19}{24} - \dfrac{7}{24} = \dfrac{19-7}{24} = \dfrac{12}{24} \qquad \dfrac{12}{24} = \dfrac{12 \div 12}{24 \div 12} = \dfrac{1}{2}$

Addition and Subtraction of Fractions with Different Denominators

To add (or subtract) $\frac{a}{b}$ and $\frac{c}{d}$, where $b \neq d$:

1. Obtain a common denominator by multiplying b and d.

2. Write $\frac{a}{b}$ and $\frac{c}{d}$ as equivalent fractions with denominator $b \cdot d$.

3. Add (or subtract) the numerators of the equivalent fractions and place the sum (or difference) over the common denominator, $b \cdot d$.

4. Reduce to lowest terms.

Example 1: Add $\frac{3}{5} + \frac{2}{7}$

Solution: A common denominator is, $5 \cdot 7 = 35$

$$\frac{3}{5} = \frac{3 \cdot 7}{5 \cdot 7} = \frac{21}{35}, \qquad \frac{2}{7} = \frac{2 \cdot 5}{7 \cdot 5} = \frac{10}{35}$$

$$\frac{3}{5} + \frac{2}{7} = \frac{21}{35} + \frac{10}{35} = \frac{31}{35}$$

$\frac{31}{35}$ is already in lowest terms.

Example 2: Subtract $\frac{5}{12} - \frac{2}{9}$

Solution: Using common denominator, $12 \cdot 9 = 108$:   Using least common denominator, 36:

$$\frac{5}{12} = \frac{5 \cdot 9}{12 \cdot 9} = \frac{45}{108}, \qquad\qquad \frac{5}{12} = \frac{5 \cdot 3}{12 \cdot 3} = \frac{15}{36},$$

$$\frac{2}{9} = \frac{2 \cdot 12}{9 \cdot 12} = \frac{24}{108} \qquad\qquad \frac{2}{9} = \frac{2 \cdot 4}{9 \cdot 4} = \frac{8}{36}$$

$$\frac{5}{12} - \frac{2}{9} = \frac{45}{108} - \frac{24}{108} = \frac{21}{108} \qquad \frac{5}{12} - \frac{2}{9} = \frac{15}{36} - \frac{8}{36} = \frac{7}{36}$$

$$\frac{21}{108} = \frac{21 \div 3}{108 \div 3} = \frac{7}{36}$$

Addition and Subtraction of Mixed Numbers

To add (or subtract) mixed numbers:

1. Add (or subtract) the whole number parts.
2. Add (or subtract) the fractional parts. In subtraction, this may require borrowing.
3. Add the resulting whole number and fractional parts to form the mixed number sum (or difference).

Example 1: Add $3\frac{2}{5} + 5\frac{3}{4}$

Solution: Whole number sum: $3 + 5 = 8$

Fractional sum:

$$\frac{2}{3} + \frac{3}{4} = \frac{2 \cdot 4}{3 \cdot 4} + \frac{3 \cdot 3}{4 \cdot 3} = \frac{8}{12} + \frac{9}{12} = \frac{17}{12} = 1\frac{5}{12}$$

Final result: $8 + 1\frac{5}{12} = 8 + 1 + \frac{5}{12} = 9\frac{5}{12}$

Example 2: Subtract $4\frac{1}{3} - 1\frac{7}{8}$

Solution: Because $\frac{1}{3}$ is smaller than $\frac{7}{8}$, you must borrow as follows:

$$4\frac{1}{3} = 3 + 1 + \frac{1}{3} = 3 + 1\frac{1}{3} = 3\frac{4}{3}$$

The original subtraction now becomes $3\frac{4}{3} - 1\frac{7}{8}$

Subtracting the whole number parts: $3 - 1 = 2$

Subtracting the fractional parts:

$$\frac{4}{3} - \frac{7}{8} = \frac{4 \cdot 8}{3 \cdot 8} - \frac{7 \cdot 3}{8 \cdot 3} = \frac{32}{24} - \frac{21}{24} = \frac{11}{24}$$

Final result is $2 + \frac{11}{24} = 2\frac{11}{24}$

EXERCISES

Add or subtract as indicated.

1. $1\frac{3}{4} + 3\frac{1}{8}$ 2. $8\frac{2}{3} - 7\frac{1}{4}$ 3. $6\frac{5}{6} + 3\frac{11}{18}$ 4. $2\frac{1}{3} - \frac{4}{5}$

5. $4\frac{3}{5} + 2\frac{3}{8}$ 6. $12\frac{1}{3} + 8\frac{7}{10}$ 7. $14\frac{3}{4} - 5\frac{7}{8}$ 8. $6\frac{5}{8} + 9\frac{7}{12}$

Multiplying Fractions

To multiply fractions, $\frac{a}{b} \cdot \frac{c}{d}$:

1. Multiply the numerators, $a \cdot c$, to form the numerator of the product fraction.
2. Multiply the denominators, $b \cdot d$, to form the denominator of the product fraction.
3. Write the product fraction by placing $a \cdot c$ over $b \cdot d$.
4. Reduce to lowest terms.

Example: Multiply $\frac{3}{4} \cdot \frac{8}{15}$.

Solution: $\frac{3}{4} \cdot \frac{8}{15} = \frac{3 \cdot 8}{4 \cdot 15} = \frac{24}{60} \qquad \frac{24}{60} = \frac{24 \div 12}{60 \div 12} = \frac{2}{5}$

NOTE: It is often simpler and more efficient to cancel any common factors of the numerators and denominators *before* multiplying.

Dividing by a Fraction

Dividing *by* a fraction, $\frac{c}{d}$, is equivalent to multiplying by its reciprocal, $\frac{d}{c}$.

To divide fraction $\frac{a}{b}$ by $\frac{c}{d}$, written $\frac{a}{b} \div \frac{c}{d}$:

1. Rewrite the division as an equivalent multiplication, $\frac{a}{b} \cdot \frac{d}{c}$:
2. Proceed by multiplying as described above.

Example: Divide $\frac{2}{3}$ by $\frac{1}{2}$.

Solution: $\frac{2}{3} \div \frac{1}{2} = \frac{2}{3} \cdot \frac{2}{1} = \frac{2 \cdot 2}{3 \cdot 1} = \frac{4}{3}$

NOTE: $\frac{4}{3}$ is already in lowest terms.

EXERCISES

Multiply or divide as indicated.

1. $\frac{11}{14} \cdot \frac{4}{5}$ 2. $\frac{3}{7} \div \frac{3}{5}$ 3. $\frac{8}{15} \cdot \frac{3}{4}$ 4. $6 \div \frac{2}{5}$

Multiplying or Dividing Mixed Numbers

To multiply (or divide) mixed numbers:

1. Change the mixed numbers to improper fractions.
2. Multiply (or divide) the fractions as described earlier.
3. If the result is an improper fraction, change to a mixed number.

Example: Divide $8\frac{2}{5}$ by 3.

Solution: $8\frac{2}{5} = \frac{42}{5}, \quad 3 = \frac{3}{1}$

$\frac{42}{5} \div \frac{3}{1} = \frac{42}{5} \cdot \frac{1}{3} = \frac{42}{15}$

$\frac{42}{15} = \frac{42 \div 3}{15 \div 3} = \frac{14}{5}$

$\frac{14}{5} = 2\frac{4}{5}$

EXERCISES

Multiply or divide as indicated.

1. $3\frac{1}{2} \cdot 4\frac{3}{4}$ 2. $10\frac{3}{5} \div 4\frac{2}{3}$ 3. $5\frac{4}{5} \cdot 6$ 4. $4\frac{3}{10} \div \frac{2}{5}$

Decimals

Reading and Writing Decimal Numbers

Decimal numbers are written numerically according to a place value system.

The following table lists the place values of digits to the *left* of the decimal point.

hundred million	ten million	million,	hundred thousand	ten thousand	thousand,	hundred	ten	one	decimal point

The following table lists the place values of digits to the *right* of the decimal point.

decimal point	tenths	hundreths	thousandths	ten thousandths	hundred thousandths	millionths

To read or write a decimal number in words:

1. Use the first place value table to read the digits to left of the decimal point.

2. Insert the word "and."

3. Read the digits to the right of the decimal point as though they were not preceded by a decimal point, and then attach the place value of its rightmost digit:

Example 1: Read and write the number 37,568.0218 in words.

			3	7	5	6	8	•
hundred million	ten million	million,	hundred thousand	ten thousand	thousand,	hundred	ten	one	decimal point

and

•	0	2	1	8		
decimal point	tenths	hundredths	thousandths	ten thousandths	hundred thousandths	millionths

Therefore, 37,568.0218 is read "thirty-seven thousand five hundred sixty-eight" "and" "two hundred eighteen" "ten-thousandths."

Example 2: Write the number "seven hundred eighty two million ninety three thousand five hundred ninety four and two thousand four hundred three millionths" numerically in standard form.

7	8	2,	0	9	3,	5	9	4	•
hundred million	ten million	million,	hundred thousand	ten thousand	thousand,	hundred	ten	one	*decimal point*

"and"

•	0	0	2	4	0	3
decimal point	tenths	hundredths	thousandths	ten thousandths	hundred thousandths	millionths

That is, 782,093,594.002403

EXERCISES

1. Write the number 9,467.00624 in words.

2. Write the number 35,454,666.007 in words.

3. Write the number numerically in standard form: four million sixty-four and seventy two ten thousandths.

4. Write the number numerically in standard form: seven and forty three thousand fifty two millionths.

Rounding a Number to a Specified Place Value

1. Locate the digit with the specified place value.

2. If the digit directly to its right is less than 5, keep the digit in step 1 and delete all of the digits to its right.

3. If the digit directly to its right is 5 or above, increase the digit in step 1 by 1 and delete all of the digits to its right.

4. If the specified place value is greater than one (i.e, to the left of the decimal point) proceed as instructed in step 2 or 3, but insert trailing zeros to the right as placeholders, and then drop the decimal point.

Example 1: Round 35,178.2649 to the nearest hundredth.

The digit in the "hundredths" place is 6. The digit to its right is 4. Therefore, keep the 6 and delete the digits to its right. The rounded value is 35,178.26

Example 2: Round 35,178.2649 to the nearest tenth.

The digit in the "tenths" place is 2. The digit to its right is 6. Therefore, increase the 2 to 3 and delete the digits to its right. The rounded value is 35,178.3

Example 3: Round 35,178.2649 to the nearest ten thousand.

The digit in the "ten thousand" place is 3. The digit to its right is 5. Therefore, increase the 3 to 4 and insert trailing zeros to its right as placeholders. The rounded value is 40,000

Note that the decimal point is not written.

EXERCISES

1. Round 7,456.975 to the nearest hundredth.

2. Round 55,568.2 to the nearest hundred.

3. Round 34.6378 to the nearest tenth.

Converting a Fraction to a Decimal

To convert a fraction to a decimal, divide the numerator by the denominator.

Example 1: Convert $\frac{4}{5}$ to a decimal.

Solution:

On a calculator:

Key in ④ ÷ ⑤ = to obtain 0.8

Using long division:
$$\begin{array}{r} 0.8 \\ 5\overline{)4.0} \\ \underline{4.0} \\ 0 \end{array}$$

Example 2: Convert $\frac{1}{3}$ to a decimal.

Solution:

On a calculator:

Key in ① ÷ ③ =
to obtain 0.3333333

Using long division:
$$\begin{array}{r} 0.333 \\ 3\overline{)1.000} \\ \underline{-9} \\ 10 \\ \underline{-9} \\ 1 \end{array}$$

Since a calculator's display is limited to a specified number of digits, it will cut off the decimal's trailing right digits.

Since this long division process will continue indefinitely, the quotient is a repeating decimal, 0.33333 . . . and is instead denoted by $0.\overline{3}$. The bar is placed above the repeating digit or above a repeating sequence of digits.

Convert the given fractions into decimals.

1. $\frac{3}{5}$ **2.** $\frac{2}{3}$ **3.** $\frac{7}{8}$ **4.** $\frac{1}{7}$ **5.** $\frac{4}{9}$

Converting a Terminating Decimal to a Fraction

1. Read the decimal.

2. The place value of the right-most nonzero digit becomes the denominator of the fraction.

3. The original numeral, with the decimal point removed, becomes the numerator. Drop all leading zeros.

Example 1: Convert 0.025 to a fraction.

Solution: The right-most digit, 5, is in the "thousandths" place.

So, as a fraction, $0.025 = \frac{25}{1000}$, which reduces to $\frac{1}{40}$

Example 2: Convert 0.0034 to a fraction.

Solution: The right-most digit, 4, is in the ten-thousandths place.

So, as a fraction, $0.034 = \frac{34}{10,000}$, which reduces to $\frac{17}{5000}$

E X E R C I S E S

Convert the given decimals to fractions.

1. 0.4   **2.** 0.125   **3.** 0.64   **4.** 0.05

Converting a Decimal to a Percent

Multiply the decimal by 100. That is, move the decimal point two places to the right and attach the percent symbol.

Example 1: 0.78 written as a percent is 78%.

Example 2: 3 written as a percent is 300%

Example 3: 0.045 written as a percent is 4.5%

E X E R C I S E S

Write the following decimals as percents.

1. 0.35   **2.** 0.076   **3.** 0.0089   **4.** 6.0

Converting a Percent to a Decimal

Divide the percent by 100. That is, move the decimal point two places to the left and drop the % symbol.

Example 1: 5% written as a decimal is 0.05

Example 2: 625% written as a decimal is 6.25

Example 3: 0.0005% written as a decimal is 0.000005

E X E R C I S E S

Write the following percents as decimals.

1. 45%   **2.** 0.0987%   **3.** 3.45%   **4.** 2000%

Comparing Decimals

1. Write the decimals one below the next, lining up their respective decimal points.
2. Read the decimals from left to right, comparing corresponding place values. The decimal with the first and largest nonzero digit is the largest number.

Example 1: Order from largest to smallest: 0.097,   0.48,   0.0356

Solution: Align by decimal point: 0.097
                                                  0.48
                                                  0.0356

Since 4 (in the second decimal) is the first nonzero digit, 0.48 is the largest number. Next, since 9 is larger than 3, 0.097 is next largest; and finally, 0.0356 is the smallest number.

Example 2: Order from largest to smallest: 0.043, 0.0043, 0.43, 0.00043

Solution: Align by decimal point: 0.043
                                                  0.0043
                                                  0.43
                                                  0.00043

Since 4 (in the third decimal), is the first nonzero digit, then 0.43 is the largest number. Similarly, next is 0.043; followed by 0.0043; and finally, 0.00043 is the smallest number.

E X E R C I S E S

Place each group of decimals in order, from largest to smallest.

1. 0.058, 0.0099, 0.105, 0.02999

2. 0.75, 1.23, 1.2323, 0.9, 0.999

3. 13.56, 13.568, 13.5068, 13.56666

Adding and Subtracting Decimals

1. Write the decimals one below the next, lining up their respective decimal points. If the decimals have differing numbers of digits to the right of the decimal point, place trailing zeros to the right in the shorter decimals.

2. Place the decimal point in the answer.

3. Add or subtract the numbers as usual.

Example 1: Add: 23.5 + 37.098 + 432.17

Solution:
```
     23.500
     37.098
 + 432.170
   _____
   492.768
```

Example 2: Subtract 72.082 from 103.07

Solution:
```
   103.070
 −  72.082
   _____
    30.988
```

EXERCISES

1. Calculate 543.785 + 43.12 + 3200.0043

2. Calculate 679.05 − 54.9973

Multiplying Decimals

1. Multiply the numbers as usual, ignoring the decimal points.

2. Sum the number of decimal places in each number; your product will contain the sum of their decimal places.

Example 1: Multiply 32.89 by 0.021

Solution:
```
    32.89      →    2 decimal places
  × 0.021      →    3 decimal places
  _____
    3289
    6578
  _____
  0.69069      →    5 decimal places
```

Example 2: Multiply 64.05 by 7.3

Solution:
```
    64.05      →    2 decimal places
  ×   7.3      →    1 decimal place
  _____
   19215
   44835
  _____
  467.565      →    3 decimal places
```

EXERCISES

Multiply the following decimals:

1. 12.53×8.2 **2.** 115.3×0.003 **3.** 14.62×0.75

Dividing Decimals

1. Write the division in long division format.

2. Move the decimal point the same number of places to the right in both divisor and dividend so that the divisor becomes a whole number.

3. Place the decimal point in the quotient directly above the decimal point in the dividend and divide as usual.

Example 1: Divide   92.4   by   0.25
                        (dividend)   *(divisor)*

$0.25\overline{)92.4}$ becomes $25\overline{)9240}$

$$
\begin{array}{r}
369.6 \\
25\overline{)9240.0} \\
-75 \\
\hline
174 \\
-150 \\
\hline
240 \\
-225 \\
\hline
150 \\
-150 \\
\hline
0
\end{array}
$$

Example 2: $0.0052 \div 0.004$
                (dividend) *(divisor)*

$0.004\overline{)0.00052}$ becomes $4\overline{)0.52}$

$$
\begin{array}{r}
.13 \\
4\overline{)0.52} \\
-4 \\
\hline
12 \\
-12 \\
\hline
0
\end{array}
$$

EXERCISES

1. Divide 12.05 by 2.5

2. Divide 18.9973 by 78

3. Divide 14.05 by 0.0002

4. Calculate $150 \div 0.03$

5. Calculate $0.00442 \div 0.017$

6. Calculate $69.115 \div 0.0023$

Algebraic Extensions

Properties of Exponents

The basic properties of exponents are summarized as follows:

If a is a real number greater than zero and n and m are rational numbers then,

1. $a^n a^m = a^{n+m}$ **2.** $\dfrac{a^n}{a^m} = a^{n-m}$ **3.** $\left(a^n\right)^m = a^{nm}$

4. $(ab)^n = a^n b^n$ **5.** $a^0 = 1$ **6.** $a^{-n} = \dfrac{1}{a^n}$

Property 1: $a^n a^m = a^{n+m}$ If you are multiplying two powers of the same base, add the exponents.

Example 1: $x^4 \cdot x^7 = x^{4+7} = x^{11}$

Note that the exponents were added and the base did not change.

Property 2: $\dfrac{a^n}{a^m} = a^{n-m}$ If you are dividing two powers of the same base, subtract the exponents.

Example 2: $\dfrac{6^6}{6^4} = 6^{6-4} = 6^2 = 36$

Note that the exponents were subtracted and the base did not change.

Property 3: $\left(a^n\right)^m = a^{nm}$ If a power is raised to a power, multiply the exponents.

Example 3: $\left(y^3\right)^4 = y^{12}$

The exponents were multiplied. The base does not change. Can you justify this result in your own mind?

Property 4: $(ab)^n = a^n b^n$ If a product is raised to a power, each factor is raised to that power.

Example 4: $(2x^2 y^3) = 2^3 \cdot (x^2)^3 \cdot (y^3)^3 = 8x^6 y^9$

Note that since the base contained three factors each of those was raised to the third power. The common mistake in an expansion such as this is not to raise the coefficient to the power.

Property 5: $a^0 = 1, a \neq 0$ Often presented as a definition, property 5 states that any nonzero base raised to the zero power is one. This property or definition is a result of property 2 of exponents as follows:

Consider $\dfrac{x^5}{x^5}$. Using property 2, $x^{5-5} = x^0$. However, you know that any fraction in which the numerator and the denominator are equal is equivalent to 1. Therefore, $x^0 = 1$.

Example 5: $\left(\dfrac{2x^3}{3yz^5}\right)^0 = 1$

Given a nonzero base, if the exponent is zero, the value is one.

Property 6: $a^{-n} = \dfrac{1}{a^n}$ Sometimes presented as a definition, property 6 states that any base raised to a negative power is equivalent to the reciprocal of the base raised to the positive power. Note that the negative exponent does not have any effect on the sign of the base. This property could also be viewed as a result of the second property of exponents as follows:

Consider $\dfrac{x^3}{x^5}$. Using property 2, $x^{3-5} = x^{-2}$. If you view this expression algebraically, you have three factors of x in the numerator and five in the denominator. If you divide out the three common factors, you are left with $\dfrac{1}{x^2}$. Therefore, if property 2 is true, then $x^{-2} = \dfrac{1}{x^2}$.

Example 6: Write each of the following without negative exponents.

a. 3^{-2} b. $\dfrac{2}{x^{-3}}$

Solution: a. $3^{-2} = \dfrac{1}{3^2} = \dfrac{1}{9}$ **b.** $\dfrac{2}{x^{-3}} = \dfrac{2}{\frac{1}{x^3}} = 2 \div \dfrac{1}{x^3} = 2 \cdot x^3 = 2x^3$

A **factor** can be moved from a numerator to denominator or from a denominator to a numerator by changing the *sign of the exponent.*

Example 7: Simplify and express your results with positive exponents only.

$$\left(\frac{x^3y^{-4}}{2x^{-3}y^{-2}z}\right) \cdot \left(\frac{4x^3y^2z}{x^5y^{-3}z^3}\right)$$

Solution: Simplify each factor by writing them with positive exponents only.

$$\left(\frac{x^6}{2y^2z}\right) \cdot \left(\frac{4y^5}{x^2z^2}\right). \text{ Now multiply and simplify. } \frac{4x^6y^5}{2x^2y^2z^3} = \frac{2x^4y^3}{z^3}$$

EXERCISES

Simplify and express your results with positive exponents only. Assume that all variables represent only nonzero values.

1. 5^{-3}

2. $\dfrac{1}{x^{-5}}$

3. $\dfrac{3x}{y^{-2}}$

4. $\dfrac{10x^2y^5}{2x^{-3}}$

5. $\dfrac{5^{-1}z}{x^{-1}z^{-2}}$

6. $5x^0$

7. $(a+b)^0$

8. $-3(x^0 - 4y^0)$

9. $x^6 \cdot x^{-3}$

10. $\dfrac{4^{-2}}{4^{-3}}$

11. $(4x^2y^3) \cdot (3x^{-3}y^{-2})$

12. $\dfrac{24x^{-2}y^3}{6x^3y^{-1}}$

13. $\dfrac{(4x^{-2}y^{-3}) \cdot (5x^3y^{-2})}{6x^2y^{-3}z^{-3}}$

14. $\left(\dfrac{2x^{-2}y^{-3}}{z^2}\right) \cdot \left(\dfrac{x^5y^3}{z^{-3}}\right)$

15. $\dfrac{(6x^4y^{-3}z^{-2})(3x^{-3}y^4)}{15x^{-3}y^{-3}z^2}$

Solving 2 × 2 Linear Systems by the Addition Method

The addition method is an algebraic alternative to the substitution method for solving a 2 × 2 linear system of the form

$$a_1x + b_1y = c_1$$
$$a_2x + b_2y = c_2,$$

where a_1, a_2, b_1, b_2, c_1, and c_2 are real numbers.

The basic strategy for the addition method is to reduce the 2 × 2 system to a single linear equation by eliminating a variable.

For example, consider the x-coefficients of the linear system

$$2x + 3y = 1$$
$$4x - y = 9.$$

The coefficients are 2 and 4. The LCM of 2 and 4 is 4. Use the multiplication principle to multiply each side of the first equation by -2. The resulting system is

$$-2(2x + 3y = 1) \qquad \text{or equivalently} \qquad -4x - 6y = -2$$
$$4x - y = 9 \qquad\qquad\qquad\qquad\qquad 4x - y = 9$$

Multiplying by -2 produces x-coefficients that are additive inverses or opposites. Now add the two equations together to eliminate the variable x.

$$-7y = 7$$

Solving for y, $y = -1$ is the y value of our solution. To find the value of x, substitute -1 for y and solve for x in any equation that involves x and y. For an alternative to determining the value of y, consider the original system and the coefficients of y, 3 and -1 in the original system. The LCM is 3. Since the signs are already opposites, multiply the second equation by 3.

$$2x + 3y = 1 \qquad \text{or equivalently} \qquad 2x + 3y = 1$$
$$3(4x - 7 = 9) \qquad\qquad\qquad\qquad 12x - 3y = 27$$

Adding the two equations will eliminate the y-variable.

$$14x = 28$$

Solving for x, $x = 2$. Therefore, the potential solution is $(2, -1)$. This should be checked to make certain that it satisfies both equations.

Depending on the coefficients of the system, you may need to change both equations when using the addition method. For example, the coefficients of x in the 2 × 2 linear system

$$5x - 2y = 11$$
$$3x - 5y = -12$$

are 5 and 3. The LCD is 15. Multiply the first equation by 3 and the second by -5 as follows:

$$3(5x - 2y = 11)$$
$$-5(3x + 5y = -12)$$

or equivalently

$$15x - 6y = 33$$
$$-15x - 25y = 60$$

Add the two equations to eliminate the x terms from the system.

$$-31y = 93$$

Solving for y, $y = -3$. Substituting this value for y in the first equation of the original system yields

$$5x - 2(-3) = 11$$
$$5x + 6 = 11$$
$$5x = 5$$
$$x = 1$$

Therefore, $(1, -3)$ is the potential solution of the system.

E X E R C I S E S

Solve the following systems using the addition method. If the system has no solution or both equations represent the same line, state this as your answer.

1. $x - y = 3$
    $x + y = -7$

2. $x + 4y = 10$
    $x + 2y = 4$

3. $-5x - y = 4$
    $-5x + 2y = 7$

4. $4x + y = 7$
    $2x + 3y = 6$

5. $3x - y = 1$
    $6x - 2y = 5$

6. $4x - 2y = 0$
    $3x + 3y = 5$

7. $x - y = 9$
     $-4x - 4y = -36$

8. $-2x + y = 6$
     $4x + y = 1$

9. $\frac{3}{2}x + \frac{2}{5}y = \frac{9}{10}$
   $\frac{1}{2}x + \frac{6}{5}y = \frac{3}{10}$

10. $0.3x - 0.8y = 1.6$
    $0.1x + 0.4y = 1.2$

Factoring Trinomials with Leading Coefficient ≠ 1

With patience, it is possible to factor many trinomials by trial and error, using the FOIL method in reverse.

FACTORING TRINOMIALS BY TRIAL AND ERROR

1. Factor out the greatest common factor.

2. Try combinations of factors for the first and last terms in the two binomials.

3. Check the outer and inner products to match the middle term of the original trinomial.

4. If the check fails, then repeat steps 2 and 3.

Example 1: Factor $6x^2 - 7x - 3$.

Solution:

Step 1   There is no common factor, so go to step 2.

Step 2   You could factor the first term as $6x(x)$ or as $2x(3x)$. The last term has factors of 3 and 1, disregarding signs. Suppose you try $(2x + 1)(3x - 3)$

Step 3   The outer product is $-6x$. The inner product is $3x$. The sum is $-3x$. The check fails.

Step 4   Suppose you try $(2x - 3)(3x + 1)$. The outer product is $2x$. The inner product is $-9x$. The sum is $-7x$. It checks. Therefore, $6x^2 - 7x - 3 = (2x - 3)(3x + 1)$.

E X E R C I S E S

Factor each of the following completely.

1. $x^2 + 7x + 12$ **2.** $6x^2 - 13x + 6$

3. $3x^2 + 7x - 6$ **4.** $6x^2 + 21x + 18$

5. $9x^2 - 6x + 1$ **6.** $2x^2 + 6x - 20$

7. $15x^2 + 2x - 1$ **8.** $4x^3 + 10x^2 + 4x$

Solving Equations by Factoring

Many quadratic and higher order polynomial equations can be solved by factoring using the Zero Product Rule, which says:

If a and b are real numbers such that $ab = 0$, then either a or b must be equal to zero.

SOLVING AN EQUATION BY FACTORING
1. Use the addition principle to move all terms to one side, so that the other side of the equation is zero.
2. Simplify and factor the nonzero side.
3. Use the Zero Product Rule to set each factor equal to zero and then solve the resulting equations.
4. Check your solutions in the original equation.

Example 1: Solve the equation $x(x + 5) = 0$.

Solution: This equation already satisfies the first two steps in our process, so you simply start at step 3.

$$x = 0 \quad \text{or} \quad x + 5 = 0$$
$$x = -5$$

Thus, you have two solutions $x = 0$ and $x = -5$. The check is left to the reader.

Example 2: Solve the equation $6x^2 = 16x$.

Solution: Subtracting $16x$ from both sides, you have $6x^2 - 16x = 0$. Since there are no like terms, factor the binomial as follows.

$$6x^2 - 16x = 0$$
$$2x(3x - 8) = 0$$

Using the Zero Product Rule,

$$2x = 0 \quad \text{or} \quad 3x - 8 = 0$$
$$x = 0 \quad \text{or} \quad 3x = 8$$
$$x = \tfrac{8}{3}$$

The two potential solutions are $x = 0$ and $x = \tfrac{8}{3}$. The check is left for the reader to do.

Example 3: Solve the equation $3x^2 - 2 = -x$.

Solution: Adding x to both sides, you obtain $3x^2 + x - 2 = 0$. Since there are no like terms, factor the trinomial.

$$(3x - 2)(x + 1) = 0$$

Using the Zero Product Rule,

$$3x - 2 = 0 \quad \text{or} \quad x + 1 = 0$$
$$3x = 2 \quad \text{or} \quad x = -1$$
$$x = \tfrac{2}{3}$$

The two potential solutions are $x = \tfrac{2}{3}$ and $x = -1$. The check is left for the reader to do.

Example 4: Solve the equation $3x^3 - 8x^2 = 3x$.

Solution: Subtracting $3x$ from both sides, you obtain, $3x^3 - 8x^2 - 3x = 0$. Since there are no like terms, factor the trinomial.

$$x(3x^2 - 8x - 3) = 0.$$
$$x(3x + 1)(x - 3) = 0$$

Using the Zero Product Rule,

$$x = 0 \quad \text{or} \quad 3x + 1 = 0 \quad \text{or} \quad x - 3 = 0$$
$$3x = -1 \quad \text{or} \quad x = 3$$
$$x = \tfrac{1}{3}$$

The three potential solutions are $x = 0$, $x = -\tfrac{1}{3}$, and $x = 3$. The check is left for the reader to do.

E X E R C I S E S

Solve each of the following equations.

1. $x(x + 7) = 0$

2. $3(x - 5)(2x + 1) = 0$

3. $12x = x^2$

4. $x^2 + 5x = 0$

5. $x^2 - x - 63 = 0$

6. $3x^2 - 9x - 30 = 0$

7. $-7x + 6x^2 = 10$

8. $3y^2 = 2 - y$

9. $-28x^2 + 15x - 2 = 0$

10. $4x^2 - 25 = 0$

11. $(x + 4)^2 - 16 = 0$ **12.** $(x + 1)^2 - 3x = 7$

13. $2(x + 2)(x - 2) = (x - 2)(x + 3) - 2$ **14.** $18x^3 = 15x^2 + 12x$

The TI–83 Graphing Calculator

Getting Started with the TI-83

ON-OFF

To turn on the TI-83 press the ⌈ON⌉ key. To turn off the TI-83, press ⌈2nd⌉ and then ⌈ON⌉.

In general, to access any of the white commands, press the black or gray key. To access the gold commands, press ⌈2nd⌉ and then the black or gray key below the desired command. Similarly, to access any of the green commands or symbols, press ⌈ALPHA⌉ followed by the appropriate black or gray key.

Contrast

To adjust the contrast on your screen, press and release the ⌈2nd⌉ key and hold ⌈▲⌉ to darken and ⌈▼⌉ to lighten.

Mode

The ⌈MODE⌉ key controls many calculator settings. The activated settings are highlighted. For most of your work in this course, the settings in the left-hand column should be highlighted.

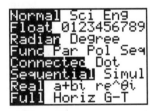

To change a setting, move the cursor to the desired setting and press ⌈ENTER⌉.

The Home Screen

The home screen is used for calculations.

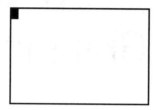

You may return to the home screen at any time by using the QUIT command. This command is accessed by pressing (2nd) (MODE). All calculations in the home screen are subject to the order of operations.

Enter all expressions as you would write them. Always observe the order of operations. Once you have typed the expression, press (ENTER) to obtain the simplified result. Before you hit (ENTER) you may edit your expression by using the arrow keys, the delete command (DEL), and the insert command (2nd) (DEL).

Three keys of special note are the reciprocal key (X⁻¹), the caret key (^), and the negative key (−).

The reciprocal command (X⁻¹) will invert the number in the home screen.

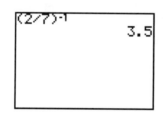

The caret key (^) is used to raise numbers to powers

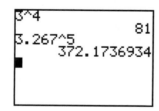

The negative key is different from the minus key. To enter a negative number use the gray key (−), not the blue (−) key.

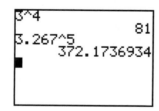

A table of keys and their functions follows.

Key	Function Description
ON	Turns calculator on or off.
CLEAR	Clears text screen.
ENTER	Executes a command.
(−)	Calculates the additive inverse.
MODE	Displays current operating settings.
DEL	Deletes the character at the cursor.
^	Symbol used for exponentiation.
ANS	Storage location of the last calculation.
ENTRY	Retrieves the previously executed expression.

ANS and ENTRY

The last two commands in the table can be real time savers. The result of your last calculation is always stored in a memory location known as ANS. It is accessed by pressing [2nd] [(−)] or it can be automatically accessed by pressing any operation button.

Suppose you want to evaluate $12.5\sqrt{1 + 0.5 \cdot (0.55)^2}$. It could be evaluated in one expression and checked with a series of calculations using ANS.

```
12.5√(1+0.5*.55^
2)
         13.41203983
▪
```
```
1+.5*.55^2
            1.15125
√(Ans)
        1.072963187
Ans*12.5
         13.41203983
▪
```

The ENTRY command recalls the last expression. Even if you have pressed [ENTER], you can edit the previous expression. The [2nd] [ENTER] sequence will recall the previous expression for editing.

Suppose you want to evaluate the compound interest expression $P\left(1 + \frac{r}{n}\right)^{nt}$, where P is the principal, r is the interest rate, n is the number of compounding periods annually, and t is the number of years, when $P = \$1000$, $r = 6.5\%$, $n = 1$, and $t = 2, 5,$ and 15 years.

Using the ENTRY command, this expression would be entered once and edited twice.

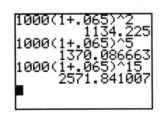

```
1000(1+.065)^2
         1134.225
1000(1+.065)^5
       1370.086663
1000(1+.065)^15
       2571.841007
▪
```

Note that there are many last expressions stored in the ENTRY memory location. You can repeat the ENTRY command as many times as you want to retrieve a previously entered expression.

Functions and Graphing with the TI-83

Y = Menu

Functions of the form $y = f(x)$ can be entered into the TI-83 using the Y = menu. To access the Y = menu press the `Y=` key. Type the expression $f(x)$ after Y1 using the `X,T,θ,n` key for the variable x and press `ENTER`.

For example, enter the function $f(x) = 3x^5 - 4x + 1$.

```
Plot1 Plot2 Plot3
\Y1∎3X^5-4X+1
\Y2=
\Y3=
\Y4=
\Y5=
\Y6=
\Y7=
```

Note the = sign after Y1 is highlighted. This shows Y1 is selected to be graphed. The highlighting may be turned on or off by using the arrow keys to move the cursor to the = sign and then pressing `ENTER`.

```
Plot1 Plot2 Plot3
\Y1=3X^5-4X+1
\Y2=
\Y3=
\Y4=
\Y5=
\Y6=
\Y7=
```

Once the function is entered in the Y = menu, function values may be evaluated in the home screen.

For example, given $f(x) = 3x^5 - 4x + 1$, evaluate $f(4)$. In the home screen press `VARS`.

```
VARS Y-VARS
1∎Window...
2:Zoom...
3:GDB...
4:Picture...
5:Statistics...
6:Table...
7:String...
```

Move the cursor to Y-VARS and press `VARS`.

Press (ENTER) again to select Y1. Y1 now appears in the home screen.

To evaluate $f(4)$, press (() (4) ()) after Y1 and press (ENTER).

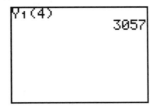

Tables of Values

If you are interested in viewing several function values for the same function, you may want to construct a table.

Before constructing the table, look at the settings in the Table Setup menu. To do this, press (2nd) (WINDOW).

TblStart tells you where the table of values will start. ΔTbl tells you the increment from one input value to the next. Set the TblStart at −2 and the ΔTbl at 0.5 then press (2nd) (GRAPH) to access the following table. (Make certain that the = sign following Y1 is highlighted.)

Use the (▲) and (▼) keys to view other values in the table.

If the input values of interest are not evenly spaced, you may want to choose the ask mode for the independent variable from the Table Setup menu.

The resulting table is blank, but you can fill it by choosing any values you like for x and pressing (ENTER) after each.

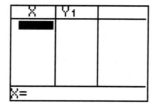
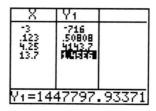

Note that the number of digits shown in the output is limited by the table width, but if you want more digits, move the cursor to the desired output and more digits appear at the bottom of the screen.

Graphing a Function

Once a function is entered in the $Y =$ menu and activated it can be displayed and analyzed. For this discussion we will use the function $f(x) = -x^2 + 10x + 12$. Enter this as Y1.

The Viewing Window

The viewing window is the portion of the rectangular coordinate system that is displayed when you graph a function.

Xmin defines the left edge of the window.

Xmax defines the right edge of the window.

Xscl defines the distance between horizontal tick marks.

Ymin defines the bottom edge of the window.

Ymax defines the top edge of the window.

Yscl defines the distance between vertical tick marks.

In the standard viewing window, Xmin $= -10$, Xmax $= 10$, Xscl $= 1$, Ymin $= -10$, Ymax $= 10$, and Yscl $= 1$.

To select the standard viewing window, press ⌈ZOOM⌉ ⌈6⌉.

If you press the ⌈GRAPH⌉ key now, you will view the following:

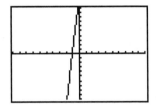

Is this an accurate and or complete picture of your function, or is the window giving you a misleading impression? You may want to use your table function to view the output values from −10 to 10.

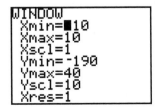
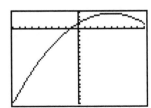

The table indicates that the minimum output value is −188 and the maximum output value is 37. Press ⌈WINDOW⌉ and reset the settings to approximately the following;

Xmin = −10, Xmax = 10, Xscl = 1, Ymin = −190, Ymax = 40, Yscl = 10

The new graph gives us a much more complete picture of the behavior of the function on the interval (−10, 10).

Specific points on the curve can be viewed by activating the trace feature. While in the graph window, press [TRACE]. A flashing cursor will appear on the curve at approximately the midpoint of the screen.

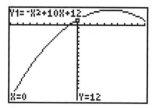

The left arrow key, [◀], will move the cursor toward smaller input values. The right arrow key, [▶], will move the cursor toward larger input values. If the cursor reaches the edge of the window and you continue to move the cursor, the window will adjust automatically.

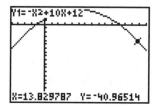

Zoom Menu

The Zoom menu offers several options for changing the window very quickly.

The features of each of the commands are summarized in the following table.

Zoom Command	Description
1:ZBox	Draws a box to define the viewing window.
2:Zoom In	Magnifies the graph near the cursor.
3:Zoom Out	Increases the viewing window around the cursor.
4:Zdecimal	Sets a window so that Xscl and Yscl are 0.1.
5:ZSquare	Sets equal size pixels on the x and y axes.
6:ZStandard	Sets the window to standard settings.
7:ZTrig	Sets built-in trig window variables.
8:ZInteger	Sets integer values on the x and y axes.
9:ZoomStat	Sets window based on the current values in the stat lists.
0:ZoomFit	Replots graph to include the max and min output values for the current Xmin and Xmax.

Solving Equations Graphically Using the TI-83

The Intersection Method

This method is based on the fact that solutions to the equation $f(x) = g(x)$ are input values of x that produce the same outputs for the functions f and g. Graphically these are the x-coordinates of the intersection points of $y = f(x)$ and $y = g(x)$.

The following procedure illustrates how to use the Intersection Method to solve $x^3 + 3 = 3x$ graphically.

Step 1   Enter the left-hand side of the equation as Y1 in the Y = editor and the right-hand side as Y2.

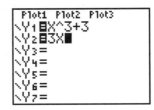

Step 2   Examine the graphs to determine the number of intersection points.

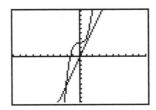

You may need a couple of windows to be certain of the number of intersection points.

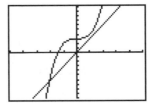

Step 3   Access the Calculate menu by pushing [2nd] [TRACE] then choose option 5: intersect

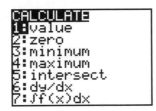 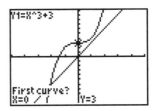

The cursor will appear on the first curve in the center of the window.

Step 4 Move the cursor close to the desired intersection point and press (ENTER).

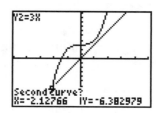

The cursor will now jump vertically to the other curve.

Step 5 Repeat step 4 for the second curve.

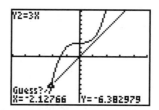

Step 6 To use the cursor's current location as your guess, press (ENTER) in response to the question on the screen that asks Guess? If you want to move to a better guess value, do so before you press (ENTER).

The coordinates of the intersection point appear below the word Intersection

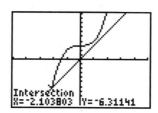

The *x*-coordinate is a solution to the equation.

If there are other intersection points, repeat the process as necessary.

Using the TI–83 to Determine the Linear Regression Equation for a Set of Data

Example:

Input	Output
2	2
3	5
4	3
5	7
6	9

Enter the data into the calculator as follows:

1. Press $\boxed{\text{STAT}}$ and choose EDIT.

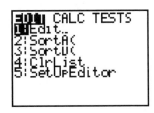

2. The TI-83 has six built-in lists, L1, L2, ..., L6. If there is data in L1, clear the list as follows:

    a. Use the arrows to place the cursor on L1 at the top of the list. Press $\boxed{\text{CLEAR}}$ followed by $\boxed{\text{ENTER}}$.

    b. Follow the same procedure to clear L2 if necessary.

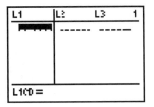

    c. Enter the input into L1 and the output into L2.

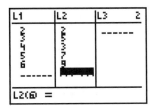

To see a scatterplot of the data proceed as follows.

1. STAT PLOT is the 2nd function of the $\boxed{\text{Y=}}$ key. You must press $\boxed{\text{2nd}}$ before pressing $\boxed{\text{Y=}}$ to access the STAT PLOT menu.

2. Make sure plot 2 and plot 3 are off and choose plot 1. Highlight On, the scatter plot symbol and the square mark as indicated below. L1 is the Xlist and L2 is the Ylist. L1 is the 2nd function of the 1 button and L2 is the 2nd function of the 2 button, and so on.

3. Press Y= and clear or deselect any functions currently stored.

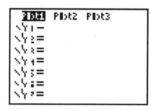

4. To display the scatterplot, have the calculator determine an appropriate window by pressing ZOOM and then 9 (ZoomStat).

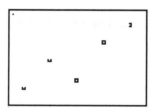

Calculate the linear regression equation as follows.

1. Press STAT and right arrow to highlight CALC.

2. Choose 4: LinReg(ax + b). LinReg(ax + b) will be pasted to the home screen. To tell the calculator where the data is, press ⌐2nd⌐ and ⌐1⌐ (for L1), then ⌐,⌐, then ⌐2nd⌐ and ⌐2⌐ (for L2) so the display looks like this:

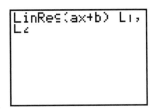

3. Press ⌐,⌐ and then press ⌐VARS⌐.

4. Right arrow to highlight Y-VARS.

5. Choose 1, FUNCTION.

6. Choose 1 for Y1 (or 2 for Y2, etc.).

7. Press ENTER .

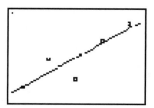

The linear regression equation for this data is $y = 1.6x - 1.2$.

8. To display the regression line, press GRAPH .

Selected Answers

Chapter 1

Activity 1.3 **Exercises: 1.** 3.143; **2.** 2.590;
3. b. 2.542.

Activity 1.4 **Exercises: 3. a.** $53.58; **b.** $3.75;
c. $8.04; **d.** $65.37; **e.** $21.79; **4 a.** $55.92; **b.** $417.56;
c. No, short $17.56.

Activity 1.5 **Exercises: 1. b.** 4.134×10^{-5};
c. 5.551405×10^{14}; **2. a.** 453,200,000,000;
d. 0.0000004532; **3. a.** 30; **d.** 12; **e.** 390625; **g.** .2;
4. b. Approximately 89,740,000,000 or eighty-nine
billion seven hundred forty million acres.

What Have I Learned? **1. b.** The total bill, including
tax, is $105.07, so $100 will not cover your purchases.

How Can I Practice? **3. b.** 13; **d.** $5.\overline{81}$;
6. approximately 8555; **7. a.** 1.7501 E 15 km;
b. 5.83 E 9 seconds $\approx$ 185 years.

Activity 1.8 **Exercises: 1.** $27.50; **4.** $\frac{11}{30}$; **6. a.** $2\frac{7}{9}$;
6. b. $\frac{17}{28}$; **6. c.** $\frac{1}{6}$.

Activity 1.9 **Exercises: 1.** 434; **3.** $\frac{5}{12}$; **5.** 22;
6. a. $2\frac{17}{21}$; **c.** $9\frac{3}{4}$; **f.** $1\frac{13}{50}$.

Activity 1.10 **Exercises: 2.** $1\frac{1}{6}$ cups.

How Can I Practice? **1.** $2\frac{7}{8}$ in.; **2.** $4\frac{9}{10}$; **4.** $1\frac{3}{4}$ ft;
6. d. $1\frac{19}{30}$; **g.** $\frac{13}{36}$; **m.** 9; **p.** 16; **t.** $1\frac{23}{26}$; **v.** $5\frac{1}{4}$.

Activity 1.11 **Exercises: 2. a.** $\frac{31}{55}$; **b.** $.56\overline{36}$;
c. $\approx$ 56%; **5.** Test 1; **6.** 0.23%; **8.** 60.5%.

Activity 1.12 **Exercises: 1. b.** 250; **e.** 15; **g.** 6;
h. 180; **2. a.** 12; **c.** 45; **3.** fifty; **4.** $36.90; **6.** 35;
8. 180; **10.** 770; **12.** $9; **13.** $125; **17.** $420,000.

Activity 1.13 **Exercises: 1.** 15; **3.** $119,600;
5. 1:36 P.M.

Activity 1.14 **Exercises: 1.** Approximately

1.94×10^{9} acres; **3.** 338,400 sq ft; **5.** 149,572,640 km;
9. 2.12 quarts; **10.** Approximately 7900 miles.

Activity 1.15 **Exercises: 1. a.** $125,000;
b. Approximately 71.4%; **3.** 12.5%; **6.** 1760; **9.** $68.04.

What Have I Learned? **Exercises: 5. a.** Growth:
35% increase; **b.** Growth: 7% increase; **d.** Decay: 14%
decrease.

How Can I Practice? **Exercises: 3.** $\approx$ 64%;
5. $21; **8.** $\approx$ 77.35%; **10. a.** 432; **b.** $29.\overline{45}$ hr; **c.** $162;
12. $8.\overline{18}$ kg, $\approx$ 2.83 ℓ; **16.** $\approx$ 29%; **17. a.** $\approx$ 6%;
20. $\approx$ 84.62.

Skills Check 1 **Exercises: 1.** 219; **2.** 5372;
3. Sixteen and seven hundred nine ten-thousandths;
4. 3402.029; **5.** 567.05; **6.** 2.599; **7.** 87.913; **8.** 28.82;
9. 0.72108; **10.** 95.7; **11.** 12%; **12.** 300%; **13.** 0.065;
14. 3.6; **15.** $0.1\overline{3}$; **16.** $\approx$ 42.9%; **17.** $\frac{1}{8}$; **18.** 3.0027,
3.0207, 3.027, 3.27; **19.** $\approx$ 41; **20.** $\approx$ 52% preferred
Coke. **21.** $\approx$ $8.67 total cost of cap; **22.** $\approx$ 32% of
monthly income is spent of food; **23.** 10,023 students are
expected this year. **24.** 37.5% decrease;
25. = 36,540,000,000 acres. **26. a.** No, the percent
reductions are based on the price at the time of the specific reduction; **b.** $450 before Thanksgiving, $270 in
December, and $135 in January; **c.** final price in January
would be $135, the same as before. Before Thanksgiving
$250; December $150; **27.** 74% of the pre-1980 scores;
28. little over $7 million; **29.** approximately $1166.

Activity 1.17 **Exercises: 3.** 43-yard gain;
6. $-11°C$; **8.** 35°C.

Activity 1.18 **Exercises: 4.** $125; **6.** total loss of $36.

Activity 1.19 **Exercises: 2. a.** 3; **b.** -19;
d. -14.4; **f.** 9; **g.** -1; **j.** $-30,000$; **k.** ≈ -11.538;
4. a. 8.8×10^{-4} lb = .00088 lb.

How Can I Practice?   **37.** 6 pounds;
38. 2616 feet apart; **39. b.** $-19°C$; **d.** 9°C drop;
e. 8° rise; **40. a.** -9; **c.** -12.625; **e.** 17; **g.** -9.

Activity 1.20   **Exercises:** **2.** 13.5 cm, 4.225 sq cm;
4. 17.2 cm, 12.25 sq cm; **6.** 12.3 cm, 7.11 sq cm;
7. 15.75 sq cm, 15 cm.

Activity 1.21   **Exercises:** **2.** $\approx$ 27.8 sq ft;
3. 66 sq ft; **5.** 1152 sq in.

Activity 1.23   **Exercises:** **1.** $\approx$ 6.28 in.;
2. $\approx$ 26.283 sq ft.

Activity 1.25   **Exercises:** **2.** No, since $80.6 < 100$,
you do not meet code. **3.** 120 ft.

Activity 1.26   **Exercises:** **1. a.** 30.62 cu in.;
b. $\approx$ \$.94; **c.** \$1.88.

How Can I Practice?   **1. a.** 40 sq ft; **b.** 95 sq m;
c. 18 sq in.; **2. b.** 7.5 sq in.; **4.** $a \approx 21.46$ sq ft;
5. $b \approx 150.8$ cu cm; **7. a.** ≈ 1703.43 sq in.; **b.** 167.12 in.

Skills Check 2   **1.** 1.001; **2.** 0.019254; **3.** 1; **4.** 9;
5. ≈ 8.49 m; **6.** 9; **7.** 2.03×10^8; **8.** ≈ 8.8 quarts;
9. 0.00000276; **10.** ≈ 9.9 ounces; **11.** 16.1 sq in.;
12. ≈ 2.95 in.; **13.** 126 sq in.; **14.** 8 students; **15.** 35
hours; **16.** 525 mistakes; **17.** 5 pieces; **18.** $\frac{4}{9}$; **19.** $2\frac{2}{7}$;
20. $\frac{34}{7}$; **21.** $\frac{37}{45}$; **22.** $1\frac{7}{10}$; **23.** $\frac{8}{27}$; **24.** 3; **25.** $4\frac{5}{6}$ cups;
26. $29\frac{5}{8}$; **27.** $2.\overline{6}$ days; **28.** $4\frac{1}{4}$ lb; **29.** 128.2; **30.** 52,800
ft; 96 oz; 80 oz; 768 oz; 48 oz; **31.** 5×10^7; 2.5×10^7;
3.5×10^7; 3.68×10^8; 7.45×10^8; 1.606×10^9;
2.674×10^9; 2.658×10^9; **32.** 1.43 lb chocolate; ≈ 6.78
oz milk; **33.** a 1-inch eraser is big enough to share;
34. $4\frac{1}{12}$; **35.** 9; **36.** 1; **37.** $9\frac{73}{144}$; **38.** $13\frac{7}{12}$; **39.** $24\frac{1}{6}$;
40. -16; **41.** 36; **42.** 24.

Gateway Review   **1.** two and two hundred two
ten-thousandths; **2.** 14.003; **3.** hundred thousandths;
4. 10.524; **5.** 12.16; **6.** 0.30015; **7.** 0.007; **8.** 2.05;
9. 450%; **10.** 0.003; **11.** $7\frac{3}{10}$ or $\frac{73}{10}$; **12.** 60%; **13.** $1\frac{1}{8}$, 1.1,
1.01, 1.001; **14.** 27; **15.** 1; **16.** $\frac{1}{4^2} = \frac{1}{16}$; **17.** 6; **18.** $\frac{2}{10} = \frac{1}{5}$;
19. $4 < \sqrt{18} < 5$; **20.** 12; **21.** 5.43×10^{-5};
22. 37,000; **23.** $\frac{25}{30} + \frac{85}{100} + \frac{60}{70} = 84.68\%$; **24.** $4\frac{1}{2}$; **25.** $\frac{23}{4}$;
26. $\frac{3}{5}$; **27.** $\frac{4}{24} + \frac{15}{24} = \frac{19}{24}$; **28.** $\frac{21}{4} - \frac{15}{4} = \frac{6}{4} = \frac{3}{2}$;
29. $\frac{\cancel{2}}{\cancel{9}} \cdot \frac{\overset{3}{\cancel{27}}}{\cancel{8}} = \frac{3}{4}$; **30.** $\frac{\overset{3}{\cancel{6}}}{\cancel{11}} \cdot \frac{\overset{2}{\cancel{22}}}{\cancel{8}} = \frac{6}{4} = \frac{3}{2} = 1\frac{1}{2}$; **31.** $\frac{\overset{3}{\cancel{21}}}{\cancel{8}} \cdot \frac{\overset{2}{\cancel{16}}}{\cancel{7}} = 6$;
32. $\frac{x}{8} = \frac{9}{4}$; $x = \frac{9 \cdot 8}{4} = 18$; **33.** $.2(80) = 16$;
34. $.25x = 50$, $\frac{25x}{100} = 50$, $x = \frac{5000}{25} = 200$; **35.** 50%;
36. ≈ 10 cm; **37.** ≈ 18.84 in.; **38.** = 14 in.; **39.** 17.5 sq
ft; **40.** ≈ 3.1 sq in.; **41.** $2.093.\overline{3} \approx 2.093$ cu cm;
42. 30% decrease; **43.** 162 students; **44.** 40 pages;
45. 1.364 lb; **46.** 0.013 in./sec; **47.** 10°C; **48.** -1;
49. -14; **50.** -13; **51.** 3; **52.** -25; **53.** -1; **54.** 2;
55. -11; **56.** 0; **57.** $\frac{1}{10}$; **58.** 19; **59.** 80; **60.** 183; **61.** 20;
62. 160 feet below surface; **63.** $-\$600$, representing a

loss of \$600 in stocks; **64.** $-\$63$, meaning you are over-
drawn and are \$63 in debt; **65. b.** $\frac{10}{120} \approx 8.\overline{3}\%$; **66.** mass
of hydrogen atom $\approx 4.4 \times 10^{-27}$ lb.

Chapter 2

Activity 2.1   **Exercises:** **1. d.** 1989; **f.** between 1967
and 1977; **h.** from 1993 to 1996.

Activity 2.2   **Exercises:** **1. b.** Yes, $n = 0$ is reasonable. You might not purchase any notebooks at the college
bookstore; **2. a.** The output is 2 times the input; **3. a.** 11,
13, 15, 17, 19, 22, 24; **4. b.** Yes. It means you don't own
any Beanie Babies; **6. a.** \$137.50 gross pay; **b.** Gross pay
is calculated by multiplying the number of hours worked
by \$6.25.

Activity 2.3   **Exercises:** **1.** d; **3.** g; **7.** a; **12.** h; **16.** f.

What Have I Learned?   **3. c.** Since input and
output values are non-negative, only the first quadrant
and positive axes are necessary.

How Can I Practice?   **1. c.** No, negative numbers
would represent distance above the surface;
2. a. Quadrant III; **d.** on horizontal axis, between quadrant I and quadrant IV; **3. c.** between 11°C and 16°C;
4. b. approximately \$1500–\$2000; **c.** approximately at
age 10.

Activity 2.6   **Exercises:** **2.** $5(x - 6)$; **5.** $-2x - 20$;
7. a. 7; **c.** -1; **e.** $\frac{2}{3}$; **8. b.** x and $\frac{1}{2}$; **d.** 2 and $(x + 3)$;
9. c. x^2, $2x$, and 5; **11. b.** Column 2: -2, -1, 1, 2; Column
3: $-.5$, -1, 1, .5; **12.** Input variable: room charge; Letter
representation: R; Output variable: cost of room and breakfast; Letter representation: C; Symbolic rule: $C = R + 10$.
Input: 75, 110, 155. Output: 85, 120, 165. **14.** Input variable: number of miles driven; Letter representation: m;
Output variable: cost of rental for 1 day; Letter representation: c; Symbolic rule: $c = 30 + 0.10m$. Input: 85, 150,
240. Output: \$38.50, 45, 54.

Activity 2.7   **Exercises:** **2. a.** $y = 0.75x$; **5.** $x = -4$.

Activity 2.8   **Exercises:** **1. d.** \$768; **f. i.** 3 credit
hours; **2. d.** \$590; **3. a.** 52.5; **b.** ≈ 41.14; **5. a.** -198.9;
b. 23; **7. a.** 1.8; **b.** 8.2.

Activity 2.9   **Exercises:** **1. a.** square the input;
6. 2, 1, 5, 26; 0, 1, 9, 36. No. They generate different
outputs; **8.** 4, 1, 13, 76; 10, 1, 37, 226. No. They
generate different outputs.

Activity 2.10   **Exercises:** **2.** $x = 16$, $y = 13$;
4. $x = 39$, $y = 27$; **6.** $x = 6$, $y = 6$; **8.** $x = 3$, $y = 6$;
13. $x = -4$; **17.** $-16 = x$; **21.** $x = \frac{8}{9}$; **24.** 70 inches or 5
feet 10 inches; **26. b.** \$144, 000; **c.** 2006.

Activity 2.11   **Exercises:** **1. c.** $y = 143x + 25$;
d. \$883; **2. b.** \$158; **d. i.** 15; **ii.** 31.

Activity 2.12   Exercises: **1. b.** $\dfrac{w + 49}{1.6} = t$;

3. b. $\dfrac{p - 2l}{2} = w$;   **4. a.** 180.989 cm.

Activity 2.13   Exercises: **2.** $\approx -29°$, $\approx 5.7°$, $\approx 15.7°$; **4. a.** $\approx -29.3°$; **5.** ≈ 1207.78 cals. The patient is being underfed. **6. c.** \$412,000; **7. b.** 70 miles; **8. c.** $I = \frac{E}{R}$; **e.** $b = y - mx$; **f.** $t = \frac{12B}{tw}$; **9. c.** 37.70 cm, 113.10 sq cm; **d.** 43.98 yd, 153.94 sq yd; **f.** 45.55 km, 165.13 sq km; **11.** 402 sq ft; **13.** 5120 cu in.

What Have I Learned?   **7. b.** Second package holds 11.4 cu in. less than the original one.

How Can I Practice?   **1. a.** There are 4 terms; **b.** 5; **c.** -1; **d.** -3; **e.** 4 and x^2; **2. b.** $10 - x$; **d.** $-4x - 8$; **f.** $\frac{1}{2}x^2 - 2$; **4. a.** $t = 2.5r + 10$; **b.** 30; **c.** -30; **5. a.** -3; **c.** 15.5; **f.** 4; **h.** $\frac{35}{3}$; **6. d.** \$250; **e.** 9666 copies; **7. a.** Multiply the number of hours labor by \$32 and add \$148 for parts to obtain the total cost of the repair; **d.** \$308; **e.** No, 3.5 hr work plus parts will cost \$260; **f.** 6 hours; **g.** $x = \dfrac{y - 148}{32}$; **9. a.** $\frac{d}{t} = r$; **c.** $\dfrac{A - P}{Pt} = r$; **e.** $\frac{7}{4}(w - 3) = h$.

Activity 2.14   Exercises: **2.** $24x - 30$; **4.** $10 - 5x$; **6.** $-3p + 17$; **8.** $-12x^2 + 9x - 21$; **10.** $\frac{5}{8}x + \frac{5}{9}$; **12. b.** x^{11} Property 1; **14.** x^6; **16.** r^9; **18.** x^{18}; **20.** $-60x^4y$; **22.** $-6s^9t^5$; **24.** $x^2 + 3x$; **26.** $5x - x^2$; **28.** $6x^4 - 3x^3$; **30.** $-20x^7 + 15x^5 + 15x^3$; **33. a.** $V = 9\pi h^3$; **b.** $V = \pi h^3$; **d.** $V = 4\pi x^3 - 5\pi x^2$; **36.** \$46.97.

Activity 2.15   Exercises: **1.** $3(x + 5)$; **3.** $5x(x - 2)$; **5.** $4(1 - 3x)$; **7.** $srt(4 - 3s + 10t^2)$; **11.** $5a + 8ab - 3b$; **15.** $2x^3 + 7y^2 + 4x^2$; **19.** $2x - 2x^2 - 5$; **23.** $24 - x$; **25.** $3x - 25$; **27.** $9 + x$; **29.** $7x - 15$; **31.** $5x - 5$; **33.** $-x^2 - 9x$; **35.** $-x^2 - 14x$; **37. c.** $368 - 26x$.

Activity 2.16   Exercises: **1. c.** The value of your stock increased sixfold; **2. a.** $3\left[\dfrac{-2n + 4}{2} - 5\right] + 6$; **b.** $-3n - 3$; **4. b.** $8x - 5y + 13$; **e.** $13x - 43$; **g.** $x - 3$.

Activity 2.17   Exercises: **1. a.** $x^2 + 8x + 7$; **e.** $10 + 9c + 2c^2$; **i.** $12w^2 + 4w - 5$; **m.** $2a^2 - 5ab + 2b^2$; **2. a.** $3x^2 - 2xw + 2x + 2w - 5$; **c.** $3x^3 - 11x^2 - 6x + 8$; **e.** $x^3 - 27$; **4. a.** 224 cu in.; **b.** $V = (7 + x)(4 + x)(8)$; **c.** $224 + 88x + 8x^2$; **d.** $\approx 92.9\%$; **e.** $V = (7 - x)(4 + x)(8)$; **f.** $224 + 24x - 8x^2$; **g.** 240 cu in.

How Can I Practice?   **1.** a and c are equivalent; **3. b.** $-x^{14}$; **e.** p^{26}; **f.** $12x^{10}y^8$; **5. b.** $2xy(3 - 4y)$; **e.** $6xy(y^2 + 3x - 2x^2)$; **6. b.** $x^2y^2 + 2xy^2$; **7. b.** 1. add 5; 2. square result; 3. subtract 15. **8. a.** $6 - x$; **c.** $3x^2 + 3x$; **e.** $-4x - 10$; **9. a.** $x^2 - x - 6$; **d.** $x^2 + 2xy - 8y^2$; **g.** $x^3 + 2x^2 + 9$; **i.** $a^3 - 4a^2b + 4ab^2 - b^3$; **11.** $30(3x - 9)$; **13. b.** $6x^2 - 9x$; **c.** $6x^2 + x - 15$; **14. c.** $198x - 45$ miles.

Activity 2.18   Exercises: **2. a.** 17; **c.** 12; **e.** 78; **g.** -7; **i.** -12; **5. b.** $B = 2A - C$; **d.** $\frac{5}{9}(F - 32)$; **g.** $P = \dfrac{A}{1 + ri}$; **6. d.** $y = 14.50x + 70$; **g.** \$215; **i.** $x = \dfrac{y - 70}{14.5}$; **8. a.** $12 - x$; **b.** $1.50x$; **c.** $0.85(12 - x)$; **d.** total cost of the dozen flowers; **e.** 14.75; **f.** 7 roses, 5 carnations.

Activity 2.19   Exercises: **1. a.** The graphs are nonintersecting parallel lines. They have no points in common. **b.** y_1 and y_2 are equivalent, so for every input, their outputs match. All real numbers are solutions; **2. a.** 9; **c.** -11; **e.** No solution. **g.** 90; **3. a.** $C = 25,600 + 200x$; **b.** $C = 5500 + 1200x$; **d.** $25,600 + 200x = 5500 + 1200x$; **e.** After 20.1 years, the cost of the electric heating system will catch up with the cost of the solar heating system. After that, the electric system costs more. **f.** \$29,620.

Activity 2.20   Exercises: **1. a.** ≈ 0.866 seconds; **c.** ≈ 1.94 seconds; **3. b.** 8 ft; **d.** 0.625 seconds; **4. a.** ≈ 49.9 mph; **c.** between 169 and 170 feet; **5. a.** ± 3; **c.** No real number solution. **e.** 49; **g.** 5.

How Can I Practice?   **1. a.** $x = 4$; **e.** $x = -3$; **g.** $y = 7$; **i.** $x = -\frac{1}{3}$; **k.** All real numbers; **m.** $x = \pm 6$; **2. a.** $y = \frac{4x - 5}{2}$; **c.** $y = 4(7 - 2x)$; **e.** $r = \sqrt[3]{\frac{3V}{4\pi}}$; **4. b.** $C = 750 + 0.25x$; **d.** 1000 booklets; **f.** 1500 booklets; **g.** 2500 booklets; **5. d.** $1.5x + 3(22 - x)$; **e.** 5 days; **f.** \$50; **6. d.** $100 + 0.30x = 150 + 0.15x$; **e.** \$333.33 in sales produces the same salary under both plans. **f.** \$200.

Gateway Review   **1. a.** $(1, 2)$; **b.** $(3, -4)$; **c.** $(-2, 5)$; **d.** $(-4, -6)$; **e.** $(0, 4)$; **f.** $(-5, 0)$; **g.** $(0, 0)$; **2. a.** $x + 5$; **b.** $18 - x$; **c.** $2x$; **d.** $\frac{4}{x}$; **e.** $3x + 17$; **f.** $12(8 + x)$; **g.** $11(14 - x)$; **h.** $\frac{x}{7} - 49$; **3. a.** $x = 9$; **b.** $y = -24$; **c.** $x = 2.5$; **d.** $x = -12$; **e.** $y = -2$; **f.** $x = 16$; **g.** $x = 144$; **h.** $y = \frac{5}{6}$; **i.** $x = 72$; **4. a.** $\dfrac{203 + x}{4}$; **5. a.** $x = 5$; **b.** $y = 42$; **c.** $x = 6$; **d.** $x = -7.75$; **e.** $y = -42$; **f.** $x = 139.5$; **g.** $x = -96$; **h.** $y = 27$; **6. a.** The number of miles driven. **b.** The cost of rental for a day. **c.** $y = 25 + 0.15x$; **d.** 40, 55, 70, 85, 100; **f.** Just under \$75; **g.** \$70.90;

i. 600 miles; **7.** ≈ $714,286; **8. a.** $100; **b.** $360; **9. a.** 12.4; **b.** $31\frac{1}{6}$; **10.** It will take a little more than 6 years for your deposit to accumulate to $20,000 in order to make a down payment on a home; **11. a.** $= 16x^2 - 24x + 9$; **c.** $16x^2 - 3$; None are equivalent; **12. a.** x^8; **b.** $6x^3y^4$; **c.** x^{28}; **d.** $8x^6y^{15}$; **e.** $x^{10}y^5$; **f.** $-s^{18}$; **g.** $4x^6y^5z^9$; **13. a.** $3x + 3$; **b.** $-12x^2 + 12x - 18$; **c.** $4x^2 - 7x$; **d.** $-8 - 2x$; **e.** $12x - 25$; **f.** $17x - 8x^2$; **14. a.** $4(x - 3)$; **b.** $x(18x + 60 - y)$; **c.** $-4(3x + 5)$; **15. a.** $6x^2 - 6x + 3$; **b.** $x^2 - 3x + 7$; **16. a.** $x^2 + 2x - 3$; **b.** $2x^2 - 7xy + 6y^2$; **c.** $x^3 - 2x^2 - 9x - 2$; **17. a.** $x = 20$; **b.** $x = 3$; **c.** $x = -13$; **d.** $x = 0$; **e.** $x = 4$; **f.** $x = 7$; **g.** $x = -3$; **h.** No solution. **i.** All real numbers are solutions. **j.** $x = \pm 2$; **k.** $x = 12$; **18. a.** $77; **e.** It is the same. **19. a.** $500 - n$; **b.** $1.50n$; **c.** $2.50(500 - n)$; **d.** $1.50n + 2.50(500 - n) = 950$; **20. a.** $C = 600 + 10x$; **b.** $I = 40x$; **c.** $P = 40x - (600 + 10x) = 30x - 600$; **d.** 20 campers; **e.** 40 campers;

g.

(20, 800)

22. a. 1.57 seconds; **b.** ≈ 7.3 ft.

Chapter 3

Activity 3.1 **Exercises: 1. a.** $C(n) = 2n + 78$; **d.** The number of students who can sign up for the trip is restricted by the capacity of the bus; **2. c.** Yes, this data represents a function because each input value is assigned a single output value; **5.** -4; **7.** 69; **9.** -13; **11.** 14; **13.** 30; **14.** a, c, d. They all pass the vertical line test.

Activity 3.2 **Exercises: 1.** As the selling price increases, the number of units sold increases slightly at first but then declines, at an increasing rate.

Activity 3.3 **Exercises: 2. a.** Domain $-25 \leq x \leq 45$; Range $-40 \leq y \leq 40$; **c.** Domain $-36 \leq x \leq 28$; Range $-20 \leq y \leq 32$; **3. a.** $0 \leq x \leq 8$; **b.** $0 \leq g(x) \leq 6$; **c.** from $x = 2$ to $x = 6$; **d.** from $x = 0$ to $x = 2$ and from $x = 6$ to $x = 8$; **e.** 6, 0; **f.** 0, 8; **g.** 2.5; **h.** $x = 1, 4$ and 7; **6. a.** -4; **b.** $(2, -4)$.

What Have I Learned? **2.** $f(1)$ represents the output value when the input is 1. Since $f(1) = -3$, the point $(1, -3)$ is on the graph of f; **3.** After 10 minutes, the ice cube weighs 4 grams; **5.** $H(5) = 100$.

How Can I Practice? **1. b.** Yes, for each number of credit hours (input) there is exactly one tuition cost (output); **c.** No, for a tuition cost (input) of $1250 there is more than one possible number of credit hours (output); **3.** 16; **5.** 3; **7. c.** We leave home and drive for 2 hours, then stop for 1 hour and finally, continue at a slower speed than before.

Activity 3.4. **Exercises: 2. e.** When age of first marriage increases, the graph rises. When age of first marriage decreases, the graph falls. When age of first marriage remains unchanged, the graph is horizontal. When the change in first marriage age is the greatest, the graph is the steepest. **3.** $-.62$ year of age per decade from 1900–1950, $-.4$ year of age per decade from 1900–1940, .2 year of age per decade from 1950–1970; **4. a.** $-.16$ degrees per hr.; **c.** The temperature dropped at an average rate of 1.5 degrees per hour from 6 P.M. to 10 P.M.

Activity 3.5 **Exercises: 1. a.** 191.4 points per year; **b.** during 1989; **c.** a decrease in the DJIA; **d.** no change in the DJIA; **e.** during 1989; **f.** during 1987; **g.** during 1981; **4. c. i.** approximately 20 million people; **ii.** 1967 through 2007; **iii.** 20 million people to approximately 42.5 million people; **e.** $P(1977) = 25$; **f.** 6.03 billion dollars per year ; **g.** .62 million people per year.

Activity 3.6 **Exercises: 2. a.** $C(t) = 45(1.07)^t$; **3. a.** increasing; **b.** decreasing; **c.** increasing; **d.** decreasing. If the base is a growth factor, greater than 1, the graph is increasing. If the base is a decay factor, less than 1, the graph is decreasing.

Activity 3.7 **Exercises: 1. b.** approximately $282 per month; **d.** (Answers will vary.) I feel more confident about the estimate for 27 months because 27 is within the boundaries of the original data. **e.** (Answers will vary.) an approximate formula is $p(m) = \frac{8000}{m}$, $24(320) = 7680$, $30(260) = 7800$, $36(230) = 8280$; **2. a.** $y = \frac{120}{x}$. Yes. y is a function of x. The domain is all real numbers except 0; **3. c.** $x = -5.5$; **d.** $x = 28$.

What Have I Learned? **2.** If the rate of change of a function is negative, then the initial value for each interval must be larger than the final value. We can conclude that the function is decreasing. **4.** No, the graph of a function cannot have more than one vertical intercept. Each input value, including $x = 0$, can be assigned at most one output value according to the definition of a function. **9.** Given a sufficient number of actual data values, the assumptions for interpolation are generally more likely to hold for the data than will the assumptions for extrapolation.

How Can I Practice? **1.** $f(-5) = 19$; **3.** -11; **5.** $R = \{-10, -7, -4, -1, 2, 5\}$; **7. a.** $f(x) = $ $2.50 + 1.50x$; **e.** $1.50 per mile; **8. b.** When the rate of change is negative, the graph decreases over that interval; **9. b.** $s(y) = 20,000(1.03)^y$; **e.** $26,878; **10. b.** We should charge $.30 for each if we wish to sell 4000 candy bars. **d.** The domain is all real numbers except zero. **e.** The practical domain is all real numbers greater than zero and less than $2 (or so); **12.** 1.5; **14.** 83; **16. b.** $f(x) = \frac{1440}{x}$; **17. c.** $-45°$F.

Gateway Review   **1.** $(-2)^2 = 4$; **2.** $-(2)^2 = -4$;
3. $(-(-2))^2 = 2^2 = 4$; **4.** $-(-2)^2 = -4$;
5. $5(-10) - 3 = -50 - 3 = -53$;

6. $4(-7)^2 - 3(-7) + 8 = 4(49) + 21 + 8 = 225$;

7. $-6(2)^2 + 2(2) - 12 = -24 + 4 - 12 = -32$;

8. $m(4) = 4^2 - 3(4) + 5 = 16 - 12 + 5 = 9$;

$n(4) = 3(4)^2 + 4(4) + 6 = 48 + 16 + 6 = 70$;

$9 + 70 = 79$;

9. $f(-3) = 2(-3)^2 - 5 = 18 - 5 = 13$;

$g(-3) = 3(-3) + 8 = -9 + 8 = -1$;

$13 - (-1) = 14$;

10. $f(3) = -(3)^2 + 3(3) + 1 = -9 + 9 + 1 = 1$

$g(3) = -8(3) + 4 = -24 + 4 = -20$

$\frac{4(3)}{g(3)} = \frac{1}{-20} = -.05$;

11. (Answers may vary.) Cost of mailing a package increases as different weight categories increase. The minimum cost is the .5 lb to 1.5 lb category and the maximum shown is the 3.5 to 4.5 category. **12.** (Answers may vary.) You traveled at a constant rate of speed for 1 hour 40 minutes, stayed at your destination for 3 hours 20 minutes, then returned home, traveling at a faster rate on the return trip, taking only 1 hour.

13. 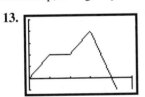 **14.**

15. 0, $3.32, $3.32, $3.32, $4.21, $5.10, $9.55, $11.33, $14.00, $18.45; **16.** The first 3 minutes cost $3.32. The next 7 minutes cost $.89(7) = $6.23. The total cost is $3.32 + $6.23 = $9.95; **17. a.** $C(0) = 0$; **b.** $C(n) = 3.32$; **c.** $C(n) = 3.32 + .89(n - 3)$; **18.** The domain of the phone call is all non-negative integers; **19.** The minimum value is 0; **20.** There is no maximum value; **21.** The function is increasing for $n \geq 3$; **22.** The function is never decreasing; **23.** The function is constant for $0 < n \leq 3$; **24.** $\frac{\Delta C}{\Delta n} = \frac{33.2 - 3.32}{3 - 1} = \frac{0}{2} = 0$; **25.** $\frac{\Delta C}{\Delta n} = \frac{14.00 - 5.10}{15 - 5} = \frac{8.90}{10} = .89$; **26.** The rate of change is positive, indicating the graph increases, as it does; **27.** $\frac{\Delta C}{\Delta n} = \frac{18.45 - 4.21}{20 - 4} = \frac{14.24}{16} = .89$; **28.** the rate of change is constant when $n \geq 3$; **29. a.** The maximum occurs in the beginning of January and the minimum at the end of January; **b.** The maximum occurs in August and the minimum occurs in May. **30. a.** The amount of electricity used for January 1, 1998 will be approximately 1150

kWh; **b.** The amount of electricity used for March 1, 1997 was approximately 960 kWh.

Chapter 4

Activity 4.1   Exercises: 1. c. For all the data points to be on a line, the rate of change between any two data points must be the same, no matter which points are used in the calculation. **3. b.** The average rate of change is not constant so the data is not linear.

Activity 4.2   Exercises: 1. Linear $\left(\text{constant rate of change} = \frac{1}{2}\right)$. Not linear (rate of change not constant); **2. d.** Every week, you hope to lose 2 pounds; **3. b.** The slope of the line for this data is $-\frac{7}{10}$ beats per minute per year of age; **5. a.** 50 feet higher; **b.** 5.2% grade; **6. a.** 50 miles per hour; **b.** The line representing the distance traveled by car B is steeper, so car B is going faster.

Activity 4.3   Exercises: 1. $p(n) = 13n - 2200$ dollars; **3.** The slope is 13. The profit increases by $13 for each additional student attending; **4.** The vertical intercept is $(0, -2200)$. The class project has a loss of $2200 if no students attend. **7.** 170 students.

Activity 4.4   Exercises: 1. Slope is 3, y-intercept is $(0, -4)$, x-intercept is $\left(\frac{4}{3}, 0\right)$; **3.** Slope is 0, y-intercept is $(0, 8)$, There is no x-intercept; **5.** Slope is 2, y-intercept is $(0, -3)$, x-intercept is $\left(\frac{3}{2}, 0\right)$ or $(1.5, 0)$; **7.** The graphs all have the same slopes but different vertical intercepts. They are parallel. **9.** $y = 12x + 3$; **11. b.** $y = 3x + 5$; **13. b.** $y = 15x + 120$; **15. a.** Horizontal intercept is $(-4, 0)$; vertical intercept is $(0, 12)$.

How Can I Practice: Exercises: 1. b. $m = 2$; **d.** $m = 2$; **f.** $m = -1.5$; **2. a.** $y = x + 3$; **c.** $y = 5x - 1$; **e.** $y = -3x + 17$; **3. a.** 7.5 feet; **4. b.** Yes, the data represents a linear function because the rate of change is constant, $28 per month; **5. d.** During the first 5 seconds, for each second that passes, your speed increases 11 mph; **7. a.** undefined; **c.** .5; **8. b.** The vertical intercept is $(0, 10)$, The horizontal intercept is $\left(\frac{20}{3}, 0\right)$; **9. a.** $y = 2.5x - 5$; **b.** $y = 7x + \frac{1}{2}$; **c.** $y = -4$; **12.** The lines all have the same vertical intercept $(0, -2)$; **13. b.** The horizontal intercept is $(3, 0)$. There is no vertical intercept.

Activity 4.5   Exercises: 1. c. $P(t) = 0.617t + 23.7$; **e.** approximately 3.6%; **f.** 40,976,000; **2. c.** The population will double in approximately 58 years' time, by 2048.

Activity 4.6   Exercises: 1. d. $t = .15i + 4$; **f.** $11,173.33; **2.** $y = 3x$; **4.** $y = 7x + 16$; **6.** $y = 2$; **8.** $y = 25x + 155$; **10.** $y = 5x - 4$; **12.** $y = 3x + 9$; **14.** $y = \frac{1}{3}x - \frac{4}{3}$.

Activity 4.7 Exercises: **7.** $g(t) = 1.29t - 29.6$; **8.** 17.35.

How Can I Practice? **1.** Slope is 2, vertical intercept is $(0, 1)$, horizontal intercept is $\left(-\frac{1}{2}, 0\right)$; **4.** Slope is $-\frac{3}{2}$, vertical intercept is $(0, -5)$, horizontal intercept is $\left(-\frac{10}{3}, 0\right)$; **7.** Slope is -2, vertical intercept is $(0, 2)$, horizontal intercept is $(1, 0)$; **10.** $y = 9x - 4$; **13.** $y = 0$; **17. a.** $N(t) = 4t - 160$; **b.** 88 chirps per minute; **c.** 87.5°F; **20. a.** $y = 3x + 6$; **d.** $y = 4x - 27$; **f.** $y = 2x - 13$; **21. a.** $y = 2x + 6$; **c.** $y = 3$; **e.** $x = 6$; **23. c.** $(-2, 0)$; **e.** $(2.6, 0)$.

Activity 4.8 Exercises: **2.** $q = 2, p = 0$; **4.** $x = 6$, $y = 1$; **6.** Examining the graphs, we see that the lines are parallel.

Activity 4.9 Exercises: **1. a.** $R(x) = 200x$; **b.** The slope of the cost function is 160. This represents a cost of $160 to produce each bundle of additional pavers. The y-intercept is $(0, 1000)$. Fixed costs of $1000 are incurred even when no pavers are produced. **e.** $(25, 5000)$.

Activity 4.10 Exercises: **1.** You need to buy 3.75 lb or more; **3.** I would catch up with my friend in 15 minutes having rollerbladed 2.5 miles.

How Can I Practice? **2.** $(-45, -40)$; **4.** 44 centerpieces; **6.** $x = -1, y = 275$; **8.** No solution. The lines are parallel; **10.** $x \approx -6.48, y \approx -15.78$.

Gateway Review **1.** The slope is -2.5, the vertical intercept is $(0, 4)$; **2.** The equation is $y = -2x + 2$; **3.** Vertical intercept is $(0, 8)$. An equation for the line is $y = \frac{-4}{3}x + 8$. **4.** row 1: c, e, b   row 2: f, a, d; **5.** Vertical intercept is $(0, 5)$. Equation is $y = 3x + 5$; **6.** The vertical intercept is $(0, -2)$, the horizontal intercept is $(4, 0)$; **7.** The vertical intercept is $(0, 3)$, the horizontal intercept is $(-4, 0)$; **8.** $y = \frac{5}{3}x - 4$; **9. a.** $y = -1.5x + 5, y = .5x + 1, x = 2, y = 2$; **10.** $a = -1; b = 6$; **11.** $y = 2, x = 2$; **12. a.** $q = -1.25p + 100$; **b.** $p = -.8q + 80$; **13.** $y = -.56x + 108.8$; **14. a.** ABC: $C(x) = 50 + .20x$; Competition: $C(x) = 60 + .15x$; **c.** The lines intersect at $(200, 90)$. The daily cost would be the same at both companies when 200 miles are traveled. **d.** It will be cheaper to rent from the competition because the distance is ≈ 235 miles and the competition's prices are cheaper if you travel over 200 miles;

15. a. $m = \dfrac{32 - 212}{0 - 100} = \dfrac{-180}{-100} = \dfrac{9}{5}$; **b.** $F = 1.8C + 32$;

c. 104° Fahrenheit corresponds with 40°C; **d.** 170°F corresponds approximately to 77°C; **16. a.** $94.97, $110.90, $116.21, $126.83; **c.** $C(k) = .1062k + 20.63$; **d.** The cost of using 875 kWh of electricity is $113.56; **e.** You used approximately 1218 kWh of electricity if your

monthly bill is $150; **17. a.** $P(t) = 2500t + 50,000$ dollars; **c.** When $t = 40$, that is, in the year 2010, the price of the house will be $150,000; **18.** $V(t) = -95t + 950$; **19. a.** The regression equation is approximately $w(h) = 6.92h - 302$; **b.** He would weigh about 252 lb; **c.** The player would be about 71.1 in. tall.

Chapter 5

Activity 5.1 Exercises: **2. c.** -4 to 5; **d.** $(0, -3)$; **e.** $(3, 0)$ and $(-1, 0)$; **f.** increasing from 1 to 4, decreasing from -2 to 1; **3. d.** Quadratic; **f.** Not quadratic; **4. a.** $a = 1, b = 0, c = -3$.

Activity 5.2 Exercises: **1. c.** The rocket is on the ground $t = 0$ seconds from launch and returns to the ground $t = 30$ seconds from the launch. **f.** It will reach its maximum height of 3600 feet 15 seconds after the launch. **2. a.** $x = 0, x = -10$; **c.** $y = 0, y = -5$; **e.** $t = 0$, $t = 1.5$; **g.** $w = 0, w = 4$; **3. a.** $x^2 - 4x + 3 = 0$; **c.** $x^2 - 25 = 0$.

Activity 5.3 Exercises: **1. a.** $x = -1$, $x = -6$; **c.** $(y + 7)(y + 4) = 0$; $y = -7, y = -4$; **2. b.** The rocket is 1760 feet above the ground 5 seconds after launch and again, on the way down, 22 seconds after launch. **e.** The rocket returns to the ground in 27 seconds. **3. a.** Not factorable; $x \approx 1.45, x = -3.45$; **c.** $x = 5, x = -3$.

Activity 5.4 Exercises: **1.** Axis of symmetry: $x = \frac{-2}{2(1)} = -1$, Vertex: $(-1, -16)$; $x = -5, x = 3$; **3.** Axis of symmetry: $x = \frac{-1}{2(-2)} = \frac{1}{4}$, vertex: $\left(\frac{1}{4}, \frac{9}{8}\right) = (0.25, 1.125)$, $x = 1, x = -\frac{1}{2}$; **5.** Axis of symmetry: $x = \frac{-10}{2(-1)} = 5$, vertex: $(5, 34)$, $x \approx 10.83$; $x \approx -0.83$; **7.** Axis of symmetry: $x = \frac{-(10)}{2(5)} = -1$, vertex: $(-1, 0)$, $x = -1, x = -1$ A double root; **11. a.** The ball reaches a maximum height of 42 feet in 1.5 seconds.

Activity 5.5 Exercises: **1.** $x = 2, x = -2$; **3.** No solution; **5.** $x = 3, x = -13$; **7.** $x = 6, x = -3$; **9. c.** The ladder must be moved approximately 7.49 feet further from the wall.

How Can I Practice? **2. a.** q; **c.** g; **e.** f; **4. b.** The y-coordinate is $y = f\left(-\frac{b}{2a}\right)$. **6. a.** $x = 4, x = 5$; **c.** $x = 8, x = 3$; **e.** $x = 3, x = -1$; **g.** $x = 5, x = 2$; **i.** $x = 0, x = -\frac{3}{2}$; **k.** $v = 0, v = -8$; **7. a.** $x = -6$, $x = 1$; **c.** $x = 9, x = -2$; **e.** $x = -8, x = 1$; **g.** $x \approx 4.414, x \approx 1.586$; **i.** $x \approx 3.414, x \approx 0.586$.

Glossary

absolute value For a non-negative number, the absolute value is that number itself; for a negative number the absolute value is its opposite.

algebraic expression A mathematical set of instructions (containing constant numbers, variables, and the operations among them) that indicates the sequence in which to perform the computations.

angle The figure formed by two rays starting from the same end point.

area The number of square units contained in a region.

average The sum of a collection of numbers, divided by how many numbers are in the collection.

axis of symmetry A vertical line that separates the graph of a parabola into two mirror images.

bar graph A diagram of parallel bars depicting data.

break-even number The number for which the total revenue equals total cost.

Cartesian (rectangular) coordinate system in the plane A two-dimensional scaled grid of equally spaced horizontal and vertical lines.

circle A collection of points that are the same distance from some given point called its center.

circumference The distance around a circle.

coefficient A number written next to a variable that multiplies the variable.

common factor A factor contained in each term of an algebraic expression.

commutative operation An operation in which you can interchange the order of the two numbers and always produce identical results. The commutative operations of arithmetic are addition and multiplication.

cone A three-dimensional surface with one circular base.

constant term A term that consists of only a number.

cylinder A three-dimensional surface with two circular bases.

delta Δ = change.

dependent variable If the input/output situation is a function, then the output variable is called the dependent variable.

diameter A line segment that goes through the center of a circle and connects two points on the circle; it measures twice the radius.

distributive property A property involving addition and multiplication that states that $a \cdot (b + c) = a \cdot b + a \cdot c$.

domain The set (collection) of all possible input values for a function.

equivalent expressions Two algebraic expressions are equivalent if they always produce identical outputs when given the same input value(s).

expand An instruction to use the distributive property to transform an algebraic expression from factored form to expanded form.

expanded form An algebraic expression in which all parenthesis have been removed using multiplication.

exponential function A function in which the variable appears in the exponent. For example, $y = 4 \cdot 5^x$.

extrapolation The assignment of values to a sequence of numbers beyond those actually given.

factor, a When two or more algebraic expressions or numbers are multiplied together to form a product, those individual expressions or numbers are called the factors of that product. For example, the product $5 \cdot a \cdot (b + 2)$ contains the three factors 5, a, and $b + 2$.

factor, to An instruction to use the distributive property to transform an algebraic expression from expanded form to factored form.

factored form An algebraic expression that has been rewritten in terms of its own factors.

fixed costs Costs (such as rent, utilities, etc.) that must be paid no matter how many units are produced or sold.

formula An algebraic statement describing the relationship between the input variables and output variable.

function terminology and notation The name of a functional relationship between input x and output y, then you say that y is a function of x and you write $y = f(x)$.

function A rule (given by tabular, graphical, or symbolic form) that relates (assigns) to any permissible input value exactly one output value.

goodness-of-fit A measure of the difference between the actual data values and the values produced by the model.

greatest common factor The product of all the common factors in the expression.

histogram A bar graph with no spaces between the bars.

horizontal (x-) axis The horizontal line, described most commonly by $y = 0$ in the Cartesian coordinate system; used to represent the input axis.

horizontal intercept The point(s) where a line or curve crosses the input (x) axis.

independent variable If the input/output situation is a function, then the input variable is called the independent variable.

input Replacement values for the variable in an algebraic expression, table, equation or function. In a cause-effect relationship, the input usually represents the explanatory variable (cause).

interpolation The insertion of intermediate values in a sequence of numbers between those actually given.

inverse operation An operation that undoes another operation. Addition and subtraction are inverses of each other as are multiplication and division. The inverse of squaring is taking the square root. Negation is its own inverse, as is taking reciprocals.

like terms Terms that contain identical variable factors (including exponents) and differ only in their coefficients.

line of best fit Line that will have as many of the data points as close to the line as possible.

linear function A function that has the form $f(x) = mx + b$ and whose graph is a straight line.

maximum value The largest output of a function.

mean The arithmetic average of a set of data.

minimum value The smallest output of a function.

operation (arithmetic) Addition, subtraction, multiplication, and division are arithmetic operations between two numbers. Exponentiation $\left(\text{raising a number} \right.$ to a power, for example, 10^4), taking roots (for example, $\sqrt{8}$), and negation (changing the sign of a number) are examples of operations on a single number.

order of operations An agreement for how to evaluate an expression with multiple operations.

ordered pair Pair of values, separated by a comma, and enclosed in a set of parenthesis. The input is given first, and the corresponding output is listed second as follows. (Input, Output). These numbers serve as an address for each point in the coordinate plane, giving a point's location with respect to the origin.

origin The point at which the horizontal and vertical axes intersect.

output Values produced by evaluating an algebraic expression, table, equation or function. In a cause-effect relationship, the output usually represents response variable (effect) corresponding to a given input.

parabola The graph of a quadratic function (second-degree polynomial function). The graph is U-shaped, opening upward or downward.

parallelogram A four-sided figure for which each pair of opposite sides are parallel.

percent(age) Parts per hundred.

percentage error The difference between an actual data value and a predicted value, divided by the actual value and then converted to a percent.

perimeter A measure of the distance around a figure; for circles, perimeter is called *circumference*.

pie chart A diagram of sectors of a circle depicting data.

piecewise defined function A function defined differently for different parts of its domain.

polynomial An expression containing the sum of a finite number of terms of the form ax^n, generally written in descending powers of the variable from left to right.

proportion An equation stating that two ratios are equal.

Pythagorean theorem The relationship among the sides of a right triangle, namely that the sum of the squares of the lengths of the two perpendicular sides (legs) is equal to the square of the length of the side opposite the right angle, called the hypotenuse.

quadrants The four regions of the plane separated by the vertical and horizontal axes. The quadrants are numbered from 1 to 4. Quadrant 1 is located in the upper right with the numbering continuing counterclockwise. Therefore, quadrant 4 is located in the lower right.

quadratic formula The formula $x = \dfrac{-b \pm \sqrt{b^2 - 4ac}}{2a}$ that represents the solutions to the quadratic equation $ax^2 + bx + c = 0$.

quadratic function A function of the form $y = ax^2 + bx + c$.

radius The distance from the center point of the circle to the outer edge of the circle.

range The set (collection) of all possible output values for a function.

rate of change The amount of change divided by the length of the interval over which the change takes place.

ratio A fraction.

regression line A line resulting from a scatterplot of data for a line of best fit.

rounding A technique to approximate numbers according to certain rules.

scatterplot A graph of the values of two variables on the coordinate plane. The values of explanatory variable (input) are plotted on the x-axis, and the corresponding values of the response variables (output) are plotted on the y-axis.

scientific notation A device for expressing very large or very small numbers as a product—(a number between 1 and 10) · (the appropriate power of 10).

signed numbers Numbers accompanied by a positive or negative sign.

slope of a line A measure of the steepness of a line.

slope-intercept form The form, $y = mx + b$, of a linear equation. The slope is denoted by m and the vertical intercept by b.

solution of a system A point that is a solution to both equations in a system of two linear equations.

sphere A three-dimensional surface for which all points are equidistant from a given point, the center.

substitution method A method used to solve a system of two linear equations. Solve one equation for one variable, then substitute for this variable in the other equation and solve for the other variable.

system of equations Two linear equations considered together.

terms Parts of an algebraic expression separated by the addition, $+$, and subtraction, $-$, symbols.

variable A quantity that takes on specific numerical values and will often have a unit of measure (e.g., dollars, years, miles) associated with it.

vertex The turning point of a parabola having coordinates $\left(\frac{-b}{2a}, f\left(\frac{-b}{2a}\right)\right)$, where a and b are determined from the equation $f(x) = ax^2 + bx + c$. The vertex is the highest or lowest point of a parabola.

vertical (y-) axis The vertical line, described most commonly by $x = 0$, in the Cartesian coordinate system that is used to represent the output axis.

vertical intercept The point where a line or curve crosses the vertical axis.

vertical line test A graph represents a function if and only if any vertical line that is drawn intersects the graph of the function in exactly one point.

volume A measure of three-dimensional space occupied by a figure.

weighted average An average in which some numbers count more heavily than others.

x-coordinate The first number of an ordered pair that indicates the horizontal directed distance from the origin. A positive value indicates the point is located to the right of the origin, a negative value indicates a location to the left of the origin.

y-coordinate The second number of an ordered pair that indicates the vertical directed distance from the origin. A positive value indicates the point is located to the above the origin, a negative value indicates a location below the origin.

Zero Product Rule The algebraic principle that says if a and b are real numbers such that $a \cdot b = 0$, then either a or b, or both, must be equal to zero.

Index